New Directions in Civil Engineering

SERIES EDITOR: W. F. CHEN *Purdue University*

Published and Forthcoming Titles

FRACTURE PROCESSES
of CONCRETE

Assessment of Material Parameters for Fracture Models

Jan G. M. van Mier

CRC Press
Boca Raton   New York   London   Tokyo

Acquiring Editor: *Navin Sullivan*
Associate Editor: *Felicia Shapiro*
Senior Project Editor: *Susan Fox*
Cover Design: *Denise Craig*
PrePress: *Kevin Luong*
Marketing Manager: *Susie Carlisle*
Direct Marketing Manager: *Becky McEldowney*

Library of Congress Cataloging-in-Publication Data

Mier, J. G. M. van.
    Fracture processes of concrete : assesment of material parameters for fracture models /
by Jan G.M. van Mier.
        p.   cm. — (New directions in civil engineering)
    Includes bibliographical references and index.
    ISBN 0-8493-9123-7 (alk. paper)
    1. Concrete—Fracture.   2. Fracture mechanics.   I. Title.   II. Series.
Ta440.M49 1996
620.1'366—dc20   96-34578
                                                                                                    CIP

No claim to original U.S. Government works
International Standard Book Number 0-8493-9123-7
Library of Congress Card Number 96-34578
Printed in the United States of America   1 2 3 4 5 6 7 8 9 0
Printed on acid-free paper

PREFACE

This is a book about concrete, but at the same time it is not. Although concrete technology and the mechanical properties of concrete are presented in Part 1, integration of experimental and numerical tools in research and applications of concrete fracture mechanics is the main topic. Fracture mechanics of concrete has grown into a mature field of research after the Fictitious Crack Model was introduced by Hillerborg and co-workers in 1976. In the two decades since then, an increasingly larger number of researchers have tried to further elaborate on the model, and to find practical applications for the new tool. Cracking is the number one durability problem. Understanding crack nucleation and growth phenomena therefore seems important when the annual amount of damage and repair on concrete structures is considered. Predicting crack growth is the goal of fracture mechanics. For successfully applying the models, however, it should have a solid physical basis, and the parameters used in the model should be determined in an easy and straightforward manner. Moreover, for the models to be accepted by practical engineers, the standard tests needed should be based on current experimental practice. In short, the demands for the new models are far reaching. They have to be "predictive" in the true sense of the word. Most unfortunately, fracture mechanics has a number of "inherently unknown parameters beyond the current unknown" (as to speak with Ralph E. Gomory, see *Scientific American*, June 1995, pp.88), which makes the application seemingly unwieldy for practical purposes. Less is true, however. The unknown has been explored extensively over the past two decades, and in spite of the inherently unknown parameters that stem from deeply founded material-structure interactions, a simple view is developing which makes the fracture models attractive for future applications. The key is, as mentioned before, a far reaching interaction between experimental and numerical modelling.

The book is separated into three major parts. After a survey of the structure and mechanical properties of concrete as usually found in standard textbooks on concrete technology in Part 1, the tools for gathering all this information are presented in Part 2. Both experimental and numerical tools are presented. Needless to say, experiments are essential to develop new views on the fracture behaviour of materials and structures. The models are not only there, however, to design the experiments, but also to critically analyze if the experiments were correctly carried out. In Part 3, such analyses are carried out and they reveal whether the model assumptions are in agreement with the experimental observations. These analyses are important for demonstrating the validity of the fracture models, but also they give confidence in the models parameters that must be used and which cannot be determined directly from experiments. Inverse analysis is therefore an essential part of applying fracture mechanics.

Thus, the methodology is important, but of course — almost unavoidable — much information is given on the structure of concrete, the mechanical properties, the fracture behaviour, testing techniques and numerical models available to date, as well as many examples.

Many of the examples and insights come from my own research work on fracture. Of course, several people were involved in "The Job". First, I would like to thank my former and present Ph.D. students, M.B. Nooru-Mohamed, Erik Schlangen, Adri Vervuurt, Jeanette Visser, Marcel van Vliet, all at Delft University, and René Vonk, Hans Bongers and Erik van Geel at Eindhoven University of Technology. Discussions with all of them have been fruitful and are greatly appreciated. The two technicians who helped with the experiments must also be mentioned: Gerard Timmers and Allaard Elgersma. Their outstanding technical skills have added considerably to the quality of the experimental work and the conclusions that could be drawn would be of a much lesser standard. For substantial help with the drawings in the book, the assistance of Theo Steijn is gratefully acknowledged. For the English language, the help of my brother Guido was essential. Finally, discussions with colleagues of the Mechanics and Structures Group at the Civil Engineering Faculty, and of the Structural Group of the Architecture Faculty at Eindhoven University, have led to new insights, and have — hopefully — added to a clear description of the fracture mechanics of concrete. And last, but certainly not least, my wife, Ria, and our two children, Ingrid and Martijn, must be mentioned. They had to endure changing tempers during the writing of the book, but managed to keep spirits high all the time. To them I dedicate this book.

GLOSSARY OF SYMBOLS AND ABBREVIATIONS

a	crack length [mm]
a_0	initial crack length or notch [mm]
b	width [mm]
$\mathbf{b}$	vector containing estimates of β_i
c	density [kg/m^3]
c	cohesion [MPa]
C	compliance [MPa^{-1}]
d	specimen depth or diameter [mm]
d_{max}	maximum aggregate size [mm]
D	diameter or specimen thickness [mm]
D	shear force
e	eccentricity
$\mathbf{e}$	error vector
E	Young's modulus [MPa]
E_a	aggregate Young's modulus [MPa]
E_b	Young's modulus of bond zone [MPa]
E_c	concrete Young's modulus [MPa]
E_m	matrix Young's modulus [MPa]
E_t	tangent Young's modulus [MPa]
f_c	uniaxial compressive strength [MPa]
f_{2c}	biaxial compressive strength [MPa]
f_p	particle strength [MPa]
f_{spl}	splitting tensile strength [MPa]
f_t	uniaxial tensile strength [MPa]
$f_{t,a}$	uniaxial tensile strength of aggregate [MPa]
$f_{t,b}$	uniaxial tensile strength of bond zone [MPa]
$f_{t,m}$	uniaxial tensile strength of matrix [MPa]
F	force [N]
G	shear modulus [MPa]
G_f	fracture energy (Hillerborg) [N/m]
G_{Ic}	critical energy release rate (Griffith)
h	specimen height [mm]
h/d	slenderness ratio [-]
I_1	first invariant of principal stresses
I_2	second invariant of principal stresses
I_3	third invariant of principal stresses
J_1	first invariant of octahedral stresses
J_2	second invariant of octahedral stresses
J_3	third invariant of octahedral stresses
k	number of variables [-]
k	spring stiffness
K	bulk modulus [MPa]
K_I	stress intensity factor in mode I (opening)

K_{II}	stress intensity factor in mode II (sliding)
K_{III}	stress intensity factor in mode III (tearing)
K_{Ic}	critical stress intensity factor, mode I [kN/m$^{3/2}$]
K_{Ic}^{s}	critical stress intensity factor, mode I, Jenq and Shah model [kN/m$^{3/2}$]
K_{M}	machine stiffness
K_{F}	loading frame stiffness
K_{r}	rotational system of testing machine
KS_{r}	sum of squares
l	length [mm]
l_{ch}	characteristic length [mm]
l_{meas}	measuring length for LVDTs [mm]
ℓ	span [mm]
L	length [mm]
M	bending moment
n	number of experimental results [-]
N	number of load-cycles [-]
N	normal force [N]
p	pressure [bar]
P	normal force [N]
P_{k}	aggregate volume
P_{s}	shear force [N]
P_{u}	ultimate force [N]
r	radius [mm]
r,ξ,θ	cylindrical coordinate system
t	time [s]
u_{i}	(i = 1,2,3) displacements in the direction of the principal stresses [mm]
v_{p}	pore volume
V	volume
w	crack opening [μm]
w_{c}	maximum crack opening at end of softening diagram [μm]
W	specimen width [mm]
W	energy
W_{e}	elastic energy
x,y,z	cartesian coordinate system
x_{i}	(i = 1,...k) experimental variables
X	design matrix
y	response variable
Y	geometrical factor
$ß_{i}$	(i = 1,...k) model parameters
γ	surface energy
δ	displacement [mm]

ε	strain [-]
ε_i	(i = 1,2,3) principal strains [-]
ε_v	volumetric strain [-]
μ	coefficient of friction [-]
ν	Poisson's ratio [-]
ξ	(=a/W) relative crack length [-]
σ	nominal stress = load divided by original area [MPa]
σ_c	confining stress (triaxial cylinder tests)
σ_i	(i = 1,2,3) principal stresses [MPa]
σ_0	octahedral normal stress [MPa]
σ_0^2	random error
σ_{peak}	strength at peak of stress-strain diagram [MPa]
$\sigma_x, \sigma_y, \sigma_z$	stresses defined in the cartesian coordinate system x,y,z
τ	shear stress [MPa]
τ_0	octahedral shear stress [MPa]
φ	angle of internal friction

AE	Acoustic Emission
CH	Calciumhydroxide
C_3S	Tricalciumsilicate
C_2S	Dicalciumsilicate
C_3A	Tricalciumaluminate
C_4AF	Tetracalciumaluminoferite
CMOD	Crack Mouth Opening Displacement
CMSD	Crack Mouth Sliding Displacement
CSH	Calcium Silicate Hydrates
$CTOD_c$	Critical Crack Opening Displacement
DEN	Double-Edge-Notched
ESEM	Environmental Scanning Electron Microscope
fpz	fracture process zone
FRC	Fibre Reinforced Concrete
LEFM	Linear Elastic Fracture Mechanics
LVDT	Linear Variable Displacement Transducer
MFSL	Multi-Fractal Scaling Law (Carpinteri)
PBFC	Portland Blast Furnace Cement
PC	Portland Cement
PFAC	Portland Fly-ash Cement
PPC	Portland Pozzolana Cement
SEL	Size Effect Law (Bazant)
SEM	Scanning Electron Microscope
SEN	Single-Edge-Notched
SF	Silica Fume
SIFCON	Slurry Infiltrated Fibre CONcrete
w/c	Water-Cement ratio

THE AUTHOR

Jan G.M. van Mier earned an Engineering degree in 1978 and a Ph.D. in 1984 at Eindhoven University, The Netherlands. In 1985 he completed a Post-Doc at University of Colorado, Boulder. In 1986 he was a Research Engineer at TNO Building and Construction Research, Rijswijk, The Netherlands. From 1987 till present, he has been an Associate Professor at Delft University of Technology, Faculty of Civil Engineering, Stevin Laboratory. From 1995 till present, he has been a Research Coordinator at The Netherlands School for Advanced Studies in Construction.

Dr. van Mier is the author or co-author of over 120 papers in refereed journals and conference and workshop proceedings. He is the editor of two conference proceedings. He is also chairman and member of numerous scientific committees of conferences and workshops on fracture of concrete and other brittle disordered materials. He is a member of the Advisory Board of *HERON, Materials & Structures (RILEM)*, and *Mechanics of Cohesive Frictional Materials* and he is a reviewer for 10 international journals. He is an elected member of the Advisory Board of the International Association on Fracture Mechanics of Concrete and Concrete Structures (IA-FraMCoS). and he is a member and secretary of numerous national and international research committees on fracture (RILEM, ESIS, CEB and ACI/SEM). Dr. van Mier is the co-applicant of two patents and he is a recipient of the 1987 RILEM Robert l'Hermite Medal.

TABLE OF CONTENTS

PART 2. EXPERIMENTAL AND MODELLING TOOLS

4 EXPERIMENTAL TOOLS

Chapter 1

INTRODUCTION: SETTING THE STAGE

1.1 CRACKING IN CONCRETE AND CONCRETE STRUCTURES

Cracks are everywhere around us. Any brittle material shows cracks at some stage of the life cycle. Humans have always tried to understand the mechanisms behind the growth of cracks and, through the decades, in particular after the pioneering work of Griffith[1]* on the brittle growth of cracks, much research has been directed towards a better and deeper understanding of the observed phenomena. Much of the research has traditionally focused on brittle materials like cast iron and glass. Interest in fracture of ductile metals emerged after Dugdale[2] and Barenblatt[3] developed a plastic crack tip model. All such models, including the very recent efforts to come to a unified approach to describe fracture phenomena in brittle disordered materials (concrete, rock and ceramics) rely heavily on experimental observations. In a time when numerical simulation is the key word in many research proposals, the role of experimentation seems however more and more under pressure and largely neglected. Of course, it is possible to keep clean hands sitting behind a computer screen. But, do the materials behave as we think they should behave? Traditionally, the role of testing has been much larger in structural and materials engineering. The analytical tools available to the engineer did not allow full computation of the behaviour of structures, and many educated guesses (engineering judgement) were necessary before new structural shapes were tried, or larger spans were introduced. With the introduction of numerical methods in engineering in the 1970s, many people believed that "prediction" of structural and material behaviour would be a matter of time. The only criterion seemed that new faster computational facilities would become available. These developments have, in other areas of science, led to speculations that eventually consciousness (a topic that could not be handled using traditional scientific methods) can be explained from numerical analogies too, e.g. Dennett.[4]

As mentioned, cracks are everywhere. In normal reinforced concrete structures cracking must occur to allow for the transfer of tensile stresses from the concrete to the steel. Such cracks, however, impair the durability and life span of the material and structures. Examples where durability becomes a leading issue is with the development of large off-shore structures. Examples are artificial islands for the recovery of oil, large bridges spanning sea arms like the Store Bælt bridge in Denmark, the Eastern Scheldt Barrier which was constructed in The Netherlands to protect the low lands from flooding during severe autumn and winter tempests, or high-rise buildings (Figure 1.1). Struc-

---

* References are numbered in order of appearance, and are collected at the end of the book.

1

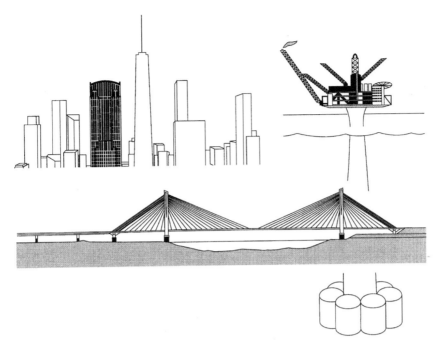

Figure 1.1 Examples of advanced, modern (reinforced) concrete structures.

tures in a marine environment are particularly susceptible to the intrusion of chlorides and sulphates from the sea-water. This may lead to corrosion of the prestressing or reinforcing steel, or to swelling of aggregates containing reactive silica. Not only is chemical attack to be considered important, but physical and mechanical loadings may cause cracking in building and civil engineering structures as well. Several strategies can be followed to try to avoid crack growth. One may try to avoid cracking completely, for example, by allowing moderate stress levels only. Needless to say, this will not lead to very practical or very economical structures. Other solutions aim at reducing the crack width by properly designing the reinforcement details, like diameter, concrete cover, etc. Again others try to solve the durability and structural problems by selecting non-corrosive reinforcement, but then other problems may have to be solved, for example the bond between the alternative reinforcement and the concrete. These are all issues that will not be dealt with in this book. Rather, we will try to come to a better and more profound understanding of crack nucleation and growth in concrete materials and structures. One of the most important issues in fracture mechanics is that the behaviour of the material and structure cannot be separated. Quite curiously, this strong interdependence between material response and structural behaviour seems to be grossly neglected. Moreover, the level of observation is very important, as will become clear in the book.

Upon considering basic fracture mechanisms of concrete under *mechanical loading* — which is one of the important limitations made in this book — one should address tensile, shear, compression and multiaxial states of stress, including any possible path dependency, rate effects and creep. Furthermore, one should address *chemico/mechanical* and *physico/mechanical* interactions. However, for a single person to comprehend all these loading cases, as well as the behaviour of structures and materials in sufficient detail, becomes more and more problematic as researchers tend to limit their activities to areas of limited scope. In the book we will aim at a much improved interaction and collaboration between different disciplines. In order to demonstrate that cooperation between disciplines is important in future research, many examples are given which show that an interaction between materials science, fracture testing and computational modelling is essential.

Interactions between disciplines, mechanical loading with a limited number of side-steps to physico/mechanical interactions, are dealt with in the book. In an effort to elucidate fracture processes in complex heterogeneous materials like concrete, the definition of the level at which the observations and models are made is of extreme importance. Therefore, in Section 1.3 we will clearly define the various levels of observation that will be used in the book. But before doing so, a very short and limited introduction of fracture criteria that have been introduced over the years, will be given in Section 1.2.

1.2 LIMIT THEORIES OF MATERIALS AND STRUCTURES

Structural engineering is generally based on strength-of-materials theories. It is assumed that failure occurs as soon as a limit strength is exceeded, and for a description of the material used in the structure, a simple determination of the strength of a material sample is considered to be sufficient. In Figure 1.2, this behaviour is indicated by means of an elastic-purely brittle stress-strain law. The determinations of the strength of the material under a wide variety of mechanical loads like uniaxial tension, uniaxial compression, and all kinds of multiaxial loads will suffice in this simplified view.

The strength could simply be determined by increasing the dead-load on a sample, and no sophisticated equipment is needed. For materials like rock, concrete and glass, in general, sudden failure was observed, but for many other materials like metals, it was found that larger deformations at a constant stress-level were needed. Ductile behaviour is observed, which is shown in Figure 1.2 as well. A perfect elastic-perfectly plastic stress-strain diagram is shown. Also, in the figure it is indicated that upon unloading not all deformation is recovered, but large irreversible deformations occur. New technical possibilities allowed for the measurement of deformations. Such measurements eventually demonstrated that in the beginning of the plastic plateau in the diagram of Figure 1.2, deformations are uniformly distributed in the sample, but later on

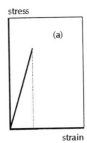
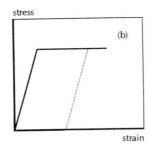
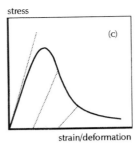

Figure 1.2 Basic behaviour laws for materials: (a) elastic-brittle, (b) elasto-plastic, and (c) softening.

they were localised in a very narrow zone of the test specimen. Moreover, it was also found that the linear elastic stress-strain behaviour up till reaching a peak stress-level, was more based on assumptions and preferred behaviour than on reality. Again, further developments of testing techniques, more important than anything else, led to the insight that some materials that were previously assumed to be brittle showed a distinct life after the peak stress level was reached. The stress-strain behaviour is shown in Figure 1.2c. Dedicated electronics nowadays allow for unrestricted testing in this so-called post-peak regime (or softening regime), and in the past two decades, much progress has been made in a further understanding of this life beyond peak.

Theoretically one can compute the ultimate strength of materials from fundamental interactions between atoms, or even from interactions between sub-atomic particles. It was the great achievement of Griffith to demonstrate, and to indicate ways to tackle the problem theoretically, that the ideal strength of materials can never be reached because stress concentrations due to impurities will cause the material to fail at a substantial lower level of stress. Later, to be more specific in the 1970s, this led to the formulation of a dedicated model for concrete fracture by Hillerborg et al.[5]

In all observations that are fundamental to mechanical behaviour laws (constitutive models) for materials, the material is considered as a black box. It is assumed that the material behaves as a continuum, i.e., as a material that has the same properties in every point. The fact that different structural features can be found in a material at different levels of observation leads to the idea that much of the behaviour observed at one level can be explained in terms of the material structure (and changes therein) at a lower level. In fact, one example was already given. Reinforced concrete can be considered as a continuum material with no internal structure, as for example proposed by Vecchio and Collins.[6] However, more widespread is the approach where reinforced concrete is regarded as a composite of the two basic materials steel and concrete. In the next section we will further explore the effect of the level of observation.

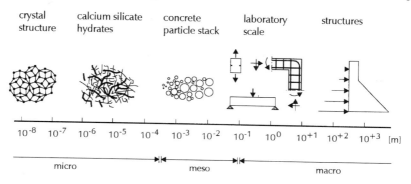

Figure 1.3 From atomic structure to large-scale building and civil engineering structures, exemplifying the various scales of observation that must be considered when studying materials and structures.

One last remark should be made. In all "classical" models, a limit strength or stress or even a critical stress intensity factor is proposed as a fundamental criterion leading to macroscopic failure. In the last decades, energy methods have been proposed as well. It should be mentioned here that strength, stress and energy are man-made quantities. In reality, a material or structure can experience deformations only.

1.3 DIFFERENT LEVELS OF OBSERVATION

Most of the classical limit models are based on the assumption that the material can be represented as a continuum. The behaviour is described independent of the structure of the material. The only geometrical ingredients are formed by the size and shape of the structure that was considered. For any material it is possible to construct a diagram as done in Figure 1.3 for concrete. Concrete is a multi-scale material. Cement, sand, aggregates, water, and sometimes a number of other additives are mixed together and harden to form a solid material. The smallest structural feature in Figure 1.3 is the atomic structure of the cement and aggregates. We are observing the material at the nanometer scale (10^{-9} m). At the micrometer scale (10^{-6} m) the individual cement grains are distinguished. Before hardening, the unhydrated cement grains can be observed; in the hardened state calcium silicate hydrates and calcium-hydroxide are visible. The complex pore structure becomes visible as well. Upon increasing the scale to 10^{-3} m, individual sand and aggregate particles can be distinguished. Larger pores can be found as well, and the interaction between the aggregate particles and the cement matrix is one of the essential features at this level of observation. At 10^{+1} m, we have reached the scale at which laboratory scale (mechanical) experiments on concrete are carried out. No internal structure is recognised at this level, and the material is assumed to have identical properties in each point of the specimen or structure. At even larger

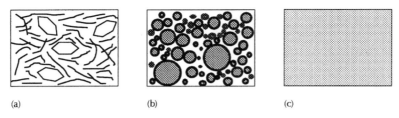

Figure 1.4 Definition of (a) micro-, (b) meso-, and (c) macro-level for cement and concrete.

scales, i.e., up to 10^{+2} to 10^{+3} m, we reach the scale of building and civil engineering structures. As mentioned before, the structures can be considered at the macro-level where reinforced concrete is considered as a continuum material, or alternatively, one might recognize some internal structure, more specifically the structure formed by the steel reinforcement. All these levels are represented in Figure 1.3. In engineering and science of concrete materials and structures, research has focused mainly on three distinct levels, namely the micro-, meso- and macro-level. By adopting familiar ideas from materials science, Wittmann[7] was the first to recognise that this so-called three-level approach could be a useful guide-line for research on concrete materials and structures. The three levels of observation are shown again in Figure 1.4, distinguishing the structural features that are important at each level. At the micro-level, the internal structure of cement and hardened cement paste is the most important structural feature. At the meso-level, the particle structure is most important. The heterogeneous nature of the material is the reason that local stress-concentrations appear, much in line with the earlier observation of Griffith that pores and impurities in materials set a limit to the strength of the material. At the macro-level, which is essentially the level at which structural engineers are working, no internal material structure is recognised, except for reinforced concrete where the reinforcement structure is normally taken into account.

In the three-level approach, it is generally assumed that behaviour at one level can be explained in terms of the material structure observed at a lower level. For example, at the micro-level, knowledge about the structure of calcium-silicate hydrates, the pore-structure of the hardened cement and the interaction between the cement and water (in different forms) helps to explain creep and shrinkage phenomena in concrete. Moreover, knowledge of and insight into the mechanical behaviour of the interfacial transition zone between the aggregate and matrix, which plays an important role when the material is observed at the meso-level, can be obtained by considering in detail the structure of hardened cement paste. At the meso-level, studying interactions between matrix and aggregates, as well as mutual interactions between aggregates, helps to better understand the mechanical behaviour of cement and concrete. In addition, large pores should be included in the analysis as well. In this book, fracturing of cement and concrete will be described both at the meso- and the macro-level. To avoid curious discussions, which are often the result from misinterpretation of the level of observation, the scale at which certain behaviour is described is always carefully defined throughout the book. Quite helpful has been an approach where alternatingly the mesoscopic and macroscopic behaviour are described and compared. The insight gained from such comparisons proves to be very helpful in coming to practical applications.

1.4 EXPERIMENTS AND (NUMERICAL) SIMULATIONS

As mentioned before, the interaction between different disciplines is considered to be very important for progress in materials science and structural engineering. Basically, three types of tools can be distinguished when research is carried out. In the first place, much of the knowledge available today is based on experimental observations. Development of new, advanced experimental techniques (loading techniques, test-control systems, deformation-measurement techniques and crack-detection techniques) is important for progress in the field. One drawback of experimental fracture research is that the fracturing of the material is highly affected by the structural environment in which it is studied. Thus, the idea of pure tests, leading to a direct measurement of, for example, the tensile strength of concrete, must be abandoned. This is not easy, in particular for those educated in a system where it was learned that mechanical properties can be measured unbiased and direct from simple experiments. In fracture research, this is certainly not true, as will be explored in depth in the book.

The second and third tools are analytical and numerical modelling. Based on experimental observations, material behaviour models and structural engineering models are constructed. The models can be either analytical or numerical. The analytical models lead to correct answers within the framework of axioms underlying the mathematics. Numerical tools can give an approximate behaviour at best. However, the great advantage of numerical analysis tools is that boundary and size-effects can be taken into account quite realistically. In particular, for fracture experiments a numerical approach seems essential because of the important interactions between crack growth and the structural environment. For many cases analytic solutions cannot be found, and a numerical approximation seems the best that can be achieved. Thus, the role of modelling should be considered in a different way than was traditionally done. Also the role of experiment is not the same. Basically, a very deep interaction is needed to come to a realistic description of the fracture process in cement composites. The modelling should not be restricted to a single level of observation. Instead, as shown in Figure 1.5, experimental research can be carried out at all levels of observation. Of course the phenomena studied at a certain level are different from properties studied at another level. Let us give two examples that clarify the deep interaction between experiments and numerical analysis.

First, let us consider structural analysis. In Figure 1.6 a basic scheme is shown that forms the basis for experimental/numerical studies at the structural level. Structural behaviour of reinforced concrete structures is studied in the

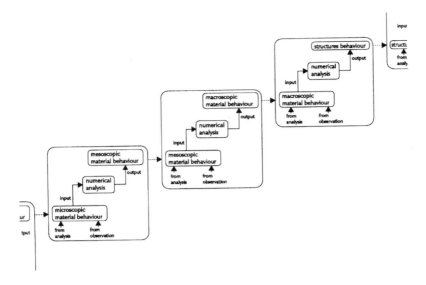

Figure 1.5 Different levels of combined experimental/numerical research of materials and structures.

laboratory by means of full-scale experiments. However, carrying out full-scale experiments becomes extremely difficult, if not impossible, if the size of the structure under consideration exceeds 10 m. Loading equipment, allowable floor-loadings and other limits, such as the capacity of a simple crane in the laboratory, have a large effect on the size of structures that can be studied. Of course for very large scale structures sometimes exceptions are made. Up till the 1960s and early 1970s, large scale testing was widespread. An example is a full scale study of pre-stressed and reinforced concrete beams with a span of 33 m for use in bridges for the Rotterdam Metro.[8] Nowadays, the behaviour of such structures can also be computed by means of numerical models. Basic properties of plain concrete, crack growth in concrete under different loading regimes, properties of the reinforcing steel and the bond-slip relations for the interaction between the steel and concrete must be determined in experiments. When all these "basic material properties" are incorporated into a finite element model, realistic structural behaviour can be computed and compared with experimental findings. Thus, the role of experimentation is twofold in the approach of Figure 1.6. First of all, experiments are needed for determining the properties of the materials, and second, experiments are needed to validate the finite element computation. Plain concrete behaviour and crack growth mechanisms are essential information in the approach of Figure 1.6. This will be the focus of the book. Experimental determination of fracture properties of plain concrete is affected by boundary and size effects from the specimen itself. In other words, the fracturing of a specimen cannot be regarded independent of its structural environment. This is the reason for placing "material properties"

OUTPUT: STRUCTURAL BEHAVIOUR   INPUT: MACRO MATERIAL PROPERTIES

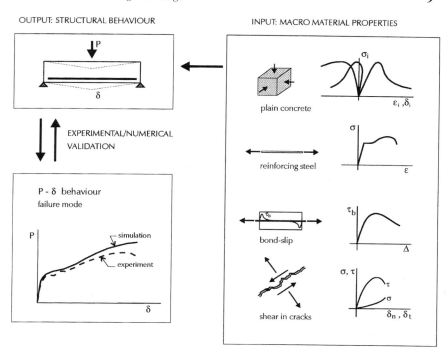

Figure 1.6 Experimental/numerical investigation of the mechanical behaviour of structures. Input parameters for the numerical model, and the behaviour of structures needed for validation of the model, are determined in experiments.

between quotation marks. When data are directly retrieved from experiments, they are a representation of the complete specimen-machine system. Therefore, it is easy to imagine that the fracture properties from material experiments in the spirit of Figure 1.6 can be analyzed by means of numerical tools as well. It is essential that boundary condition and size effects are incorporated in the analysis. Moreover, in such an approach, it becomes essential to generate different material characteristics. Figure 1.7 applies to the new situation. The macroscopic material experiment is analyzed by means of numerical tools at the meso-level. Boundary conditions in the experiment are mimicked in great detail. Moreover, information from the meso-structure of the material is incorporated into the model. Input data for the analysis include the strength and stiffness of the constituents of the material, as well as the properties of the interface between cement and aggregate. Such properties must be determined from other, lower-level, material properties experiments. Thus, again, the role of experiment is twofold. In the first place, experiments are needed to determine the new, lower-level, material properties. And second, macroscopic fracture experiments are needed to validate the outcome from the numerical simulations. It will be obvious that the sequence can be continued to increas-

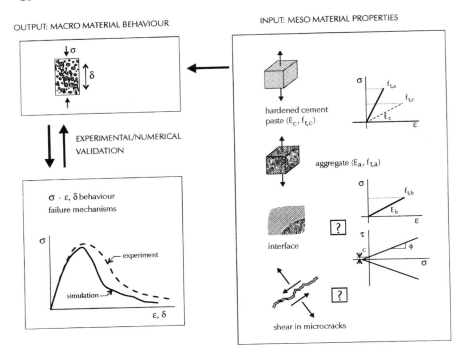

Figure 1.7 Experimental/numerical research in materials science. The input parameters for the numerical models, as well as the macroscopic material behaviour needed for validation of the model should be derived from experiments.

ingly lower structural levels, until we reach the smallest level presently reached in particle physics.

One important observation should be made. The role of the numerical tool has changed also when Figures 1.6 and 1.7 are combined. Numerical simulations are applied to further understand the behaviour observed in experiments, and this opens the road to a mutual beneficial state of symbiosis. The numerical simulations cannot be separated from experiments and vice-versa. The final outcome will be a reliable numerical tool for structural engineering, but also for materials engineering.

The last remark deserves a little more attention. The behaviour of materials and structures is closely related. The meso-level numerical tools needed for the analyses following the approach of Figure 1.7 can be used as a tool — in combination with experiments — for designing new materials for structural applications. In materials engineering, it is attempted to develop an integrated approach where materials for structural applications are designed in such a way that the material behaves most optimal in the structure for which it was developed. Thus, in engineering new materials for specific applications one might attempt to control the micro-/meso-structure of the material, as well as

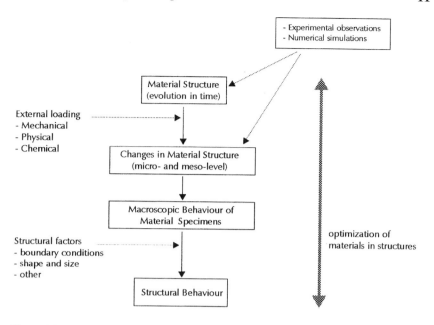

Figure 1.8 Optimizing materials for application in structures.

changes in the micro-/meso-structure, to obtain the best possible behaviour of the material in a structure. The process is elucidated in Figure 1.8.

1.5 ORGANISATION OF THE BOOK

Let us now explain the organisation of the book. From the above, it may be clear that in an integrated approach it is not always possible to make a strict hierarchical scheme for the book. We are dealing with material properties at different levels of observation, with experimental and numerical tools that should be explained to at least some degree and, of course, all the material should then be used in an integrated manner.

The book is separated into three main parts; Part 1: Structure and Mechanical Behaviour, Part 2: Experimental and Modelling Tools, and Part 3: Synthesis. Part 1 is comprised of Chapters 2 and 3. In Chapter 2, the hydration of portland cement, the pore structure of hardened cement paste, and the structure of hardened concrete are discussed. Because the book is mainly about fracture mechanisms, the part on cement chemistry is very limited. The main purpose of presenting this material is because the structure of cement largely determines the properties of the interfacial transition zone. For the meso-level modelling that will be presented later on in Part 3, the properties of the interfacial transition zone are of prime importance. The determination of the properties of the interfacial transition zone is not straightforward. As a matter of fact, fracture properties of the interfacial transition zone depend to a large extent on the scale of observation and the type of model adopted. In Chapter 2, the discussion is taking place at the micro- and meso-levels.

In Chapter 3, the mechanical properties of concrete are discussed, again with emphasis on the fracture properties. Basically, a conventional overview of mechanical properties as it is usually found in classical texts on concrete technology and properties is given. At places, however, some of the ideas that are worked out in the book are surfacing already. Many of the results are placed in the experimental context, which seems the only way of properly explaining them. The discussion in Chapter 3 is at the meso- and macro-levels. Mechanical behaviour of concrete subjected to uniaxial compression (Section 3.2), uniaxial tension (mode I in classical fracture mechanics terminology, in Section 3.3), shear (Section 3.4) and multiaxial states of stress (Section 3.5) is presented at both levels. Phenomena observed at the macro-level, at least qualitatively, in terms of structural features appearing at the meso-level is explained. More detailed description of the meso-level mechanisms will follow in Part 3.

Part 2 focuses on the experimental and modelling tools that can be used to obtain the information presented in Chapters 2 and 3. Moreover, the modelling tools are needed for Part 3. Part 2 is comprised of Chapter 4 on experimental tools and Chapter 5 on modelling tools. The tool kit is very important in fracture mechanics. It is not permitted to study fracture properties independent from the experimental environment. Moreover, the definition of the level of observation and the type of model used are of importance, as they indicate how the obtained results should be interpreted. So, here we break with the classical notion that material properties can be defined independent of their structural context. The consequence of this is rather shocking. It means that properties determined under some structural conditions cannot be used unrestrictedly under different structural conditions. In Chapter 4, methods for performing stable fracture experiments and important influence factors will be discussed. The goal is to isolate the real properties of the material from the experimental environment. It will be shown that this can be done by combining different experimental techniques, but also by applying the different types of fracture models that are presented in Chapter 5. The models are the final goal of many researchers, as they should be used for practical applications. In this book, however, they are merely a tool for explaining experimental observations. On the other hand, if a model is capable of describing the outcome of laboratory experiments, it should also be capable of explaining practical situations. The key issue is to obtain a sound knowledge of the conditions under which cracks nucleate and propagate.

In Part 3, numerical models are applied for the analysis of laboratory scale experiments. Models are used both at the meso- and macro-level, and depending on the level of observation more or less detail is included in the model, and the type of fracture model may change. The definition of the level of observation is crucial to come to proper interpretation of fracture processes. Part 3 contains three chapters. In Chapter 6, first inverse modelling tools for the determination of model parameters will be given. Fracture properties of concrete cannot be determined directly, but rather inverse modelling from close comparison of simulations and experimental findings seems the only way to define parameters for numerical models. This is of course a weakness of numerical models. It will never be possible to exactly "predict" structural behaviour outside the range of experimental observations. This is true both for the meso-level and macro-level models. It seems questionable whether fracturing can ever be described in terms of first principles starting from a lower level of observation as done in the present book. This means that applications will only start to emerge when the theories and attitudes needed for applying fracture mechanics have become sufficiently widespread. The attitude is particularly important. It is difficult to learn fracture mechanics from textbooks. Although in this book I tried to sketch how to apply fracture mechanics, a certain critical attitude is needed, which cannot be learned from books only. The hands-on method, both in experimentation and simulation is essential to appreciate the value of the new tools. However, one warning should be made: never believe your result until it has been confirmed by different approaches.

In Chapter 7, examples of simulations at laboratory scale are presented. Attention is given to tensile (mode I) fracture, macroscopic shear fracture, and compressive failure. Results from meso- and macro-level simulations are shown and compared to experimental findings presented in earlier chapters. One of the examples that returns repeatedly throughout the book is the non-uniform opening in uniaxial tension. Due to the heterogeneous nature of concrete it is difficult (if not impossible) to maintain uniform deformation distributions around the circumference of a tensile specimen. Rather, the nature of the material will force a crack to grow from one side of the specimen and propagate through the specimens cross-section. Depending on the specific boundary conditions adopted in the experiments, more than one crack zone may develop. This problem has been analyzed by means of all the different models, namely linear elastic fracture mechanics, non-linear fracture mechanics, experimentation, and numerical modelling both at the meso- and macro-level. In this way, the example becomes instrumental in showing differences between the various approaches. Also, the example helps us to appreciate the necessary interaction between experiment and (numerical) model.

Finally, in order to reach out to practice, two examples of structural details are discussed in Chapter 8. The first example shows the bond between steel and concrete; the second example focuses on the pull-out of steel anchors from concrete. Both problems can be solved by means of fracture mechanics. Next to these two numerical examples, the optimisation of materials in structures on the basis of so-called brittleness numbers will be discussed. The role of fracture mechanics is however limited. It should be regarded as a rational tool for developing new design rules in model codes. It is hoped that the present book gives insight, not only to freshman in the field, but also to practical engineers, who will not apply the new models directly in practice, but who will be confronted with the new theories as they spread out over the engineering community.

PART 1

STRUCTURE AND MECHANICAL BEHAVIOUR

Chapter 2

STRUCTURE OF CEMENT AND CONCRETE

Hydraulic cement, in the form as it is used today, has been known under the name of portland cement since 1824 when it was first used by Joseph Aspdin in England. Cement has an extremely complex microstructure, which is still not completely understood. In order to describe, for example, creep and shrinkage phenomena in cement and concrete a detailed knowledge of the cement structure is essential (see for example in Xi and Jennings[9]). For the purpose of this book, however, we can remain at a slightly higher level, although, when it comes to the interface between cement and aggregate, the structure of the cement becomes increasingly more important. In this chapter, a necessarily limited description of the structure of cement and concrete is given. Rather limited is the paragraph on hydraulic cements, somewhat more elaborate is the section on hardened concrete. In several chapters of this book, concrete is modelled at the "meso-level" (10^{-4} to 10^{-2} m) which means that the aggregative structure of the material as well as the larger pores or air voids are explicitly modeled. For more detailed accounts, the reader is referred to text books on cement and concrete.[10-14]

2.1 STRUCTURE OF CEMENT

Cement is the binding agent in concrete. Normally about 250 to 350 kg is added to 1 m^3 of concrete, which is sufficient to bind all aggregates together to form a solid material. As mentioned, cement as it is currently used has been known since 1824 when it was patented by Joseph Aspdin. Aspdin produced hydraulic cement by calcining lime containing argillaceous (clayey) impurities in a kiln. It was found that the burnt material hardened when water was added. For the hardened cement he proposed the name portland, because of the similarity of the hardened cement to a natural stone that was quarried near Portland in Dorset. It carries too far to give a detailed account of the history of cement here. An excellent survey can be found elsewhere.[15] Instead we will limit ourselves to a description of modern cements, including some of the modified cements, in which ordinary portland cement is mixed with other (sometimes latent) hydraulic materials. In a book on concrete technology, the complete structure of cement and concrete would have to be described, i.e., from the structure of the anhydrous cement particles and other reactive additives to the process of hydration and the structure of the hardened paste. For the purpose of this book it suffices to know the structure of the hardened cement paste, more specifically how it is contained within the structure of concrete where the cement binds together a complex heterogeneous conglomerate of aggregates. Therefore, in this section we will give only a short account of the

TABLE 2.1
Blended (or Modified) Cements on Basis of Portland Cement

Cement type	Portland clinker (%)	Other components (%)	Abbreviation
Portland cement	100		PC
Portland blast furnace cement	65–90 15–65	10–35 slag 35–85 slag	PBFC
Pozzolana cement	60–80	20–40 pozzolana	PPC
Portland fly-ash cement	75	25 fly-ash	PFAC

Note: Percentages are by weight.

raw materials of which portland cement and other modified cements are composed. The hydration process is shortly explained, followed by a description of the effect of w/c ratio on porosity and strength of the hardened cement paste.

2.1.1 RAW MATERIALS AND CLINKERS

As mentioned, portland cement, the most widely applied hydraulic cement, consists mainly of lime, but contains argillaceous impurities. Nowadays, the composition of the raw materials is carefully monitored, especially when high-quality cement must be produced. Next to the ordinary portland cement other so-called modified cements are used. These modified cements consist for the major part of portland cement clinker, but other (latently hydraulic) materials are added, such as blast furnace slag (a by-product from the production of iron), natural pozzolana's (e.g., volcanic ash, the word comes from the Italian "pozzuolano", after Pozzuoli near Naples) or fly ash (a rest product from coal-burning power plants). The blast furnace slag and fly ash must have the right composition. Both materials are residues from industrial processes, and in this way cement production helps to limit the amount of waste. In Table 2.1 several of the modified cements are gathered. The major composition is shown, and the abbreviation that will be used in the book is given as well.

Today, the name portland cement is used to describe a cement containing calcareous (lime) and argillaceous materials or other silica-, alumina- and iron-oxide bearing materials. Thus, the raw material basically consists of limestone ($CaCO_3$), silica (SiO_2), alumina (Al_2O_3) and iron oxide (Fe_2O_3), as well as some minor compounds (MgO, TiO_2, Mn_2O_3 and the alkali oxides K_2O and Na_2O). Of these minor compounds, especially the alkali oxides are of importance, as they may react with some types of aggregates. The reaction products have a larger volume than the original materials, and may cause severe damage in hardened concrete. This so-called alkali-aggregate reaction will not be discussed here; other sources contain detailed descriptions.[10,16]

TABLE 2.2
Portland Cement Clinkers

Name	Chemical composition	Abbreviation
Tricalcium silicate	$3CaO.SiO_2$	C_3S
Dicalcium silicate	$2CaO.SiO_2$	C_2S
Tricalcium aluminate	$3CaO.Al_2O_3$	C_3A
Tetracalcium aluminoferrite	$4CaO.Al_2O_3.Fe_2O_3$	C_4AF

After the materials are ground and mixed in the required proportions, the raw material is fed into a rotary kiln where it is burned. The product from the kiln is the portland cement clinker, which hardens as soon as water is added. In general the kiln is a long rotary tube, typically in the range of 100 m with a diameter of 6 m (in the case of the so-called "dry process", in contrast to the "wet process" that will be descibed later). The temperature increases gradually from the front where the raw materials are fed into the kiln to the end where sintering temperatures exist. At a temperature of approximately 1450°C, the materials sinter and recombine into balls (of size up to 25 mm) known as clinker. The clinker is cooled, and subsequently ground to a fine powder. At this stage some gypsum ($CaSO_4.2H_2O$, typically around 4%) is added to the cement clinker. The reason for adding the gypsum is that it retards the hydration. The processes taking place will be elucidated further in the next section. Today, in western countries, the dry burning process is used mostly. The main advantage of this process is that the amount of energy consumed is substantially lower than in the older "wet process". In this latter process, the raw materials were fed into the kiln as a slurry (typical size of the particles in the slurry is smaller than 90 μm). Obviously, the water contained in the slurry must be removed in the kiln during the clinkering process, of course at additional energy costs. For this purpose, longer rotary kilns are needed than in the dry process. More modern kilns have a pre-heating facility or pre-calciner before the material is fed into the rotary kiln, thereby even further reducing the amount of energy needed. In the pre-calciner, the $CaCO_3$ disintegrates before it is fed into the rotary kiln.

The type and amount of clinkers that develop in the kiln typically depends on the ratio between the raw materials and the temperature in the kiln. The reactions that take place are fairly complicated and are still not completely understood. It carries too far for the purpose of this book to explain all the details. Again, the interested reader is referred to more specialised books on the subject, for example Skalny.[12] The main clinkers that develop are given in Table 2.2, together with the chemical composition and the common abbreviation that is used by cement chemists. The size of the cement particles is in the order of 1 μm to 100 μm,[17] with a specific surface area (Blaine) of around 300 m^2/kg cement for ordinary portland cement.

In portland blast furnace cement and fly-ash cement, the portland clinker is mixed with slag particles and fly-ash, respectively. The fly-ash particles are

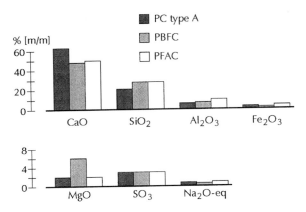

Figure 2.1 Chemical composition of three Dutch cements: Portland Cement type A (PC), Portland Blast Furnace Cement (PBFC) and Portland Fly-Ash Cement (PFAC).

small, glassy spheres with a size distribution that is similar to the anhydrous portland cement. The larger particles are hollow. Fly ash consists mainly of reactive silica glass containing several oxides as shown in Figure 2.1. Essentially the composition can be compared to that of ordinary portland cement. The same is true for blast furnace slag which is a by-product from the production of iron. The slag particles are latently hydraulic, but have almost the same composition as the portland cement clinker.

The chemical composition of a typical Dutch portland cement is given in Figure 2.1. This is only used as an illustration. The cement shown here is comparable to the German PZ35F and the American type I cement. Moreover, a comparison is made with cements containing blast furnace slag (65% PBFC) and fly-ash (25% PFAC). Clearly the major compound is CaO (63% m/m) in the example of Figure 2.1, followed by SiO_2 (21%), Al_2O_3 (6%) and Fe_2O_3 (3%). The two modified cements both have a reduced amount of CaO and an increased proportion of SiO_2. The result of these differences in composition is that both modified cements react at a slower rate with water than the ordinary portland cement.

It is quite common to show the chemical composition of cements as in Figure 2.1. However, for many purposes the breakdown of the cement composition into major compounds is preferred. The problem is, however, that the experimental methods to determine the compound composition are more elaborate than the simple analysis needed to obtain an overview of the chemical composition of cements as shown in Figure 2.1. A way to solve this problem is to compute from the oxide composition the major compounds (clinkers) by using, for example, the formulas derived by Bogue.[18] The numbers from such analyses are approximations only. For example, the compound composition of the portland cement of Figure 2.1 is shown in Figure 2.2. From the computation using the Bogue formulas it would be found that the amount of C_3S is equal to 52%, C_2S would be 21%, the C_3A content 11% and C_4AF 9%. For comparison,

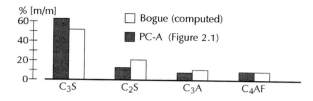

Figure 2.2 Compound composition of the Dutch Portland Cement, type A.

these numbers have been included in Figure 2.2 as well. Obviously there are deviations. The sum is not equal to 100% because some minor components are formed, whereas also some material is lost during the analysis.

After the production of the cement clinker, the clinker balls are ground to a fine powder. The size of the particles varies. In The Netherlands it is common to refer to the fineness of the cement through the addition of the letter A, B or C. The A-cements are rather coarse, whereas the C-cements are much finer. The fineness of the cement is characterised through the specific surface area (Blaine), which is equal to 300, 440 and 550 m^2/kg for portland cement type A, B, and C, respectively. The hardening of the cements proceeds faster when a finer cement is used, simply because more surface area is exposed to water.

The properties of portland cement depend to a large extent on the clinker composition, i.e., the relative abundance of the four major compounds. For example, a rapid hardening cement is obtained when the cement contains more C_3S and C_3A, whereas a low heat cement is obtained (for use in mass concrete structures) when both the C_3S and C_3A are reduced and the amount of C_2S is increased. In Table 2.3, the composition of three cements used in the U.S. are shown. Type I is an ordinary portland cement, Type III is a rapid hardening cement, and type IV is a low-heat cement. The above trends can be recognised. It should be mentioned that the blended (or modified) cements like PBFC and PFAC also produce less heat in the hardening phase because the fly-ash or blast furnace slag that is added hydrate much slower than the portland cement clinker. We will discuss these matters in somewhat more detail in the next section.

2.1.2 HYDRATION OF CEMENT

When water is added to the portland cement, several chemical reactions occur. The end product of these reactions is the hardened cement paste. It will be obvious that the composition of the final paste and the details of the chemical process depend to a large extent on the compound composition of the cement. The reaction with water is usually referred to as hydration process, and the final solid material is called hydration product. The structure of these hydration products, or calcium-silicate-

TABLE 2.3
Clinker Composition of Three
U.S. Cements[19]

	C_3S	C_2S	C_3A	C_4AF
Type I	49	25	12	8
Type III	56	15	12	8
Type IV	30	46	5	13

hydrates (abbreviated as CSH) explains much of the behaviour of hardened cement paste.

For the hydration to initiate and proceed, a certain amount of water is needed. The amount of water added is usually expressed as the water/cement ratio (m/m). The water/cement ratio is important as it affects the porosity of the cement paste, and thus, has a direct influence on the mechanical behaviour of the concrete. The amount of water has a major effect on creep and shrinkage mechanisms in cement and concrete as well.[9,20] Therefore, it is important to give a short description of the hydration process and the effect of the water content on the mechanical properties of hardened cement paste.

Another point of interest is the heat of hydration. The reactions between the cement compounds and water are exothermic, which means that heat is released during the reactions. Depending on the original compound composition, more or less heat is produced. Since this heat is produced in a relatively short time span (from a few minutes up to some hours or several days), it has some effect on the initial microstructure of the concrete, and thus may even affect the properties of the hardened concrete under mechanical load. Therefore, some information will also be included on the heat of hydration produced during the hardening of the concrete. Numerical models are developed at present,[21] in which the hardening process of cement is modelled in great detail. Others approach these matters by looking at the chemical reactions only, and by trying to mimic these reactions in a cellular automata type of computation.[22] We will return to this cement modelling as it promises to be an excellent tool for elucidating the structure of cement (and in concrete, in particular, phenomena happening at the matrix-aggregate interface), at least when the models are used in relation of direct observations. For this, the situation is not much different from the application of numerical models for studying fracture phenomena. Now let us proceed with the hydration process.

When the cement grains come in contact with water, the outer areas of the particles start to react. The reactions initially are limited to this outer surface. Gypsum plays a prominent role in this. The first reactions to occur are between water, C_3A and gypsum. They lead to the formation of a relatively hard shell of ettringite crystals at the surface of the cement grains.

$$C_3A + 3CaSO_4.2H_2O + 26H \rightarrow C_6A(SO_4)_3H_{32}$$

The ettringite layer is more or less impermeable to water. This means that the hydration stops, or at least is slowed down, as soon as this layer has formed. Only after one or two hours the shell is broken due to further chemical reactions in the shell, and due to pressure building up in the cement grains, which cause the hard shell to break. It should be mentioned here that if no gypsum was added to the cement, the initial hard ettringite shell would not develop. Instead, very rapidly tetra calcium aluminate hydrates (C_4AH_{13}) would form, which are not stable and which disintegrate after some time.

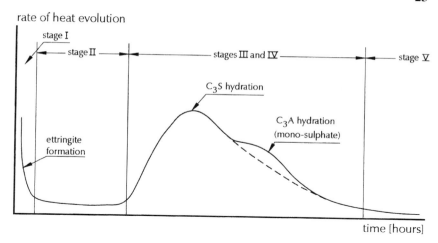

Figure 2.3 Rate of heat evolution during the hydration of C_3S and C_3A (schematic).

The hydration process can now proceed and the C_3S and C_2S hydrate to form calcium silicate hydrates. The hydration reactions are quite similar for both calcium silicates:

$$2C_3S + 6H \rightarrow C_3S_2H_3 + 3CH$$

$$2C_2S + 4H \rightarrow C_3S_2H_3 + CH$$

Thus, in both reactions calcium silicate hydrates (abbreviated as CSH) plus calcium hydroxide (CH) are formed. Due to the formation of CH, the alkalinity of the cement is affected. The pH of the resulting pore water rapidly increases to 12.5, thereby protecting reinforcement steel against corrosion. The afore-mentioned reactions of C_3A, C_3S and C_2S with water are all exothermic, i.e., heat is liberated during the reactions. Both for the calcium aluminate (C_3A) and for the calcium silicates, the heat evolution peaks at the beginning of the reaction, drops off, and eventually shows a second peak. The time lapse between the first and second peak, as shown in Figure 2.3 is called the dormant period. The dormant period typically lasts for 2 to 4 hours, during which the cement paste remains plastic and can be casted into moulds. In other words, the paste sets after some 2 to 4 hours.

The heat-evolution curve of C_3A and C_3S/C_2S shows five characteristic stages, which are described in detail by Mindess and Young,[10] and Gardner and Gaidis.[23] The latter authors call them: *initial reaction, induction period or dormant period, acceloratory period, deceleratory period* and *final slow reaction.* Mindess and Young refer to the last stage as *steady state.* The duration of all these stages can be read from Figure 2.4. The reactions are explained by several competing mechanisms and will not be discussed here; for further

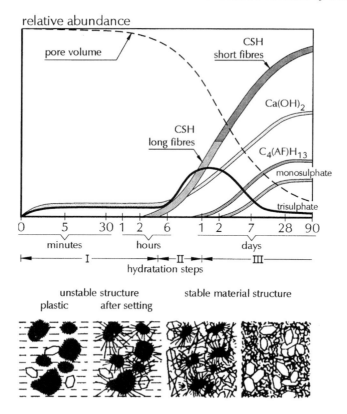

Figure 2.4 Hydration of portland cement, schematic after Locher et al.[24]

information see for example Van Breugel[21] and Gardner and Gaidis.[23] In the C_3A reactions, the second peak is attributed to the breakdown of the ettringite layer when all the sulphate from the gypsum has been used. The ettringite is transformed in so-called monosulfoaluminate (or sometimes simply monosulphate), and further hydration of the C_3A may proceed.

The hydration process has been visualised clearly by Locher et al.[24] Their figure is reproduced here (see Figure 2.4). Along the x-axis, the time is shown; along the y-axis, the amount of each product formed. The figure shows the rapid reaction of C_3A leading to ettringite, and after 2 hours the onset of the C_3S reactions. During these last reactions, needle-like structures develop from the cement particles. The needles are the calcium silicate hydrates, which are the final product. As mentioned, both the C_2S reaction and the C_3S reaction lead to CSH. However, a small difference exists, namely the CSH fibres produced from the C_3S reaction are longer than those from the C_2S reaction. This is visualised in the figures below the main graph of Figure 2.4. First, the longer fibres develop, and at a later stage the short fibres appear, filling up the open

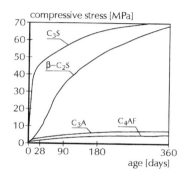

Figure 2.5 Strength development of the individual clinkers.[25]

space further. The first contacts between the cement grains are caused by the CSH from the C_3S hydration; subsequently the void space is reduced by the CSH from the C_2S. Throughout this fibrous cement gel structure, CH crystals appear, which have a hexagonal shape. The amount of CH can be considerable, up to 25% of the cement gel, but it appears that the CH has a low strength itself. Considering the strength of hardened cement paste it would be favourable to limit the amount of CH, even more so because the material can easily dissolve in water, which may eventually lead to an increase of porosity (and thus an even larger decrease of strength). In some of the blended cements, however, CH is an essential ingredient. For example, when fly-ash or blast furnace slag is added to the portland cement, the CH is needed to initiate the hydration of the fly-ash or slag, respectively. It should be mentioned that the hydration of blast furnace slag and fly-ash proceeds much slower than that of portland cement. On the average, blast furnace slag hydrates three times slower than PC, whereas fly-ash is about seven times slower than PC. Consequently, as we will see, the amount of heat liberated in PBFC and PFAC is much lower than in ordinary PC.

The hydration of C_4AF proceeds similar to that of C_3A, yet it is much slower and produces less heat. It has been shown that cement with a low C_3A content but with a high C_4AF proportion is more resistant against sulphate attack, which may for example occur in a marine environment.

It may be obvious from the foregoing that CSH contributes most to the strength of cement paste. Both the hydrates from C_2S and C_3S are essential for the formation of a strong material. In Figure 2.5 the strength development of the individual clinkers is shown. Clearly, the strength of the reaction products from C_3A and C_4AF is considerably lower than that of the C_2S and C_3S hydration products. Again, by selecting the ratio between the various clinkers in the cement, one may affect the strength characteristics of the material, as well as the increase of strength in time. However, not only is the composition of importance, but also the particle size and the amount of water available for the hydration are of major importance.

TABLE 2.4
Heat Production from Hydration of the Four Clinkers
in PC[11]

Clinkers	→	Hydrates	+	Heat (kJ)	[J/g]
$2C_3S + 6H$	→	$C_3S_2H_3 + 3CH$	+	230	500
$2C_2S + 4H$	→	$C_3S_2H_3 + CH$	+	86	250
$C_3A + 12H + CH$	→	C_4AH_{13}	+	362	1340[a]
$C_4AF + 9H + 4CH$	→	$C_4(AF)H_{13}$	+	204	400
$C + H$	→	CH	+	66	1170

[a]900 J/g for the formation of C_3AH_6

The amount of heat liberated from the hydration of the different clinkers is summarised in Table 2.4. Most heat is produced in the rather violent reaction between water and C_3A and C_3S. In particular, the first reaction leading to the formation of ettringite gives a rapid increase of heat production. The effect on the cement is marginal because the material is still plastic in the first few hours. More important is the effect of heat of hydration in the first few days of hardening. The strength of the cement has not developed to its full capacity, and due to differential temperatures (which may be very large indeed in mass concrete structures), tensile stress concentrations may appear, leading to early-age thermal cracking. An example is the growth of surface cracks in plain concrete breakwater elements. The effects can be minimised by selecting a cement that produces less heat, or alternatively by minimizing the temperature gradients in the structure either by insulating the outer surface so no heat can escape, or by cooling the interior part of the structure.

Depending on the relative abundance of the various clinkers, the cement will produce more or less heat. This was already pointed out in Table 2.3, where the composition of three different types of U.S. cements was shown. The type IV cement contains much less C_3A and C_3S, and therefore produces considerably less heat. After 7 days, about 220 J/g is produced for the type IV cement, which is considerably lower than the 360 J/g for the type I cement. After 365 days, these values are 355 and 430 J/g, respectively. Also, the blended cements containing either fly-ash, blast furnace slag or pozzolana's produce much less heat because part of the portland cement has been replaced by a much slower reacting material. In the Dutch portland cements type A, B and C, the heat production after 7 days of hardening varies between 320 and 380 J/g. For a type A portland blast furnace cement, the heat production is 190 J/g after 7 days of hardening at 20°C (isothermic measurement), and this material can be regarded as a truly low-heat cement. Type A portland fly-ash cement will produce about 310 J/g, which is higher than the PFBC but still lower compared to ordinary PC.

2.1.3 WATER-CEMENT RATIO AND POROSITY

For the hydration to proceed smoothly, a certain amount of water is needed. The amount of water used is commonly described by the water/cement ratio (abreviated to w/c ratio), which indicates the mass of the water added to the mass of the cement used. Full hydration of the cement is obtained when 25% of the weight of cement is added as water. This amount of water is chemically bound to the cement gel. However, quite a bit of water is also physically absorbed and as such is not available for hydration. The physically absorbed water mounts up to 15% of the cement weight. Thus, in order to hydrate all the cement, 40% of the cement weight has to be added as water, or in other words, a w/c ratio of 0.4 should theoretically lead to full hydration of the cement. In reality a w/c ratio of 0.4 does not guarantee full hydration because the water will not always reach the core of all cement particles. In that case, pockets of unhydrated cement remain in the hardened cement paste structure, which however are not affecting the strength of the material. The amount of hydrated cement is usually expressed by the degree of hydration. The w/c ratio used in practice normally deviates from the theoretical value of 0.4. In order to obtain a good workability (plasticity) of the fresh concrete mix, quite often higher values are used. However, when special additives like superplasticizers are added, the amount of water is reduced to values as low as 18 to 20%. Superplasticizers reduce the surface tension of the particles. However, in the context of this book we will not delve too deeply into these matters. The trend in low w/c ratio is in particular visible in the development of new very high strength concretes. The major effect of excess water on the structure of hardened cement paste is on the porosity. If more water is added, the surplus is not used in the chemical reactions and remains as free water in the cement structure to form capillary pores. This is shown schematically in Figure 2.6.

On the other hand, when the amount of water is decreased, i.e., when a w/c ratio smaller than 0.4 is selected, not enough water is available for all the cement to hydrate. The cores of the cement particles do not react. However, no free water remains in the cement structure, and the total porosity decreases substantially. In this context it is worthwhile to mention that due to chemical shrinkage, i.e., the volume of the reaction products is smaller than the total volume of water and solid cement particles, already an increase of porosity will occur. The porosity of cement paste is a very important factor. Both strength and durability of the cement are directly affected by the pore structure. Pores of different size are found in hardened cement paste. Very small pores (in the nanometre range) exist in the cement gel itself, whereas larger pores (of micrometre size) develop as capillary pores between the CSH particles. Even larger air voids may occur during mixing. Because of a poor compaction of the cement paste or concrete these larger air voids (of millimetre size) will become an integral part of the material structure. In particular the amount of larger pores (and air-voids) has a substantial effect on the mechanical properties of

(a)   w/c ratio < 0.4
       unhydrated cement
       particles remain

(b)   w/c ratio = 0.4
       all cement hydrated

(c)   w/c ratio > 0.4
       all cement hydrated
       large capillary pores remain

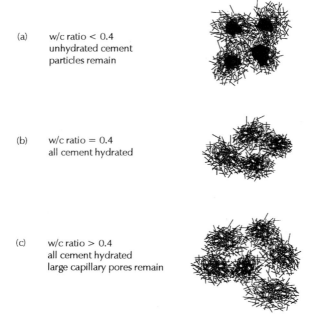

Figure 2.6 Effect of water/cement ratio on the structure of hardened cement.

cement and concrete. This is perhaps best illustrated in Figure 2.7. Here results
are shown of flexural tests on small notched beams made of ordinary portland
cement and so-called "Macro-Defect-Free" (MDF) cement. In MDF cement,
the amount of large pores and air-voids are minimized by optimally squeezing
and vibrating the material (Figure 2.7b). A comparison between the structure
of hardened portland cement and MDF cement is shown in Figure 2.7. The
larger black spots in the portland cement structure (Figure 2.7c) are the larger
air-voids and pores.

After hardening of the material, beams are prepared with different notch
depth. The notches are small saw cuts made in the middle bottom side of the
beams. Subsequently the flexural strength of the beams is determined and
plotted as a function of the notch depth, see Figure 2.7a. For the portland
cement, the effect of the notch cannot be recognised before the depth has been
increased up to 1.2 mm. In contrast, for the MDF cement, creating a 0.2 mm
deep notch already leads to an enormous drop in flexural strength as compared
to a beam with a 0.1 mm deep notch. Quite clearly, the MDF cement is much
more "notch-sensitive" than the hardened portland cement. This can be easily
explained from the presence of the larger air-voids in the PC structure. Stress-
concentrations around these air-voids, as will be explained in Chapter 5, lead
to a substantial decrease of the overall strength. One might say that the size of
the initial notches in the portland cement is directly determined by the size of

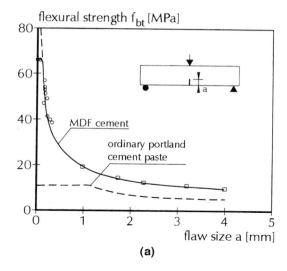

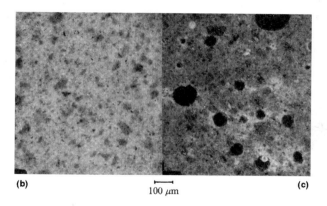

Figure 2.7 Effect of porosity on the flexural strength of ordinary portland cement compared to "Macro-Defect-Free" cement (MDF cement), after Birchall et al.[26] (From *Nature*, 1981, Macmillan Magazines, Ltd. With permission.)

the air-voids, and sets a limit to the flexural strength that can be achieved. By selecting a very balanced particle composition, one might reach a rather large flexural and compressive strength of concrete. For example, Richard and Cherezy[27] claim to have reached a flexural strength up to 140 MPa for a so-called Reactive Powder Concrete. Also Bache[28] reached very high flexural and compressive strength for a material called compact reinforced composite. In such materials, silica fume, a very small grained silica powder, is used to further increase the density of the concrete (see Section 2.2.1). Moreover, care should be taken to obtain a very balanced grading of all the particles, i.e., the cement, silica fume and sand and gravel, in order to obtain these exceptionally

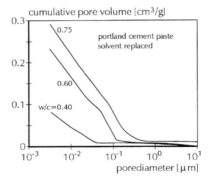

Figure 2.8 Effect of w/c ratio on the pore-size distribution and pore volume of hardened cement paste, after Hansen et al.[29]

dense material structures. The MDF cement used by Birchall et al.[26] could only be produced at a relatively small scale. Richard and Cherezy indicate that their Reactive Powder Concrete can be used for very large structural applications. In general the ultra high strength materials are extremely brittle. In some applications this disadvantage is overcome by adding fibres to the concrete.[27,28]

As mentioned, the pore-size distribution is directly dependent on the w/c-ratio used. In Figure 2.8, the effect of w/c ratio on the pore-size distribution and total pore volume of portland cement is shown. Clearly, the total porosity increases with increasing w/c ratio, and it seems that the increase is limited mainly to an increase of the amount of smaller pores. Porosity can be measured using different types of test methods, for example gas or fluid absorption. Pore-size distributions can be obtained using mercury intrusion under pressure. An interesting new development is the application of optical or SEM microscopy-based image analysis (see, for example, Diamond and Leeman[30] and Scrivener[17]). Image analysis techniques will be discussed in Section 4.4.3.

2.1.4 STRUCTURAL MODELS FOR HARDENED CEMENT PASTE

In Figure 2.9 several photographs of hardened cement paste are presented. The scale at which the hardened cement paste is shown here is in the order of 10^{-7} to 10^{-5} m. Thus, following the scale range of Figure 1.4, we are now looking at the material at the micro-scale. In Figure 2.9a the growth of ettringite is shown, in Figure 2.9b the structure of CSH is visible, whereas in Figures 2.9c and 2.9d a blast furnace slag particle and fly-ash particles covered by CSH are shown, respectively. From the last two figures it can clearly be deduced that the development of CSH from the portland cement clinker in these blended cements precedes the hydration of the slag and fly-ash particles. Due to the fact that specimens should be dried first before they can be examined in conventional Scanning Electron Microscopes, the free water cannot be shown. Many researchers believe that the structure of the material is damaged during the SEM preparation. Therefore, several of the structural models of hardened

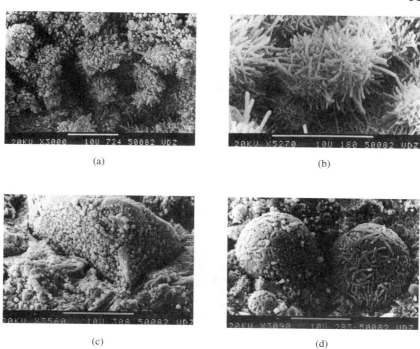

(a)

(b)

(c)

(d)

Figure 2.9 Electron microscopy images of (a) ettringite, (b) Calcium Silicate Hydrates (CSH), (c) a blast furnace slag particle embedded in CSH, and (d) fly-ash particles in CSH, after Wischers and Richartz.[31] (Reprinted with kind permission of the publisher, Beton-Verlag GmbH.)

cement paste have not been derived from direct observation only, but from other chemical and physical analysis techniques as well.

The most important product in hardened cement is the cement gel (CSH). Next to this, CH is quite abundant, whereas free water in capillary pores plays an important role as well. These three materials can be considered as the three most important components making up the structure of hardened cement paste. Of course, other components are present as well, for example residues of alumina's. Some of these minor constituents can be quite hazardous, such as these alumina's, which may react with sulphates from the environment to which the cement (or concrete) has been exposed. In that case, formation of ettringite is observed, which swells substantially, and which might impair the coherence in the hardened cement (or concrete) structure. As far as structural models for hardened cement are concerned, the interaction between water in the structure and the CSH needles seems of major importance. Several structural models have been proposed in the past, and the models by Powers,[32] Feldman and Sereda[33] and Wittmann[20] are best known. The three models are depicted in Figure 2.10.

In most of these models, the interaction between the solid CSH and water plays an eminent role. Two types of bonds are distinguished, namely primary

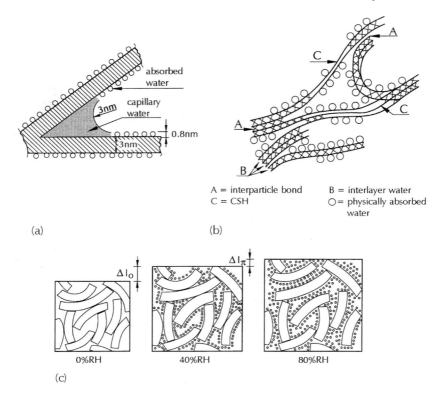

Figure 2.10 Structural models for hardened cement paste. (a) Powers' model,[32] (b) structure after Feldman and Sereda,[33] and (c) the Münchener model of Wittmann.[20] (Figure (a) and (b) are reprinted by kind permission of RILEM, Figure (c) by permission of Deutscher Ausschuss für Stahlbeton.)

bonds between rest-valencies of molecules at the CSH surface, and secondary bonds. The secondary bonds are VanderWaals forces, a physical attraction between solid surfaces. Depending on the distance of the solid surfaces the VanderWaals forces are stronger or weaker. In the cement structure, water plays an important role, as under wetting the mutual distance between the CSH particles may increase, thereby loosening the secondary bonds. According to the model developed by Wittmann,[20] i.e., the Münchener model, the primary bonds always remain intact, whereas under drying or wetting the secondary bonds are affected as described, see also Figure 2.10c. Important are the wedging forces in wedge-shaped slits between the CSH particles. In these wedges water is contained through capillary forces and adhesion. Powers suggested that at the tip of a wedge, disjoining pressure is active, which is counteracted by VanderWaals forces. In the Feldman and Sereda model (Figure 2.10b), water is absorbed at the surface of the CSH particles, but also appears as interlayer water in the CSH "needles". The various ways in which

Figure 2.11 Rolled-up plate of CSH.

water is bound in the hardened cement paste is depicted in Figure 2.10 using different symbols. The CSH "needles" are in this model rolled-up plates with small mutual distance as shown in Figure 2.11. It should be mentioned that creep and shrinkage can be modelled quite realistically using the Münchener model.

Further details of the structure of cement will be given in Section 2.2.2 where the structure and properties of the interfacial transition zone between aggregates and matrix is discussed.

2.2 STRUCTURE OF CONCRETE

Now that to some extent the structure and properties of the hardened cement paste have been described, we will focus on the structure of hardened concrete. Of interest here are the aggregates, specifically their properties and grading, as well as the interface between the aggregates and the cement paste. This interface is of prime importance if the mechanical behaviour of hardened concrete is modelled at the meso-level. For some time it has been understood that the interface is the weakest link in the structure of concrete.[34–36] Some of the physical processes during the hydration of the cement have a direct effect on the interface. These effects will be discussed in the forthcoming sections. The scale at which the concrete is studied is in the order of 10^{-4} to 10^{-2} m. An exception is made for the interfacial zone between aggregate and cement matrix, where the structure of the cement paste in direct contact with the aggregates must be considered at the micro-level.

2.2.1 AGGREGATES, TYPES AND PROPERTIES, PARTICLE DISTRIBUTIONS

Aggregates are normally about 75% of the total volume of concrete. Because of this large volume fraction much of the properties of concrete depend on the type of aggregate chosen. Most common is to use natural rocks as aggregates in concretes. Demands are made regarding the size distribution of particles that can be used. In addition to natural aggregates, alternative materials are used sometimes also, for example rest-products from a number of industrial processes. An example of using residues from other industrial pro-

TABLE 2.5
Examples of Natural Aggregates for Concrete

Material	c [kg/m³]	V_p [%]	f_c [MPa]	f_t [MPa]	E [GPa]	v [-]	α [10⁻⁶K]	G_f [N/m]
Basalt	2900	0.3–4	250–400	10–35	48–105	0.25	6–8	—
Diabase	3000	0.2–8	140–330	15–40	60–115	0.23	5–7	128.7
Granite	2700	0.4–5	100–270	5–18	35–70	0.18	8–12	124–159
Quartz	2600	0.4–3	100–325	9-21	50–75	0.19	—	162.8
Magnetite	4300	—	80–200	6–35	80–200	0.23	—	—
Limestone	2600	—	50–160	4–7	50–58	—	—	50.6–74.3

Values taken from Reinhardt,[11] except the fracture energy values G_f, and the properties of limestone, which are from Hassanzadeh.[42] f_t is splitting tensile strength; c = density; V_p = porosity, G_f = fracture energy according to RILEM.[67]

TABLE 2.6
Examples of Alternative Aggregates for Concrete

Aggregate type	c [kg/m³]	s [m/m]	f_p [MPa]
Lytag sintered fly-ash	1410	18	>5
Aardelite pelletised fly-ash	1430	25	>3
Argex expanded clay	1140	19	2
River gravel	2650	<1	90

c = dry density, s = water absorption after 24 hours immersion in water, and f_p is the particle strength.

cesses was already given in the previous section, namely blast furnace slag is mixed through the cement. This slag, when it has been properly cooled down, and also when it has the right chemical composition, has latently hydraulic properties. Blast furnace slag is sometimes also used as normal aggregates in concrete, but then less emphasis is on the chemical composition and the particles are larger of course. When blast furnace slag is used as aggregates the material structure can be different from the slag used in the cement. For cement, the slag must be cooled down very rapidly in order to obtain an amorphous slag structure. This demand is dropped when larger slag particles are used as aggregative material. In some countries, like for example in the U.K. and The Netherlands, lightweight by-products from industrial processes are sometimes used as aggregate material. Generally the mechanical and other properties of such aggregates deviate from those of natural aggregates as shown in Tables 2.5 and 2.6. The main difference is in the density, which is generally very low because of the high porosity of such materials. In general this means that when such aggregates are used in concrete, one must take care that the relevant corrections for the w/c ratio are made. For lytag, for example,

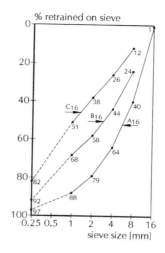

Figure 2.12 Example of practical grading for concrete, after the Dutch concrete codes.[37]

it is quite customary to soak the particles in water before using them in concrete. The different porosity of these lightweight materials has a significant effect on the flow of water at the cement-aggregate interface as will be discussed in the next section. Also quite important is the particle strength. For the lightweight materials the strength is lower because of the high porosity (see also the discussion in the previous section, specifically Figure 2.7).

For high strength concrete, selected strong aggregates are recommended, such as basalt or good quality lime stone. Selecting strong aggregates is not the only measure that must be taken. One should also strive towards a very balanced aggregate grading. The goal is to limit the porosity of the material as much as possible. Model codes usually contain detailed descriptions of aggregate gradings that must be used in practice. For example, in Figure 2.12 practical gradings from the Dutch model code are shown. The requirement is that the grading used in a given concrete is situated between the A and C curves, or alternatively when higher demands to the concrete are made, between the A and B curves.

The choice for a given particle grading not only affects the total porosity of the concrete, but in addition, the grading has a significant influence on the amount of water that must be mixed into the concrete for obtaining a certain workability. Workability is a measure for the plasticity of the concrete mixture in the first two hours after mixing the cement, aggregates and water together. More water will be absorbed to the surface of small sized aggregates. This means that for a fine grading (for example the C-curve in Figure 2.12), much more water must be added for obtaining a given workability than for a coarse grained aggregate mixture (for example the A-curve in Figure 2.12). Also the aggregate shape and texture have an important effect on the workability.

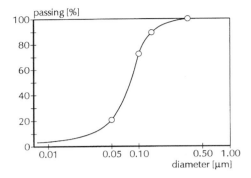

Figure 2.13 Example of particle distribution of silica fume.[38]

(a)   (b)   (c)

Figure 2.14 Effect of superplasticizer and silica fume on the density of cement paste: (a) cement without additives, (b) with superplasticizer, and (c) with silica fume.

Rounded, smooth aggregates will generally give an improved workability. The water demand from the aggregate grading used directly affects the amount of cement that must be used in the concrete mixture for a prescribed w/c ratio.

For the high strength concrete, however, it is not sufficient to limit the porosity originating from the particle grading, but also the fine porosity in the cement matrix should be reduced as much as possible. Using superplasticizer is one measure that can be taken; however, the addition of an ultra-fine filler like silica fume will further reduce the porosity of the hardened cement paste. The silica fume consists of very fine particles, as shown in Figure 2.13. These particles tend to fill the pore space in the cement paste, but an additional advantage is that the silica fume is latently hydraulic. Eventually the material also starts to hydrate, causing a very dense material structure with very good strength properties. The increased density of the cement structure is shown schematically in Figure 2.14 in three steps. The highest density is obtained when silica fume is added.

The development of high strength concretes receives some international attention these days. This is quite surprising because the principles of making strong materials have been known for quite some time, e.g., Kelly and MacMillan.[39]

The use of silica fume has different effects on the strength of the concrete. First of all, there is the small particle effect. Silica fume will reduce the pore-

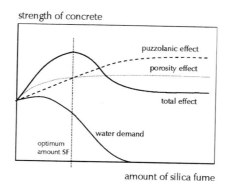

Figure 2.15 Effects of silica fume dosage on strength. The effect can be separated in a small particle effect, increased water absorption, and hydration effect (puzzolanic effect).

space, which has a positive effect on strength as shown schematically in Figure 2.15. Second, with increasing amount of silica fume, the amount of mixing water must increase because the specific surface of the silica fume is very high indeed (i.e., up to 20,000 m^2/kg (see for example in ACI[40]), compared to a Blaine of 300 to 600 m^2/kg for ordinary portland cement and up to 700 m^2/kg for fly-ash). The increased water demand tends to increase the w/c ratio, which has a negative effect on the strength of the concrete. Third, the hydraulic properties of the silica fume will have a positive effect on the strength, given sufficient time to hydrate. The three major effects are shown schematically in Figure 2.15. When the three effects are combined, an optimum amount of silica fume will be found. In practice, 10 to 20% of silica fume will normally be added as a slurry to obtain high strength concrete. The technology is still under development. An interesting side-effect is that silica-fume concrete, made with superplasticizer, has very good viscous properties. To obtain a dense concrete, it is generally not necessary to compact the material by vibrating because the material flows very easily into the moulds. For some applications, not only is the improved strength the criterium to select high strength concrete, but quite often the improved workability or the reduced permeability (which is important for durability of concrete) is the main reason.

One last remark should be made regarding the structure of concrete. Up till now it has been assumed that cement and aggregate particles are uniformly distributed throughout the concrete volume. However, this is not true. Concrete is generally cast in moulds, and wall effects appear near the sides of the moulds. Moreover, microstructural gradients appear at the casting surface where water is more free to evaporate from the concrete. Apart from that effect, the wall effect causes a rather different size distribution of aggregate particles in the parts of the concrete that are in contact with the mould. Many small particles will fill the space between the larger aggregates and the mould. Usually the skin is also rich in cement. This cannot be avoided. Kreijger,[41] has pointed at these matters, and has tried to measure the microstructural gradients in the skin of concrete. Because this part of the concrete is in contact with the environment it should perhaps receive more attention than it did to date.

We will not discuss further the structure of concrete here. Of main interest for the book is the interface between aggregate and matrix, as has been mentioned before. Therefore, in the remainder of this chapter attention will be given to the structure and properties of the interface.

2.2.2 INTERFACE BETWEEN AGGREGATES AND CEMENT MATRIX

The interface between cement and aggregate is normally regarded as the weakest link in concrete. For normal concrete this has been confirmed over and over, but for new concrete-like materials containing alternative aggregates, or improved cement matrices, the interface might have different properties too. In this section we will limit ourselves to the structure and properties of the aggregate-cement interface. Other interfaces appear in concrete as well, such as the interface between steel (fibres) and cement, the interface between unhydrated cement particles and the hydrated cement paste, or the interfacial transition zone between slag particles and cement. In Chapter 8 an example will be given of the pull-out of a steel bar from concrete, where the properties of the interface are of extreme importance. They are shown to have a major effect on the behaviour of such structural details. The strength of the interfacial transition zone between cement matrix and aggregate has a significant effect on the global strength of the concrete. However, as will be discussed in Chapter 7, the increase of strength of concrete is not as large as one would expect on the basis of pull-off tests of chunks of aggregate from cement paste.

Let us first consider the structure of the interfacial transition zone in normal strength concrete containing rounded river gravel. As shown in Table 2.5 and 2.6, the porosity of such aggregates is very low indeed, and one may assume that no moisture exchange will take place between the aggregate and the cement matrix in the hardening stage of the concrete. Similar as described in the previous section, wall effects appear where the cement is in contact with the aggregate. Severe microstructural gradients are present. The structure of the cement layer near the interface is shown schematically in Figure 2.16. The model is quite generally accepted, and is based on research carried out by, among others, Rehm et al.[35] Many overviews have been and are still being published (see, for example, Mindess[36] and Maso,[43] and more recently Diamond et al.[44]). Generally a layer of CH precipitates at the physical boundary between aggregate and cement matrix. Next follows a rather open layer containing oriented CH crystals, ettringite, and CSH. This so-called intermediate layer, or contact layer, or interfacial transition layer has a very high porosity as will be discussed later. The thickness of the porous transition zone is generally between 20 and 60 µm for concrete containing river gravel. As we will see, the structure of the interfacial transition zone depends to a large extent on the type of aggregate used in the concrete.

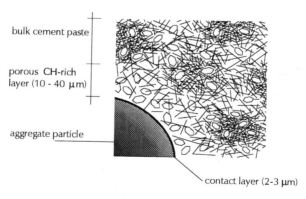

bulk cement paste

porous CH-rich
layer (10 - 40 μm)

aggregate particle

contact layer (2-3 μm)

Figure 2.16 Structure of the interfacial transition zone between rounded, dense, natural aggregate and portland cement matrix.

The bond strength between aggregate and matrix depends on a number of different mechanisms. Basically, mechanical interlock, physical attraction and/ or direct chemical bonding between cement and aggregate can occur. Mechanical interlock occurs, for example, between the different constituents in the cement structure (interlocking between CSH particles), but also at a larger scale due to the roughness of the aggregate surface which is in contact to the cement matrix. The exact mechanical behaviour of the interfacial transition zone is still poorly understood, although recent research points to a number of important phenomena that might influence the strength of the highly porous interfacial transition zone. Close observation of fracture surfaces generally shows that fracture occurs not directly at the physical boundary between aggregate and matrix, but rather slightly away from the interface in the porous transition zone.[45] The question is of course why this extreme porous zone can develop. Moreover, oriented large hexagonal CH crystals are quite abundant in the porous zone. This material is known to be very weak. A view of the CH crystals in the transition zone is given in Figure 2.17a, whereas in Figure 2.17b a view through the physical contact layer into the transition zone is shown.

Let us first consider the relatively large porosity at the interface. Scrivener[17] studied the structure of the interfacial transition zone using a Scanning Electron Microscope (SEM), in combination with image processing techniques. Microstructural gradients were measured by simply counting the number of elements along lines that were drawn parallel to the aggregate surface as shown in Figure 2.18b. The original image is shown in Figure 2.18a. Using image analysis techniques, the porosity in the interfacial zone is revealed (Figure 2.18c), and the counting procedure can be used to show the porosity gradient near the aggregate-cement interface. Scrivener and Gartner[46] used specimens containing isolated aggregate particles, and some of their results have been reproduced in Figure 2.19. In Figure 2.19a, the amount of anhydrous material is shown, whereas in Figure 2.19b the porosity gradients are presented. Results are

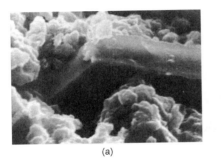
(a)

(b)

Figure 2.17 Hexagonal calcium hydroxide crystal in the porous interfacial transition zone (a), and view into the transition zone through the physical contact layer (b), after Zimbelmann.[45] (Reprinted with kind permission from Elsevier Science Limited.)

(a)
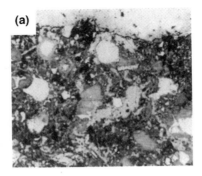

(b)
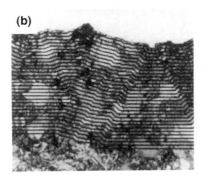

(c)
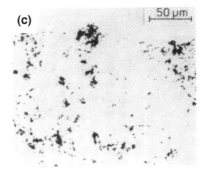

Figure 2.18 Measurement of microstructural gradients in the interfacial transition zone. (a) SEM image of aggregate and interfacial transition zone, (b) black lines parallel to the aggregate interface along which microstructural features are counted, and (c) porosity in the cement paste extracted from (a) through image analysis, after Scrivener.[17] (Reprinted by kind permission of the American Ceramic Society.)

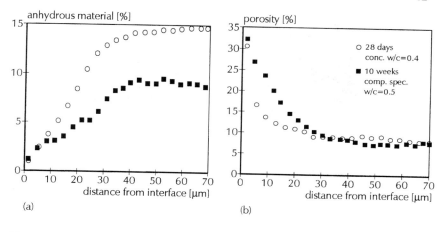

Figure 2.19 Microstructural gradients in a 10-week old composite system containing a single isolated aggregate particle, and in 28-day old concrete with a w/c ratio of 0.4. In (a) the distribution of anhydrous material is shown, in (b) the porosity distribution as a function of the distance to the interface, after Scrivener and Gartner.[46]

shown both for the composite specimen at an age of 10 weeks, and for a normal concrete with a water-cement ratio of 0.4 at an age of 28 days. Clearly there are a number of important details that must be considered in the interpretation of such images, but the general trend is an increased porosity close to the aggregate surface, and a reduced amount of hydrated cement. Thus, such measurements clearly reveal the open structure of the interfacial transition zone. Note that in the representation of the results, it is assumed that porosity is uniformly distributed around the circumference of the aggregate particle. Figure 2.18c reveals that the porosity is not evenly distributed, and moreover the size of the pores may vary significantly.

The increased porosity can perhaps be explained as follows. During mixing of the concrete, water is physically absorbed at the surface of the aggregate particles. This might result in an increase of the effective w/c ratio in the layer of cement in contact with the aggregates. The increased w/c ratio will lead to an increase of porosity as discussed in the previous section. Moreover, it is generally accepted that relatively large weak open areas may develop under the larger aggregates because of flow of bleed water in the early stages of hardening of concrete. The third reason might be the wall-effect that was explained quite clearly in a number of computer simulations of the hydration process by Garboczi and Bentz.[47] The hydration of cement was simulated using an idealized model shown in Figure 2.20. The big square in the centre is an idealized aggregate particle, the white circles represent cement, and water is shown as black. Using a cellular automata type of simulation technique, the cement and water reacted to form CSH, CH and pore space. Using the same counting technique for characterizing microstructural gradients as shown in Figure 2.18, the porosity gradients were computed for two different cases.

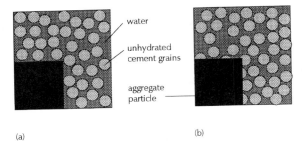

(a)   (b)

Figure 2.20 Idealized particle model to study structure development in the interfacial transition zone by means of a numerical model, after Garboczi and Bentz.[47]

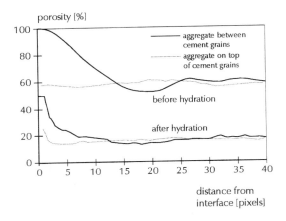

Figure 2.21 Porosity plotted against the distance from the interface. Results from the computation of the two situations shown in Figure 2.20. (a) The combined wall effect and one side growth mechanism of cement is simulated and (b) corresponding to Figure 2.20b, one side growth is simulated only, after Bentz et al.[48]

In the first situation, mixing of concrete was simulated, and the cement particles were placed at random locations around the large aggregate. Because of the wall effect, the cement particles could only touch the edge of the aggregate. This means that at the edge of the aggregate, in the beginning of the simulation, 100% porosity is measured (note that before hydration this "pore" space is filled with water of course). The second model is possible only in a numerical environment; it cannot be made in the laboratory. Here the cement particles were placed randomly over the entire image, and on top of which the aggregate particle was placed. This means that some cement particles intersect with the aggregate, but now the porosity is evenly distributed over the complete image and no wall effect exists. The initial porosities can be seen in Figures 2.21a and b for the two different starting positions described above. In the first case a relatively large porosity is found near the interface; in the second case,

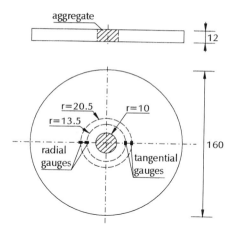

Figure 2.22 Single particle model for studying eigenstresses in concrete, after Torrenti et al.[50]

porosity is uniformly distributed. After hydration, the total porosity has decreased substantially in both situations. Only near the interface, a steep gradient remains, which extends over a larger distance in the first case as compared to the second situation, where the cement particles were allowed to overlap with the aggregate. The second simulation indicates that a significant "one-sided" growth mechanism of the hydration products occurs in the interfacial transition zone. In the first starting position, the wall-effect plays an important role as well, as cement particles can in reality never overlap with an aggregate. Distinguishing between the wall-effect and the one-sided growth mechanism can only be done in computer simulations. However, similar as with the fracture analyses that will be discussed in Chapters 6 through 8, uncertainties always remain regarding the model parameters. This means that simulations must *always* be accompanied by experiments.

The flow of water around the aggregate particles is very important, and perhaps the effect has been neglected by researchers too much. One of the important properties of hardened cement paste is that substantial hygral shrinkage may occur. At places where shrinkage is restrained, cracks may develop. The shrinkage of the cement paste around stiff aggregate particles may cause eigenstresses in the cement matrix, and if the situation is severe enough, interface cracking might occur. The situation was simulated in simple experiments carried out by Torrenti et al.[50] and Acker et al.[51] A single cylindrical aggregate particle was placed in a thin disk of cement paste. Two radial strain gauges and two tangential strain gauges were attached to the specimen immediately after casting. The disk specimen with the strain gauge locations is shown in Figure 2.22.

Specimens were stored in two different manners, i.e., either a specimen was covered by a coating to prevent drying, or a specimen was allowed to dry freely. In the second case, drying shrinkage will occur, but around the aggre-

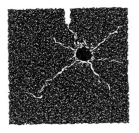 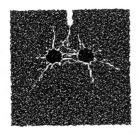 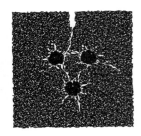

Figure 2.23 Simulation of shrinkage cracking in a numerical lattice model for fracture. Uniform shrinkage of the cement matrix around (a) one, (b) two and (c) three aggregates.[52]

gate this is restrained. In another context, the drying shrinkage mechanism is visualised in Figure 3.56, but the same principle applies here as well. After some time the specimens were split into two halves, and the strains were measured. It was found that under free shrinkage microstrains up to 260 were measured, indicating that substantial eigenstresses may develop. In the radial direction compressive strains were measured, whereas in the tangential direction tensile strains developed. Depending on the strength of the material, shrinkage cracking may appear. Recently, shrinkage cracking around cylindrical aggregates was simulated using the lattice model described in Section 5.2.1. In Figure 2.23 the simulated crack patterns are shown for the situation with one, two or three aggregate particles. Uniform shrinkage was assumed in the matrix only, and the crack patterns that were found are qualitatively in good agreement with experiments on simple composite systems.[53]

Often arguments are raised against the single particle geometries, e.g., Mindess[36] and Scrivener.[17] The particles are fixed into a mould before the cement matrix is cast. Therefore the conditions are different from those in an ordinary concrete where the aggregates and cement and water are mixed together. Also in real concrete, the interfacial transition zones might touch, and bulk matrix areas may be very small indeed, if not non-existent. Therefore, the interface regions might develop in different ways. For example, moisture movement might be completely different in composite systems and in real concrete. Nevertheless, single particle systems can be used because the geometry is more easily accessible to (numerical) computation. However, results from such experiments should always be interpreted in comparison to real concrete.[54]

Up till now, the structure of the interfacial transition zone was considered for concrete containing dense aggregates like many natural rocks. The situation at the interface may change drastically when porous aggregates are used, where exchange of water with the cement matrix may occur. When water from the cement matrix is sucked into dry porous aggregates, the interfacial transition zone tends to become more dense. Depending on the size of the pores of the aggregates, cement hydration may even occur in the skin of such aggregates. This means that in contrast to concrete with natural dense aggregates (but also

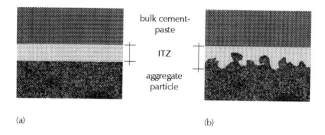

Figure 2.24 Differences in interfacial transition zone for concrete containing aggregates with a dense outer shell (a), and for concretes containing aggregates with porous outer surface (b).

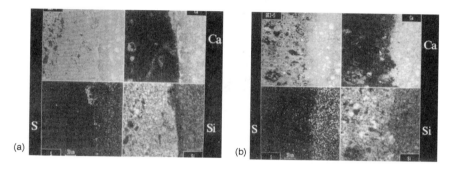

Figure 2.25 Differences in microstructure of the interfacial transition zone between aggregates with a hard, dense outer shell and cement matrix (a) and between a porous aggregate and cement paste (b), after Zhang and Gjørv.[55] (Reprinted with kind permission from Elsevier Science Limited.)

when lightweight aggregates are used with a dense impermeable skin), the interfacial transition zone becomes rather diffuse, and no sharp straight boundary between aggregate and matrix can be found. The difference is shown schematically in Figure 2.24. This difference in microstructure was first recognised by Zhang and Gjørv.[55] Using X-ray analysis, they measured the relative abundance of Ca, S and Si in the interfacial transition zone, and were thus able to distinguish between the microstructure of aggregate particles with a dense outer shell and porous, lightweight aggregates like Lytag. A comparison between the microstructure between Liapor 8 aggregate and cement paste, and between a Lytag aggregate particle and cement paste is shown in Figure 2.25. Quite pronounced is the boundary between paste and aggregate in Figure 2.25a, whereas the transition is much more diffuse in case of Lytag (Figure 2.25b). Zhang and Gjørv suggested that the bond might be improved in the case of aggregates with a porous outer shell because of the better mechanical interlock between the aggregate and the matrix.

The interfacial transition zone might also be densified by adding silica fume to concrete. Not only does the silica fume fill the larger pores in the hardened cement paste, but the silica fume reacts with CH, thereby limiting the amount of the weak CH crystals in the interfacial transition zone.[56] Consequently, a much denser interface zone is found, with much improved mechanical properties. Other authors also refer to this effect, and show SEM images that suggest a much denser interfacial transition zone.[36] Silica fume is used in high strength concrete technology, but even in high strength concrete quite often failure of the interfaces is observed.[57] It has been suggested that failure shifts from the porous interfacial transition layer to the contact layer, which is the next weak zone. In this respect it is perhaps important to realize that with increasing bond strength, the cracks tend to grow *through* the aggregate particles rather than *around* them. This has an important influence on fracture energy and brittleness of concrete. Indeed, normally it is accepted that both lightweight concrete and high strength concrete behave more brittle than normal gravel concrete. We will return to these matters in Sections 3.3 and 7.1. So, improvement of bond might lead to changes in other properties of the material that might have more severe consequences.

2.2.3 INTERFACE PROPERTIES: STRENGTH AND FRACTURE ENERGY

The determination of interface properties is not straightforward. As has become clear in the previous section, the structure is very small, and the behaviour seems very much affected by moisture transport in the cement matrix itself, but also by moisture transport between the aggregate and matrix. In the latter case the porosity of the aggregate plays an important role. Some insight in the properties of the interface between aggregate and cement matrix can be obtained from a sensible combination of macroscopic tests, microstructural observation and characterization, and numerical simulation (and/or analytical modelling). As has been mentioned before, insight in the interfacial transition zone can be obtained from idealized single-particle composites, or from real concrete. Testing can be done both on idealized composite systems or directly on concrete. In the first case, relatively large specimens are tested in tension or shear; in the second case, one might perform micro-indentation tests to determined the fracture toughness.[58] Modelling can be used in combination with the experiments. Essentially this is the role of modelling as it is used throughout this book (see Chapter 5 and further). The model is not necessarily perfect, but rather a tool to come to a better understanding of experiments. It is not our intention to give an exhaustive overview of all possible test methods and a critical comparison of test results in this section. Rather, we will limit ourselves to a global description of some of the macroscopic experiments carried out to date, including a very limited survey of some of the test results. Much work still has to be done in this area.

Aggregate-cement bond tests have been carried out for a long time. From the older tests, perhaps those by K.M. Alexander et al.[59] are most widely known. In general, in these early experiments, strength was the main parameter searched for, whereas in more recent research emphasis is shifting towards the

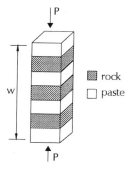

Figure 2.26 Composite compression prism for determination of the stiffness of the interfacial transition zone, as proposed by Alexander and Mindess.[62]

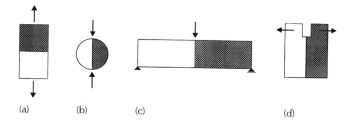

Figure 2.27 Different composite test geometries for measuring the tensile or flexural bond strength between aggregate and cement matrix: (a) uniaxial tensile test, (b) splitting tensile test, (c) three-point-bend test, and (d) wedge splitting test.

determination of fracture mechanics parameters like fracture energy and fracture toughness.[42, 60] M.G. Alexander[61] and M.G. Alexander and Mindess[62] proposed the use of two different geometries, namely the ISRM test method, i.e., the Chevron notched bend specimen (ISRM = International Society for Rock Mechanics, the specimen geometry is presented in Figure 4.80b), and composite compression cylinders. In the first experiment it is tried to measure the strength and fracture toughness; in the second test it is attempted to determine the material stiffness. The latter test is based on the series model for the elastic modulus of composite systems, see also Figure 3.2. A three-phase material is considered, i.e., a series coupling of bond-, matrix- and aggregate-phases. The proposed test specimen is shown in Figure 2.26. As far as the determination of the fracture strength and fracture energy is concerned, not only flexural tests can be performed, but also uniaxial tensile tests (e.g., Zimbelmann[45]), Brazilian tests (e.g., Hassanzadeh[42]) and wedge splitting tests (e.g., Tschegg et al.[63]).

A number of test geometries are summarised in Figure 2.27. In most of such tests half of the specimen consists of the matrix material, whereas the other part is the specific rock or other aggregative material under consideration. Some-

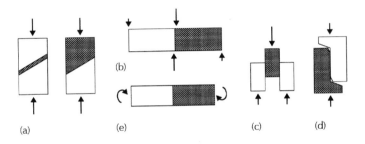

Figure 2.28 Different composite test geometries for the determination of the shear strength of the interfacial transition zone between aggregate and matrix: (a) specimen used by Taylor and Broms,[64] (b) four-point-shear beam, (c) push-through cube, (d) compact shear specimen and (e) cylinder subjected to torsion.

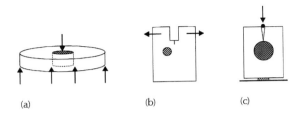

Figure 2.29 Examples of single-particle composite specimens for studying the mechanical properties and crack growth in the interfacial transition zone: (a) shear test designed by Mitsui et al.,[65] (b) single-particle geometry of Vervuurt et al.,[54] and (c) two-phase composite model adopted by Lee et al.[66]

times the aggregate interface is roughened prior to casting, but in other cases, smoothly ground interfaces are tested. Natural weathered rock is used sometimes as well.

Testing shear properties can also be done, for example by using bi-material specimens with the interface located exactly on the line where the shear force is applied, see Figure 2.28. Alternative tests are single-particle experiments for studying tensile and shear fracture of the interface zone, as shown in Figure 2.29. The objections to all such experiments have been raised in the previous section. The conditions during manufacturing are different from those during casting of ordinary concrete, which might affect the structure and properties of the interfacial transition zone. Remedy against this criticism is to perform both tests on simple composite systems and on real concrete made with the same type of aggregative material, as for example attempted by Vervuurt et al.[54] In the last example, next to testing, numerical simulations form an integral part of the investigation. The different experiments summarized in Figures 2.27 and 2.28 are quite complicated, as will be discussed in depth in Chapter 4 (and further worked out later in Chapters 5 and 7). In particular in fracture experiments, boundary and size effects have an important influence on the experimental outcome.

Now let us consider some of the test results. First we will look at the interface stiffness. Using the ISRM test method (Figure 4.80b), Alexander and Mindess[62] determined the stiffness from the initial portion of the load-deflection curves of the test. Results were obtained for ordinary portland cement paste OPC (ASTM Type I cement, w/c ratio 0.3, with 15% silica fume replacement in some of the cement pastes [which are designated OPC+SF]), three different types of rocks (andesite, dolomite and granite), and the various interfacial transition zones. The stiffness of the paste/rock interfaces were generally between the values for OPC and the type of rock used. For example, for OPC a Young's modulus of 24.1 GPa was measured, for andesite 99.7 GPa and for dolomite 112.6 GPa. The interfaces andesite/OPC and dolomite/OPC gave values of 45.6 and 46.8 GPa, respectively. When OPC+SF was used as matrix (with E = 24.4 GPa), the interface stiffness andesite/OPC+SF and dolomite/OPC+SF changed to 38.2 and 41.7 GPa, respectively, i.e., slightly lower than the values found for OPC. Results of the composite cylinder test (Figure 2.26) are scarce. Quite some experimental difficulties are encountered in this test, namely bleeding water under the aggregate discs will lead to an interfacial zone with different properties than the interfaces in ordinary concrete.

Strength values for the paste/aggregate bond are quite abundant. Results were systematically gathered by Alexander et al.[59] Composite specimens as shown in Figures 2.27 and 2.28 were used to determine the tensile and shear bond strengths. For example, the flexural strength of the specimen shown in Figure 2.27c yielded a lower value at 7 days of 4.62 MPa for reef quartz having a broken surface and 4.82 MPa for basalt with a sawn surface. The highest bond strength (flexural strength) at 7 days was measured for dacite, namely 10.4 MPa. The hardened cement paste was made of ordinary portland cement with a w/c ratio of 0.35. At 28 days, the flexural strength varied between 4.6 MPa and 6.9 MPa for reef quartz and basalt (both with broken surface), and up to 14.7 MPa for toscanite having a sawn surface. The treatment of the rock before the cement paste is casted has a significant effect on the test result. The use of natural weathered rock usually leads to a decrease of bond strength. Sawn, polished or broken surfaces may have different effects. Coarse grained rock with a broken surface before casting the cement paste may contain some microcracks in the interfacial transition zone, which might lead to a lower strength. On the other hand, the increase of mechanical interlock at the interface of a broken rock might lead to an increase of the bond strength. Polished aggregate surfaces may cause an increase of strength, in particular if the type of aggregate used tends to react with the cement. K.M. Alexander et al.[59] give the following flexural strengths for limestone/cement interfaces: 9.1 MPa for a laboratory produced smooth interface surface, 8.0 MPa for a natural water worn surface, and 7.4 MPa for a broken surface. Other examples were pub-

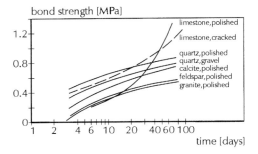

Figure 2.30 Tensile bond strength between hardened portland cement paste and different types of rock with broken or polished surfaces, after Zimbelmann.[45] (Reprinted with kind permission from Elsevier Science Limited.)

lished by Zimbelmann,[45] who determined the paste-aggregate bond strength from uniaxial tensile tests. An ordinary portland cement matrix (PC, w/c ratio = 0.35, f_t = 4.8 MPa at 28 days) or a portland blastfurnace cement matrix (PBFC, w/c ratio = 0.35, f_t = 5.0 MPa at 28 days) were cast against different types of rocks. The rock surfaces had different textures, i.e., broken or polished material was used. The bond strength was determined at different ages, viz. from 3 to 80 days. For all rocks a rather low tensile bond strength was measured as can be seen in Figure 2.30. Slight differences were found for different types of rock, but in most cases the bond strength did not exceed the 1 MPa. Obviously this is much lower than the flexural bond strength values reported by K.M. Alexander.[59] Note however that the conditions at failure are different in a uniaxial tensile test and in a three-point-bend test. Using non-linear fracture mechanics tools, such differences can be explained, but we will return to these matters later on. It should also be mentioned that the uniaxial tensile test is very difficult to perform. The measured peak strength depends on possible eccentricities (see Figure 5.36) and also the boundary conditions in the test (i.e., the rotational freedom of the loading platens) has a substantial effect (see Sections 4.1.3.2 and 7.2.2). From a careful analysis of the results of Figure 2.30, Zimbelmann showed that the bond strength depended on physical attraction between rock and cement paste and on mechanical interlock. Chemical reactions played a minor role only.

As mentioned, the current trend is to determine the interfacial fracture properties rather than the bond strength. A uniaxial tensile test or a three-point-bend test can be used to determine the fracture energy G_f as defined in the Fictitious Crack Model (see Sections 3.3 [Figure 3.32] and 4.6.1). Alternatively, one might follow an approach based on linear elastic fracture mechanics and try to determine the critical values of the stress intensity factor (see Section 5.3.2). Here we will only summarize some of the test data found in literature. Further comments on the underlying theoretical models and the validity of such models will be discussed in depth in Chapters 3 through 7.

Let us consider the fracture energy-based approach first. Hassanzadeh[42] and Tschegg et al.[63] both determined the fracture energy G_f of the interface between rock and cement paste. Hassanzadeh used cylinders with a length of 300 mm and diameter of 60 mm. Half of the cylinder was made of rock, and cement paste (with and without silica fume) was cast against the rock half. A simple straight notch with a length of 30 mm was sawn in the cylinder exactly at the location of the interfacial transition zone. Next to these displacement controlled three-point-bend cylinder tests, splitting tensile tests (using cylinders as shown in Figure 2.27b) were performed. These last specimens had a length of 120 mm and a diameter of 60 mm. Nine different types of rocks were studied, ranging from fine grained granite to gabbro, diabase and limestone. Not only the interfacial splitting tensile strength was measured, but also the fracture energy G_f of the interface and the different types of rock used according to the RILEM recommendation (RILEM,[67] see Section 4.6.1). A bond strength of 3.7 MPa was measured for all rocks and cements with silica fume but with different w/c ratio's, whereas the average decreased to 3.4 MPa when silica fume was not added. The differences between the nine rocks studied were not very prominent. It should be mentioned that the rock surface was roughened before casting the cement paste. The fracture energies were also quite similar for all types of rocks, except for the limestone which showed a slightly higher value. The mean values were $G_f = 26.9$ N/m for paste with silica fume, and 23.4 for paste without silica fume. For the different types of rocks the fracture energy G_f was measured as well. Some of the values are included in Table 2.5, which shows a variation between 50.6 N/m for limestone to 162.8 N/m for quartzite. Unfortunately, G_f was not measured for the different cement paste used in these tests. However, the weakness of the interface as compared to the rock is quite clear.

Fracture energies were also measured by Tschegg et al.,[63] but instead of the three-point-bend test a wedge splitting test was used (see Section 4.6.1 and Figure 2.27d and 4.77). A fracture energy of 15 N/m was measured for a sandstone/cement paste interface, and only 6 N/m for a limestone/paste interface. The fracture energy of the cement matrix (with a water-cement ratio of 0.60) itself was 80 N/m. Unfortunately, Tschegg et al. do not give any information on the properties of the rock that was used.

Second, instead of using the fracture energy approach, critical stress intensity factors can be measured using, for example, the ISRM chevron notched bend beam (Figure 4.80b).[60,61] The fracture toughness K_{CB} (Equation 4.17) for the cement paste with (OPC+SF) and without (OPC) silica fume was measured, $K_{CB} = 0.54$ MN/m$^{1.5}$ for OPC and 0.50 MN/m$^{1.5}$ for OPC+SF. The rocks gave the following values: 2.23 and 3.49 MN/m$^{1.5}$ for dolomite and andesite, respectively. For the interface fracture toughness values of 0.80 and 0.71 MN/m$^{1.5}$ were measured for andesite against OPC and OPC+SF, respectively, whereas for dolomite 0.57 and 0.64 MN/m$^{1.5}$ were measured for the OPC and OPC+SF systems, respectively. A comparison with test results obtained by others showed that these values are relatively high. M.G. Alexander and Mindess[62] explained these deviations from differences in rock surface prior to casting. The surface of the rock was roughened in the tests shown here, whereas, according to M.G. Alexander and Mindess, smooth surfaces were used in the other tests.

The above results are quite contradictory. Very likely the adopted test methods have a significant effect on the test result, quite similar to what will be discussed for macroscopic fracture tests on concrete in Chapters 4 through 7. The above results were all obtained from tests on composite systems, which many researchers regard as not representative of real concrete, where aggregates and matrix are mixed together. This criticism may be valid, although the use of simple geometries seems the only way to get at least some grip on the properties of the interfacial transition zone. A quite powerful tool available today is numerical simulation. Examples of the development of the cement structure around aggregate particles were given before (Figure 2.21). Using such models, the flow of water around porous aggregates and the resulting differences in interface structure can be computed as well (see for example Bentz et al.[48]). Such effects seem to have a large effect on the properties of the interfacial transition zone, and might force cracks to grow through aggregates rather than along the interface. This was shown in splitting tensile test on cement plates containing a single cylindrical aggregate of either dense granite or porous sandstone[54] (see also Figures 4.66 and 4.67). Looking at the fracture behaviour, composite systems containing a single cylindrical or rectangular aggregate may be quite interesting. Tests and analyses have been (and are currently being) carried out.[54, 66, 68] In such combined numerical (or analytical) and experimental studies it is important to get a better understanding of the properties of the interfacial transition zone. It should be mentioned that such combined studies are usually aimed at the determination of parameters needed in the model that is used to analyze the experiment.

Finally, it should be mentioned that because of the mismatch in Young's moduli of the aggregate and matrix materials, stress concentrations appear in the interfacial transition zone. Also, as was shown in Section 2.2.2, eigenstresses from differential shrinkage of aggregate and matrix may be the cause of a substantial number of micro-cracks in concrete at an early age, i.e., already before any mechanical load has been applied. Such effects are the cause of the extreme low tensile strength of concrete composites. By controlling the relevant factors, much stronger materials can be developed.

2.3 FINAL REMARKS

In this chapter, the structure of cement and concrete was discussed. From the material presented, it may have become clear that we are dealing with a complex, multi-scale material. For cement and silica fume, the material must be considered at the micro-scale (10^{-9} to 10^{-6} m), for concrete at the particle

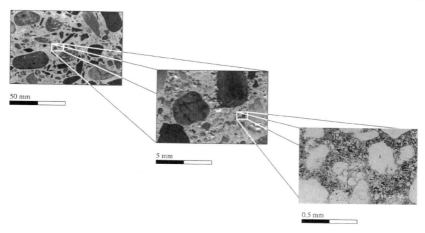

Figure 2.31 Concrete as a multi-scale material, as it will be regarded in this book. The photograph of the cement structure has been kindly provided by J. Larbi.

level, the characteristic size is between 10^{-2} and 10 mm, whereas at the macro-scale, no internal structure is recognised at all. For a detailed understanding of the properties of the interfacial transition zone, which is considered as a key factor in the behaviour of concrete, the properties and structure of hardened cement paste and moisture flow in the cement paste, but also the interaction with porous aggregate particles seems of great importance. In this book, we will not descend to the micro-level. Instead, we will remain at the meso-level to explain fracture behaviour of concrete composites as affected by the material structure. Once more it is emphasized that at the meso-level, we are dealing with a complex multi-scale material as perhaps is best illustrated with Figure 2.31. At this level, the interfacial transition zone will be considered as a porous weak continuous layer between aggregates and matrix. Also, the matrix and aggregates will (in general) be regarded as continuum materials at the meso-level. Of course this is not completely true, but for the experiments and analyses to come this is a necessary (but sufficient) assumption.

Chapter 3

MECHANICAL BEHAVIOUR OF CONCRETE

3.1 INTRODUCTION

In this chapter, experimental observations of the mechanical behaviour of concrete will be given. The discussion will take place at the meso- and macro-level, i.e., at the particle and continuum level. The size scales range from 10^{-3} to 10^{-1} m or, in other words, the discussion is restricted to the behaviour observed at the laboratory scale. Of course, as argued before in Chapter 1 and later on in Section 4.2.3, the laboratory scale results have to be extended to larger scales before they become interesting for real practical applications. However, the present book is much more focused on deriving a sound physical understanding of the observed fracture phenomena.

In the paragraphs below, a number of definitions are given. Most important are those regarding the representative volume and the definitions for stress and strain. In the past, much attention has been given to the definition of the representative volume for concrete.[69–71] The representative volume is defined as the minimum volume of a laboratory scale specimen, such that the results obtained from this specimen can still be regarded as representative for a continuum. Thus, the size of the characteristic material dimensions such as the particle size, the void (or pore) size and so on, should be small with regard to the size of the specimen volume under consideration. Moreover, when the material is prone to cracking, like concrete, the size of the (micro-) cracks should be small enough as well. Here we come directly to one of the major aspects in fracture studies of concrete, rock and ceramics. As soon as larger (localized) cracks start to grow in the material (specimen), structural changes in the order of the specimen size occur. Therefore, the above definition of a representative volume is violated, even though the ratio between the character-istic material size and specimen size was chosen correctly. Thus, the inherent problem in fracture mechanics is that the structural changes are at a scale of the same order as the size of the specimen (or structure), which means that it is not possible anymore to define a representative volume. The consequence is that size/scale effects and boundary condition effects start to play a major role. How to solve this problem is the main issue addressed in this and the coming chapters.

The other definitions needed are those of nominal stress and strain. The nominal stress σ is defined as the quotient of totally applied load and the total area of the (unloaded/undeformed) specimen. Thus, effects of area decrease due to plastic deformation (yielding), cracking and so on, are not accounted for in the computation of the nominal stress. For convenience we will drop the term "nominal stress" and use the terminology "stress" instead throughout this book.

Compressive stresses are negative and tensile stresses are positive. However, when it comes to defining the concrete quality by means of the compressive strength, the minus sign is omitted (for example B45 denotes a concrete with a uniaxial compressive strength of -45 MPa).

Experimental information on the stress-strain (or rather stress-deformation) response of concrete is important input for constitutive models. Essentially, a stress-deformation curve is the fingerprint of the concrete and reflects the cracking process taking place in the material structure. Recent findings about fracture response limit the applicability of continuum-based constitutive models. In particular, the specimen geometry and boundary condition effects observed in fracture experiments suggest that the constitutive laws as they were developed in the sixties, seventies and early eighties when finite element analysis of concrete (and other geo-materials) became popular, are limited to application in the pre-peak regime of the stress-deformation curve only. Because the state variables are stress and strain, constitutive modelling is per definition at the macro level. As we will see in this chapter, because of localization of cracking and shear-sliding in the post-peak regime, structural response is measured rather than material behaviour. This violates the implicit assumptions made in the choice of state variables in macroscopic constitutive models.

In view of the above it is always important to consider the test data in the context in which they were obtained. Thus, the experimental result will always reflect the total specimen-machine behaviour, and effects related to specimen-machine interactions must be eliminated before we really can define the observed characteristics as material properties. In the paragraphs below these matters will be addressed in detail.

The specimen response is (at least) partly governed by processes taking place at the particle level. Therefore, we will try to give a qualitative description of the fracture processes taking place at the meso-level. They are all based on experimental observations, that will be further analyzed in Chapter 7 using the models presented in Chapter 5. It should be mentioned here that the discussion is restricted to static loading only, i.e., loading rates in the range of 10^{-5}/sec (strain controlled) or 0.05 MPa/sec (load controlled). Moreover, only steadily increasing loadings will be addressed. Only occasionally, when needed or for presenting phenomena in a wider scope, effects due to creep (long-term loading), cyclic un- and reloading, fatigue and impact will be included.

3.2 UNIAXIAL COMPRESSION

3.2.1 INITIAL STIFFNESS

When a concrete specimen (larger than the representative volume) is loaded in compression, a load-deformation response of Figure 3.1 is observed. Typically, an ascending branch is measured, followed by a peak that is often called the compressive strength of the concrete, and finally a descending or falling

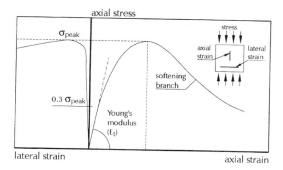

Figure 3.1 Stress-strain curve for concrete under uniaxial compression.

branch, sometimes also referred to as softening branch. The initial part of the ascending branch is more or less straight (i.e., up to stress levels of 30% of peak stress), and the slope is called Young's modulus (the tangent modulus). For a composite material like concrete, the initial slope E_t depends on the composition of the material. As we have already seen in the previous chapter, concrete is a multiscale material consisting of aggregates embedded in a matrix of binder (Figure 2.31). The composite properties depend of course on the properties of the constituents and the properties of the interface between these constituents. Due to increased temperatures during the hydration process, the interface between aggregates and matrix is already loaded, and some initial microcracks may have developed in the porous and weak interfacial zone.

Chemical and drying shrinkage in the cement paste in the composite material structure may have some effect too. Initial de-bonding cracks have been observed by many investigators, but among the first were Hsu et al.[34,49] At relatively low loads, these cracks will not propagate, and the microstructure can be considered stationary. Using the rules of mixtures, the composite Young's modulus can be estimated from the moduli and volume fractions of the constituting phases in the concrete. Basically, the Young's modulus of the composite is situated between the Hill bounds[72]. Much research on the analysis of the composite Young's modulus was done in the 1950s and 1960s.[73-75] In concrete textbooks, often reference is made to the so-called Voigt and Reuss upper and lower bounds for the composite Young's modulus. Essentially, the upper (Voigt) bound is a volume weighting of the phase stiffnesses. The assumption is made that the strain field is uniform. The upper bound can be visualised as a parallel spring model as shown in Figure 3.2a. The composite stiffness is then calculated as

$$E_c = V_a E_a + V_m E_m \tag{3.1}$$

where V_a and V_m are the aggregate and matrix volume, respectively, and E_a and E_m are the respective Young's moduli. Similarly, the lower (Reuss) bound can

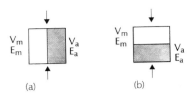

Figure 3.2 Upper and lower bounds for the composite stiffness: (a) series and (b) parallel spring models.

be regarded as a series model, see Figure 3.2b. Again, the stress field is assumed to be constant. The lower bound is essentially a volume weighting of the phase compliances and can be computed following

$$\frac{1}{E_c} = \frac{V_a}{E_a} + \frac{V_m}{E_m} \tag{3.2}$$

An example of the application of the series model was mentioned in Chapter 2. There it was applied to determine the Young's modulus of the interfacial transition zone between aggregate and cement matrix (Figure 2.26). In that example, concrete is considered as a three-phase material, and the interface is considered as an independent material with its own strength and stiffness.

Several combinations of parallel and series arrangements have been proposed, for example the models of Counto and Hirsch (see the overview in Newman[76]). We will not delve into these matters much further here. The effective modulus computed with these "advanced models" lies between the upper and lower bound. Depending on the scatter in test results, either of these models will do. It should be mentioned that these upper and lower bounds require a minimum of information about the actual microstructure of the material. Hashin and Shtrikman[77] also made proposals for an upper and lower bound, which are widely used too.[70,78] Important in the application of the upper and lower bound models is that for a particle composite like concrete, at some moment a ratio between matrix and aggregate volume is reached such that not all aggregates are surrounded by matrix anymore, or that the layers of matrix are so thin that only "interfacial properties" exist (see Section 2.2.2). Thus, the bulk properties of the matrix material change: they are related to the size of the available space between adjoining sand particles. Consequently deviations from the theories start to emerge.[78] For example, for mortars with different aggregate volume fractions, a deviation from the theoretically expected stiffness occurs when the volume fraction of aggregates becomes higher than 50%, see Figure 3.3. It is shown clearly that a deviation from the models, in this case the Hashin-Shtrikman bounds, occurs near the threshold of 50% for mortar. Wittmann and co-workers calculated the cement layer thickness between the sand particles in the mortar, and found that it was smaller than the average cement grain size. Consequently a higher porosity exists in mortars with high aggregate fraction, which was subsequently shown by porosity measurements.

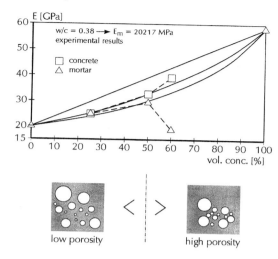

Figure 3.3 Effect of aggregate content on composite stiffness of concrete and mortar with a w/c ratio of 0.38, after Wittmann et al.[78] (Top diagram reprinted with kind permission of Presses Polytechniques et Universitaires Romandes, Lausanne, Switzerland.)

Thus, it is not sufficient to consider concrete as a two-phase material, in particular when extreme ratios of matrix to aggregate volumes are considered. This would imply that the (upper- and lower-) bound models (such as Equations 3.1 and 3.2) are applicable only to a small range of situations, and the generality that is normally assumed, is not allowed. This example also supports the idea that a close relation exists between material microstructure, limits imposed by the manufacturing method, and the response of the material. Another example of this manufacture-microstructure-response-relation will be given in Section 3.4.

The initial stiffness is also affected when the direction of loading is changed with respect to the direction of casting.[79,80] Initial bleeding, segregation and shrinkage cracking occurs preferably under the larger aggregate particles in the concrete. Consequently, weak zones will develop in the material. These weak layers have preferential orientations. Figure 3.9 shows the decrease of initial stiffness when loading is applied parallel to the direction of casting, rather than perpendicular.

Because the Young's modulus of aggregate and matrix are included in the formulation, different values of the composite modulus will emerge when different types of aggregate, binder and/or different volumes of the constituting phases are used. For example, lightweight concrete containing porous low modulus aggregates has a lower overall Young's modulus as normal weight concrete with natural aggregates (see for example Wang et al.[81] for compressive loading, and Cornelissen et al.[82] for tensile loading). The situation is shown schematically in Figure 3.4. The composite stiffness will be lower or

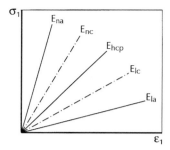

Figure 3.4 Composite stiffness of different two-phase materials. Subscripts *na*, *hcp* and *la* stand for natural aggregate, hardened cement paste and lightweight aggregate, respectively. Concretes with natural aggregates (subscript *nc*) will have a Young's modulus between the values for natural aggregate and hardened cement paste, respectively. For lightweight concrete (*lc*), the stiffness is between cement paste and lightweight aggregate.

higher than the matrix stiffness, depending on the stiffness of the aggregates used. As shown in the example of Figure 3.3, porosity may play an important role as well. In the extreme case, no aggregates are present, but the second phase consists of air-filled voids. An example of such material is foamed concrete. The Young's modulus will then depend mainly on the stiffness of the cement matrix, which will be lower depending on the porosity of the material. Essentially increasing the w/c ratio of paste has an influence on porosity already, and tends to lower the Young's modulus. For foamed concrete initial Young's moduli of 1.5 GPa have been reported,[83] which is extremely low in comparison to values of 30 GPa and 18 GPa for normal concrete and light-weight concrete, respectively.[84]

As we will discuss in Chapter 4, the measurement of strain in the initial part of the stress-strain curve is not straightforward. In particular, it is rather questionable if the strain measured at the surface of a specimen is representative for the complete volume. Normally, when Young's moduli obtained from different experiments using different strain measurement techniques are compared, relatively large deviations are found, with a scatter sometimes up to 30% (see for example Figure 4.5 in Mihashi et al.[85]).

3.2.2. MICROCRACKING AND MICROMECHANISMS

Up till now the discussion has been limited to the initial Young's modulus. As mentioned, when external loading is imposed, i.e., a steadily increasing compressive stress, which is assumed to be uniformly distributed over the specimens cross-section, a curvilinear stress-strain diagram is found as was shown in Figure 3.1. Although the initial part of the curve seems almost straight, detailed measurements have shown that a slight curvature exists from the very beginning of loading.[80,86] The explanation for the curved stress-strain response is that microcracking takes place in the microstructure as soon as the external compression is applied.

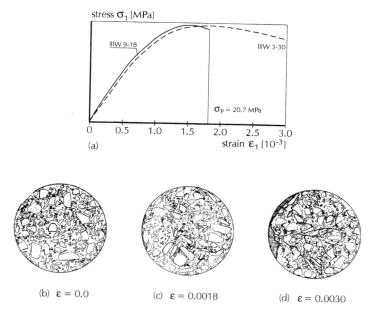

Figure 3.5 Relation between microcracking and stress-strain behaviour. (a) Stress-strain curves and (b) crack patterns in a section perpendicular to the loading direction at $\varepsilon = 0$, (c) 0.0018 and (d) 0.0030. After Hsu et al.[34] (Reprinted with permission of ACI, Detroit, MI.)

It was mentioned in the previous chapter that due to differential shrinkage and thermal mismatch between aggregates and matrix, tensile stress-concentrations will appear at the interface between aggregate and matrix. Hsu et al.[34] were the first to recognise that these cracks might propagate under load. They showed the relation between microcracking and the shape of the stress-strain curve. In Figure 3.5 some of their results are shown. Different cylinders were loaded up to a prescribed level of axial strain, unloaded, and cut into pieces. Cracking was visualised using a coloured dye. The results showed the presence of initial microcracks at the interface between aggregate and matrix (although this will always be debatable when sections cut from specimens are considered, because the cutting itself may cause microcracks), the growth of microcracks into the matrix and the development of macrocracks at large strains (Figure 3.5d, $\varepsilon = 0.003$). Unfortunately, crack growth cannot be followed in a single microstructure because of the way the specimens are cut. The conclusion drawn from their work was that microcracks initiate at the interface, subsequently propagate into the matrix and form continuous macrocracks around the peak of the stress-strain curve.

The most important question seems to be: what is the mechanism underlying the growth of the (micro-) cracks? In Chapter 5, an introduction to fracture mechanics applied to concrete will be given, followed by detailed analyses of cracking in concrete. Here we will limit the discussion to a qualitative over-

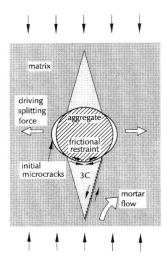

Figure 3.6 Stresses acting at a single particle embedded in a continuous matrix and hypothesized failure mode, after Vile.[87]

view. Cracks can propagate under three different modes, i.e., tensile opening mode, in-plane sliding mode and out-of-plane sliding mode (see Figure 5.20). Combinations of these three modes are also possible. When concrete is considered as a continuum, no reason for crack growth exists, unless disturbances in the stress field are made, for example by introducing a geometrical imperfection. In concrete, however, as discussed in Chapter 2, the (micro-) structure of the material is far from the ideal continuum. In fact, tensile stress concentrations will appear around any type of heterogeneity in the material structure. The simplest case is the single void or aggregate embedded in a continuous matrix. A plausible explanation of the processes taking place around a single aggregate particle was discussed in depth by Vile.[87] The key figure of his detailed explanation is shown in Figure 3.6. The situation is a stiff aggregate particle embedded in a soft matrix. Two interfacial cracks have developed, likely due to differential shrinkage as described in Chapter 2. Because the matrix has a lower stiffness than the aggregate, the material tends to "flow" around the particle. The microcracks in the interfacial zone between cement and aggregate are stable: they can propagate only when the external load is increased further. Because of the mismatch in Poisson's ratio, the matrix material above and below the particle is confined, i.e., a state of triaxial compression develops above and below the aggregate particle. Shear forces exist at the interface as indicated in the figure. As will be shown in Section 3.5, the strength of concrete increases under triaxial confinement. In fact, the situation above and below the aggregate particle is completely identical to the boundary sliding effects in uniaxial compressive tests as will be described in Sections 3.2.4 and 4.1.3.1. Because a triaxially stressed region exists above and below the aggregate particle, failure has to occur away from the aggregate matrix interface. It has been observed that "en-echelon" cracks join to form a shear zone along the conical triaxially stressed region as indicated in Figure 3.6. It is an accepted fact that shear failure in concrete and rock under triaxial confinement is initiated by tensile crack growth parallel to the direction of the (major) compressive stress. Subsequently, a shear plane will develop through the array of "en-echelon" cracks. Therefore, it is argued by researchers in rock mechanics, for example Gramberg,[88] that failure of heterogeneous materials like concretes and rocks is essentially due to tensile failure at the level of the microstructure of the material. It should be appreciated that rocks are also conglomerates of grains and particles, joined together under high pressure, sometimes cemented by another agent. It carries to far to explore the structure of rock here as well.

Returning to Figure 3.6, we can now qualitatively identify a sequence of fracture mechanisms. First there are the de-bonding cracks, most likely nucle-ated during the hardening of the concrete. Next, depending on the flow of paste along the aggregate particle, the de-bonding cracks might be subjected to lateral splitting loads and propagate. The propagation stage stops as soon as the triaxially loaded cones are reached, and failure can proceed only through extension of "en-echelon" shear cracks through the paste. The existence of these shear cones on top and below aggregate particles was shown in experi-ments.[70,80,89]

In the model of Figure 3.6, tensile splitting is induced by the flow of the soft mortar around the aggregate particle. Several parameters are of importance in this model of concrete fracture. First of all it is important to have a sufficiently large mismatch between the Young's moduli of the aggregate and matrix, respectively. As a matter of fact, the Young's modulus of the aggregate particle should be larger than that of the matrix for the above mechanism to develop. When the aggregate particle is less stiff, different mechanisms occur as will be shown below. Furthermore, the behaviour is strongly affected by the bond strength of the matrix to the aggregate particle, the "shear" strength of the matrix material which defines the failure of the shear cones, and the roughness of the aggregate surface which defines the amount of confinement in the shear cones. The amount of matrix material surrounding the aggregate affects di-rectly the deformability of the matrix around the particle. The tensile strength of the matrix affects the behaviour of the shear cone. Note that the tensile strength of the matrix material depends directly on the w/c ratio. With increas-ing w/c ratio, the porosity increases and the strength will decrease (see also Section 3.3).

The above model is rather limited as it considers the behaviour of a single stiff particle in a soft matrix only. The situation changes when the complete composite is considered. Evidently interactions between the individual aggre-gate particles must occur. In Figure 3.7 a conglomerate of interacting aggregate particles is shown. The concrete can be considered as a stack of spherical

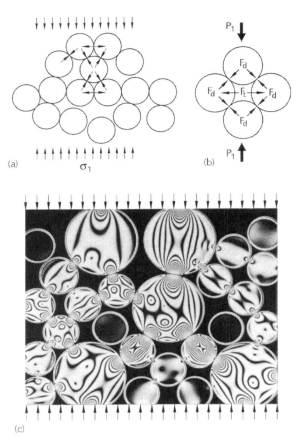

Figure 3.7 (a) Load transfer in a particle stack, and (b) splitting forces between four particles, representing the behaviour of normal weight concrete. (c) shows a result of photo-elastic experiments carried out by Wischers and Lusche.[90] (Figure (c) reproduced by kind permission of the publisher, Beton-Verlag GmbH.)

particles (an ideal shape is considered here), and when an external compressive stress is applied, splitting (tensile) forces will develop in the material structure. The splitting forces are assumed to be the driving force in the fracture process, and explain, at least partly, the cracks that occur. Experiments on discs of photo-elastic material have shown that indeed large stress concentrations occur at the contacts between the various particles. In Figure 3.7c, an example of a photo-elastic experiment of Wischers and Lusche[90] is shown. Clearly visible are the stress-concentrations at the contact points between the discs. Some of the discs don't seem to carry any load at all; they are situated under arches formed by other discs. Gallagher et al.[91] carried out similar photo-elastic experiments on a model with highly irregular photo-elastic particles. Their

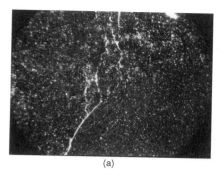

(a)

(b)

Figure 3.8 Experimental evidence for shear fracture near large aggregates, after Stroeven.[70] Example of microcracks near a shear cone (a) from fluorescenting spray experiments and (b) rupture element found after the completion of the test.

model was developed for simulating the behaviour of clastic rock, which seems essentially the same as the behaviour of concrete. However, the arches that could build up in the regular particle model of Figure 3.7c were not so distinct in the model of Gallagher et al. Perhaps this should be attributed to the fact that the irregular shaped particles fill up space better than the circular discs of uniform size. In principle, this model can explain the splitting forces, and also the large lateral deformations that may occur in uniaxial compressive tests on concrete. However, the previously reported mechanism will probably also develop. Also, the shear cones have been found in experiments, see Figure 3.8. Therefore, most likely crack initiation may be caused by the splitting forces generated through the interaction of the various aggregate particles, whereas global failure may eventually occur when the "en-echelon" shear zones start to develop.

Vonk et al.[92] carried out a number of crack-detection tests on prisms loaded in uniaxial compression. In these tests, specimens loaded up to a prescribed axial deformation were vacuum impregnated with a fluorescenting epoxy and cut into parts as soon as the epoxy had hardened (the procedure is described in somewhat more detail in Section 4.4.3). Thus, the crack patterns were fixated and internal crack patterns could be studied in detail. However, again the same disadvantage remains as mentioned in Hsu's experiments referred to earlier: crack growth is not followed in a single specimen, but a number of specimens must be loaded to different levels of damage. The material microstructure changes from specimen to specimen. However, in contrast to Hsu's experiments, in the tests of Vonk et al., not a section perpendicular to the loading axis was made, but rather a section parallel to the direction of loading. The tests

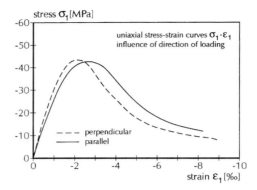

Figure 3.9 Effect of direction of casting with respect to the direction of loading on the uniaxial stress-strain diagram in compression.[80]

showed some of the conical crack shapes near aggregates (see Plate 9).* At the surface of the specimen the cones are hard to identify. Either the aggregates cannot be seen as they are covered by a layer of cement (the skin of concrete, due to the wall effect when concrete is cast in a mould), or in the case of sawn specimens, the aggregates have been cut and different splitting mechanisms may appear as will be discussed in Section 4.2.

As mentioned before, initial microcracking may occur due to differential shrinkage, bleeding and segregation during casting. As a result weak layers develop preferentially under the larger aggregate particles. Different initial stiffness will be measured when the loading direction is rotated with respect to the direction of casting. In addition, the initial and preferentially oriented weak zones will undoubtedly affect the failure mechanism. The weak spots are more or less important depending on their orientation with respect to the direction of loading as will be discussed in Chapter 5. In terms of stress-strain diagrams, it was already mentioned that the initial stiffness decreases when a specimen is loaded parallel to the direction of casting. In Figure 3.9 complete compressive stress-strain curves for a medium strength concrete loaded in uniaxial compression are shown. The differences in initial stiffness are as described above. Another difference is in the pre-peak behaviour, which shows a larger curvature when loading is applied parallel to the direction of casting. This suggests that crack growth is delayed in that case. At the meso-level the mechanism shown in Figure 3.10 is hypothesized. When loading is applied parallel to the direction of casting, the orientation of the initial weak layers or microcracks is very unfavourable to propagation in a compressive stress field in the direction of casting, see Section 5.3.2. In contrast, when the loading is applied perpendicular to the direction of casting, the cracks are oriented more favourable, and crack propagation may proceed much faster. In the experiments, the delayed and accelerated crack growth for parallel and perpendicular loading, respectively, seems to lead to differences in peak strain and not so much in differences

*All plates follow Page 98.

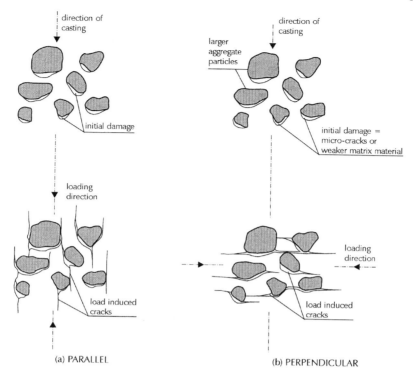

Figure 3.10 Hypothesized micromechanisms, explaining the effect of initial anisotropy in concrete.[80]

in peak stress. The concrete quality is probably of importance. For example, Hughes and Ash[79] also reported an effect on strength for a low strength concrete.

The situation changes dramatically when low strength aggregates are mixed in the concrete, for example, as done in lightweight concrete. In that case, the cracks tend to run through the aggregate particles which are now loaded as in a brazilian splitting test, see Figure 3.11. However, through the lateral extension of the aggregates, the particles will force splitting cracks to develop in the mortar matrix. In the composite, most of the compressive load will be carried by the matrix material, which has a higher stiffness now.[90] In Section 3.3.1.1 (Figure 3.42) and Chapter 7, examples of crack growth in lightweight concrete under externally applied uniaxial tension will be shown.

In high strength concrete, cracks also tend to run through the aggregates. In this type of material the differences in stiffness and strength between matrix and aggregates is relatively small. However, interfacial fracture may still occur as in general the increased compressive strength is derived from a well-balanced particle distribution. The increased amount of cement in high strength

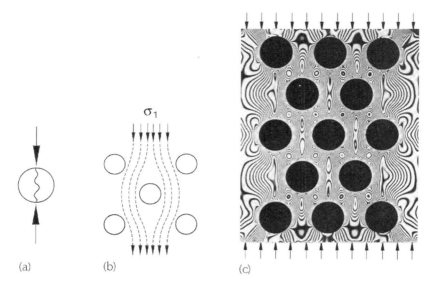

Figure 3.11 Load transfer in lightweight concrete. (a) Splitting cracking in individual particles, (b) load transfer in a particle-matrix composite. (c) shows the result of a photo-elastic experiment by Wischers and Lusche.[90] (Figure (c) reprinted with permission from the publisher, Beton-Verlag GmbH.)

concrete does not generally lead to an increased bond strength between aggregates and matrix.[57] The stress-strain behaviour of lightweight and high strength concrete in compression will be addressed in Section 3.2.5.

3.2.3 STRESS-DEFORMATION RESPONSE

Let us now consider the stress-deformation behaviour of concrete loaded in compression. There are a number of effects that should be mentioned here. First of all, results of uniaxial compression tests on a medium strength concrete are shown in Figure 3.12. In this figure the results of four cube tests between brushes are shown (see explanation of loading systems in Section 4.1.3.1, Figures 4.23 to 4.27). Both the axial and lateral deformations were measured and plotted against the axial compressive stress. All the stages mentioned with the schematic diagram of Figure 3.1 can be recognised. The ascending branch is curved, leads to a peak (the uniaxial compressive strength), and after that a descending branch is observed. The mechanisms discussed in the previous section are all active in the pre-peak stress-strain curve. The energy released by the nucleation and growth of microcracks is still relatively small compared to the total amount of energy contained in the specimen. These matters are clarified in detail in Section 5.3.2.4. It is interesting to look at the lateral deformations. They are plotted at the left side of the diagram. The lateral deformations are initially smaller than the axial deformations. Basically, the Poisson's ratio (defined as $\nu = \varepsilon_1/\varepsilon_2 = \varepsilon_1/\varepsilon_3$, where ε_1 is the axial strain and ε_2

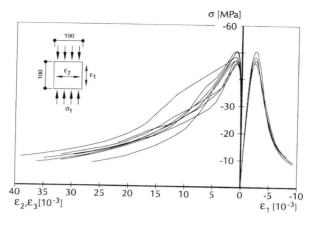

Figure 3.12 Stress-strain curves of concrete in uniaxial compression.[80]

and ε_3 are the strains in the unloaded directions) of normal weight concrete is in the order of 0.15 to 0.20, at least when the stress levels are relatively low (i.e., smaller than 30% of peak stress). At a stress-level of approximately 70% of peak-strength, the lateral deformation suddenly starts to increase significantly. From that point on to the peak stress level, the Poisson's ratio increases from 0.20 to 0.50. The value at peak is equal to 0.50, which is the same value found for elastomers. If we continue to use the same definition for Poisson's ratio in the descending branch, the value increases even more significantly, namely up to 4 or 5. The reason for the highly non-linear Poisson's ratio must be sought in the crack-growth processes referred to in the previous section. Because the lateral deformation is initially smaller than the two lateral deformations, the specimens volume will decrease at low stress levels. Note that the volume change is normally defined as $\varepsilon_v = \varepsilon_1 + \varepsilon_2 + \varepsilon_3$ (i.e., under the assumption of small strains). As a result of the fast increasing lateral deformations at higher stress levels, the specimen volume starts to increase. The change from volume decrease to volume increase is found near stress levels of approximately 70% of peak stress. The volumetric strain is plotted in Figure 3.13 for the same experiments that were shown in Figure 3.12. The minimum volume is clearly visible, and also it can be seen that after the peak stress has been reached a fast volume increase or dilatancy is measured. For many years, the minimum volume bound has been interpreted as the onset of (macroscopic) failure of a specimen loaded in uniaxial compression.[34,93,94] At this level, crack growth becomes unstable, and when the loading would be halted and kept constant, spontaneous fracture of the specimen would occur after some time. Therefore, the minimum volume bound is sometimes related to the long-term strength of the concrete as well.

Beyond the peak a descending branch is measured. In terms of strain: axial strains up to 1.0% are measured, and lateral strains easily increase beyond 10%. Knowledge of the descending branch in uniaxial (but also multiaxial)

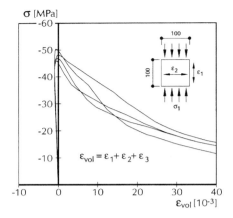

Figure 3.13 Stress-volumetric strain curves for concrete in uniaxial compression.[80] The curves were derived from the experiments in Figure 3.12.

compression is needed for the analysis of the rotational capacity of concrete beams.[95,96] In the next section we will discuss in detail the post-peak response of concrete in uniaxial compression.

3.2.4 POST-PEAK BEHAVIOUR AND LOCALIZATION OF DEFORMATIONS

After the compressive peak a gradual descending branch can be measured. The experimental techniques for measuring softening will be described in detail in Chapter 4. In the past years, the interest for describing the post-peak behaviour in more detail has increased considerably. Notably the development of numerical models has been an important factor. First we will show some of the mechanisms. In the next section we will discuss the effect of concrete composition and other factors.

In Figure 3.14 complete stress-strain curves for a medium strength concrete (40 MPa) are shown. Prisms of various slenderness ratios (h/d = 0.5, 1.0 and 2.0, where h is the specimen height, and diameter d = 100 mm) were loaded in uniaxial compression between brushes (Figure 4.26) at a moderate displacement rate, i.e., 1 μm/sec. The stress-strain curves are almost the same in the pre-peak regime, but beyond peak the slope of the descending branch decreases with decreasing specimen height.

In 1984, this was quite a surprising result, and an explanation of the phenomenon can be found if the post-peak curves are plotted as stress-post peak displacement curves. Essentially the displacements are calculated through

$$\delta = \varepsilon h - \varepsilon_{peak} h \qquad (3.3)$$

These diagrams are shown in Figure 3.15.

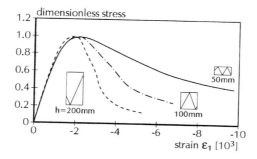

Figure 3.14 Stress-strain curves for medium strength concrete in uniaxial compression: effect of slenderness ratio h/d ($d = 100$ mm for all tests).[80]

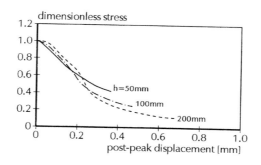

Figure 3.15 Dimensionless stress-post peak deformation diagrams for prisms with different height.[80]

The diagrams now more or less fall on top of one another, indicating that the same displacement is needed to fracture the specimens. In different words, the post peak deformation must be localised in a small zone, and cannot be interpreted as a strain. The phenomenon can best be explained from a series coupling model for localization as done by Bažant,[97] and later also by Hillerborg,[95] both on the basis of Van Mier's test results. The results clearly indicate that fracture of concrete in uniaxial compression is a localized phenomenon. This would imply that strain cannot be used anymore as state variable in constitutive laws for concrete.

The localization of deformations can be explained when failure occurs through the growth and development of inclined shear fracture planes in the specimens. Normally this is not easy to verify for the lower specimens. In Figure 3.16, the failure modes observed in specimens of three different heights are shown. Although often first surface splitting occurred, internally inclined fracture planes were found as shown in this figure. The inclined fractures were later confirmed in a number of impregnation experiments.[89] The development of these shear fracture planes has a direct effect on the lateral deformations. In

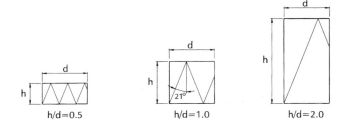

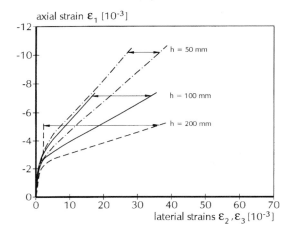

Figure 3.16 Failure modes for prisms with varying height.

Figure 3.17 Lateral strains ε_2 and ε_3 plotted against the axial strain ε_1 for specimens of different height.[80]

Figure 3.17 the lateral strains ε_2 and ε_3 are plotted against the axial strain ε_1. The range of extreme lateral strains is shown for the three specimen heights investigated, and the results show an increase of variability with increasing specimen height. This increased variability depends on the discrete strain measurement method that was adopted, and will be further elucidated in Chapter 4. The discrete shear band formation is the direct cause for the large variability: a measuring point may be situated such that it either crosses a shear plane, or that it is fixed on a part of the specimen where no fractures develop. The probability for a measuring point being located at an intact part of the specimen decreases with decreasing specimen height. This can be directly seen from the fracture patterns plotted in Figure 3.16.

For rock, essentially the same problems are encountered. In fact, long before it was recognised in the field of concrete, localization of deformations in shear bands was observed in rocks.[98,99] The localization in uniaxial compression was,

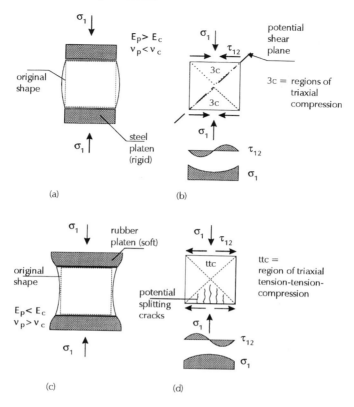

Figure 3.18 Effect of loading platen on specimen response: (a) deformed specimen and (b) load distribution with rigid platens; (c) deformed specimen shape and (d) load distribution with soft loading platens.

however, not as clear as in the above results because different loading platens were used. In the experiments of Figures 3.12 through 3.15 brushes were used to reduce the amount of boundary shear, but in rock tests it seems not very customary to reduce the boundary friction between the loading platen and specimen. Therefore, in the experiments of Hudson et al.[99] in which rigid steel platens were used, both the peak stress and the post-peak behaviour were affected when the specimen slenderness and size were varied, and a clear conclusion could not be drawn.

Let us now investigate the effect of the boundary friction. When a concrete (or rock) specimen is loaded between rigid loading platens, like for example steel, the lateral deformation of the concrete is restricted at the specimen ends. This is shown schematically in Figure 3.18a. The bulged shape of the specimen develops because the specimen ends are forced to have the same lateral deformation as the rigid steel platens. In this figure, full frictional restraint is assumed, i.e., no sliding can occur in the interface between loading platen and specimen. Thus, shear stresses will develop between the specimen and loading platen, causing a three-dimensional state of stress in the specimen ends as shown in Figure 3.18b. In fact, the situation is similar to the triaxially stressed cones that develop on top of the aggregate particles in a normal strength cement composite (see Figure 3.6). A completely different situation emerges when loading platens with a low stiffness and high Poisson's ratio are selected, for example rubber. In this case, the platens will give rise to large lateral deformations, resembling outward-directed shear forces at the interface as shown in Figure 3.18c,d. In Figure 3.18c, it is assumed that the upper part of the "rubber" platen can deform unrestrictedly in the lateral direction, which of course is not easy to accomplish in a real experiment. Because the frictional forces are now directed outward, a triaxial tension/tension/compressive state of stress may develop in the specimen ends, possibly leading to splitting cracks originating from the specimen ends as indicated in Figure 3.18d. Obviously, the only correct way of loading concrete specimens is by means of loading platens that have the same material constants throughout the loading process. This would imply that only platens made of the same concrete as the test specimen are suitable. Normally, this is the step taken in practice: higher specimens are selected, thereby reducing the end effects caused by the loading system. We will come back to these end-effects in more detail in Sections 4.1.3.1 and 7.4.2.

Let us confine ourselves here to a number of experimental observations. In triaxial testing, specifically in true triaxial tests on cubical specimens, the use of increased specimen height is impossible. The specimen will always be a cube with end effects in the order of the rib size of the cube. Therefore, investigators have tried to mimic the lateral deformability of concrete by introducing specially designed loading platens. Examples of loading platens are described exhaustively in Chapter 4. In general we can distinguish between rigid steel platens without friction-reducing measures and platens with friction-reducing measures. In Figure 3.19, the effect of type of loading platen on the compressive stress-strain behaviour of a low and medium strength concrete are shown. The results shown here were obtained by Kotsovos,[100] and were later confirmed by Vonk et al.[101] Kotsovos performed experiments on cylinders with an aspect ratio of 2.5; Vonk et al. carried out experiments on 100 mm cubes, but the effect was essentially the same as shown in Figure 3.19. In his experiments, Kotsovos used the following systems (in order of decreasing frictional restraint): active restraint, rigid steel platens, brushes, rubber layers, and MGA pads (a special design, low-friction interlayer). For both concretes investigated, the pre-peak behaviour (in dimensionless form) was found independent of the choice of loading system. However, beyond peak the dimensionless stress-displacement curves were more steep when a system with a very low coefficient of friction was used. This led Kotsovos to the conclusion that the best post-peak model would be a purely brittle model, i.e., with a steep falling branch. However, in view of the models of Figure 3.18, this is not completely certain.

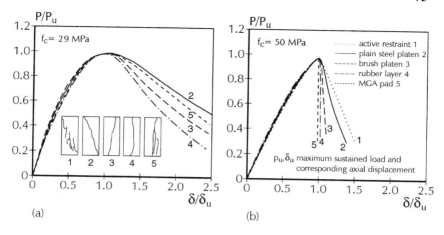

Figure 3.19 Dimensionless load-displacement curves for various load application systems and for two different concretes: (a) $f_c = 29$ MPa and (b) $f_c = 50$ MPa, after Kotsovos.[100] In the inset of (a), the change of failure mode is shown with decreasing boundary restraint from left to right. (Reprinted with kind permission from RILEM, Paris.)

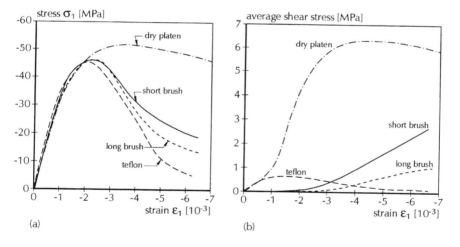

Figure 3.20 Effect of loading system on uniaxial compressive stress-strain response (a), and average shear stress-strain curves for the four different loading systems used (b), after Vonk et al.[101,102]

As mentioned, the results of Vonk et al. showed a similar tendency and they are reproduced in Figure 3.20a. Cubical specimens are loaded between four different loading systems. With increasing frictional restraint they are: teflon platens, long brushes, short bushes and dry steel platens. In separate experiments Vonk determined the frictional behaviour of these loading platens, see Figure 3.20b.

From these "friction tests" it was found that both types of brushes had a low friction at smaller deformations. However, at larger deformations, the restraint caused by the brushes would increase due to lateral resistance from bending of the brush rods. For dry steel platens, for any deformation, large frictional restraint was measured. For teflon, relatively high shear stresses could develop at low deformations (stick-slip behaviour: it is quite well known that sliding between teflon and steel only will start after a certain resistance has been overcome, see Figure 4.25). Beyond strains of 0.0025 the specimen is well into the softening regime, as can be seen from comparing the axial strains in Figure 3.20a. Note that the frictional restraint caused by the brushes decreases with increasing brush length. It should be mentioned here that the inclination of the dominant cracks in the specimen decreases when boundary restraint decreases, as can be seen in the inset of Figure 3.19a. Obviously failure changes from a shear mode under high boundary restraint to a splitting mode under low boundary restraint.

Now what do we conclude from these results? Obviously, the ductility of the specimen changes when different loading platens are applied. In all of the above cases, the specimens were loaded in machines with a hinge above the upper loading platen, and a fixed platen at the bottom of the specimen. In general, the hinge is used only for placing the specimen such that good contact with the upper and lower loading platen exists. Thus, specimens can be slightly skewed: the upper platen will adjust itself. This point will be further elucidated in Chapter 4. The size effect shown in Figures 3.14, 3.15 and 3.17 was obtained from experiments with brushes. On the basis of the results of Figures 3.19 and 3.20 it must be concluded that brushes must have an effect on the post-peak behaviour in the size tests. However, it is difficult to assess the effect. Recently, the experiments of Figure 3.14 were reproduced in about 11 laboratories. This was achieved by RILEM Technical Committee 148-SSC "Strain-Softening of Concrete", for the purpose of defining a standardised test method for measuring compressive softening in concrete. The problem is that the loading platen not only affects the post-peak behaviour, but also the peak strength as will be discussed indepth in Section 4.1.3.1. In Figure 3.21 the contribution from Delft University to the Round Robin test are shown. The slenderness tests were carried out by using a newly developed teflon layer (comprising two 100 µm thick sheets of teflon foil, with 50 µm of bearing grease in between), or rigid steel platens as loading system. Experiments were carried out on a normal strength concrete (approximately 54 MPa compressive strength measured from tests on 150 mm cubes loaded between rigid steel platens) and a concrete with a higher strength (approximately 85 MPa). In Figure 3.21a and b, the results for the normal strength concrete are shown. In Figure 3.21c and d the results for the high strength concrete are shown. The "low-friction" results are shown in Figures 3.21b and d, whereas the high-friction results are shown in Figures 3.21a and c. The tendency is the same for both concretes. The application of a teflon interlayer results in an almost constant peak strength, irrespective of

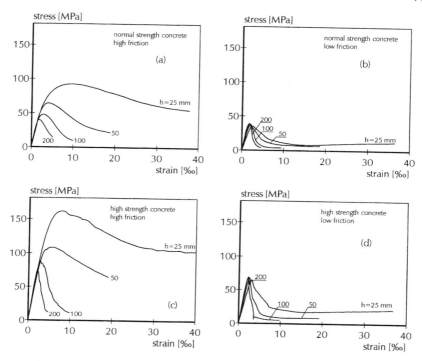

Figure 3.21 Stress-axial strain curves for normal strength concrete (a,b) and high strength concrete (c,d) tested between rigid steel platens (a,c) and between teflon loading platens (b,d).[103]

the specimen height. In this respect the results are quite comparable to those of Figure 3.14. The frictional restraint of the teflon interlayer, however, is much smaller than in the brush tests of Figure 3.14. Effectively, this results in steeper softening curves. For the high strength concrete, for some of the high slenderness specimens it was not possible to obtain a stable test result because to much elastic energy was released during softening (see Chapter 4). In contrast, an increase of peak strength is observed when specimens are loaded between rigid steel platens. The lowest specimens gave the highest strength. Note that the slenderness range in these tests was extended from $h/d = 0.25$, 0.50, 1.0 to 2.0 (with $d = 100$ mm). The results confirm the observation made in Figure 3.14, namely, the softening branch becomes steeper with increasing specimen height. Again, when the results are plotted as stress-post-peak deformation curves, the same unique softening behaviour is found independent of the specimen slenderness, as shown in Figure 3.22. However, the effect of the loading system remains quite notable. The "bundle" of curves from experiments between steel platens gives a more gradual softening curve than the teflon tests. Comparison with the data in Figure 3.14 indicates that the results from the brush tests are falling somewhere in between. Clearly three "bundles"

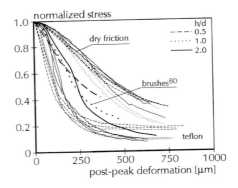

Figure 3.22 Post-peak, normalized stress-displacement curves for normal strength concrete specimens of different slenderness and loaded between different boundary conditions.[105]

of test results can be distinguished, exemplifying the influence of boundary shear. However, the effect of specimen slenderness has disappeared when the softening curves are presented as dimensionless stress-post peak displacement curves. However, the pre-peak behaviour is unique in terms of stress and strain. This all suggests that around peak a transition from distributed to localized failure must occur.

At this point a remark regarding the peak strength is in place. The matter will be more deeply discussed in Section 4.1.3.1, but it is interesting to show the effect of specimen slenderness on the peak-stress and strain at peak-stress for the tests between steel and teflon platens. In Figure 3.23, the results from all experiments are gathered. Figures 3.23a and b clearly show that for tests between rigid steel platens, an increase of strength is observed when the specimen slenderness decreases. This is observed both for the normal strength concrete (Figure 3.23a) and for the high strength concrete (Figure 3.23b). The teflon results, in contrast, show almost no effect of slenderness on strength. Even, quite surprisingly, the peak stress seems to decrease with decreasing slenderness. The reason for this is not clear at present. Also visible from Figures 3.23a and b is the decrease of scatter when teflon loading platens are used. This is an important observation, as it shows clearly how a standardised test for compressive softening should be defined (see Section 4.6). The strains at peak follow the same tendencies as mentioned above for the peak-stress (Figure 3.23c and d). Again, the strain at peak seems independent of the specimen slenderness when teflon is used, whereas the strain increases for low slenderness when steel platens are used. Moreover, the scatter is substantially larger when steel platens are used.

Let us finally address the modes of failure. The fractured specimens were carefully photographed and drawn. In Figure 3.24, the observed modes of failure for prims with three slenderness ratios $h/d = 0.5$, 1.0 and 2.0 are shown. The three specimens at the top row were loaded between rigid steel platens, whereas those at the lower row were loaded between teflon platens. The

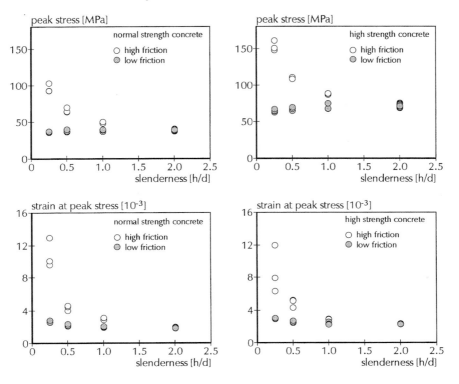

Figure 3.23 Peak-stresses (a,b) and strains at peak-stress (c,d) for the normal strength concrete (a,c) and high strength concrete (b,d) tests of Figure 3.21.[103]

experiments between steel platens clearly showed the well known "hour-glass" failure mode, whereas a more splitting type of failure occurred in the teflon tests. Note that the specimen surface in contact with the loading platen remained almost uncracked in the case of steel loading platen, whereas a more distributed failure occurred, with many cracks developing in the top and bottom surface as well, when teflon interlayers were applied. Quite clearly, the loading system has an important effect on the behaviour observed in the uniaxial compression test.

The global impression of all these results is that the compressive stress-strain behaviour is a system behaviour. Everything seems to depend on the actual loading conditions. In his thesis, Vonk also showed that the compressive softening curve is affected by the ability of the loading platen to rotate or not. In tensile fracture the effects of loading platen rotations are even more significant, and the point will be further elucidated in Sections 3.3, 5.3 and 7.2.

The above results are confirmed by the work of the other participants to the RILEM round robin test.[106-113] Also recent experiments by Taerwe[114] and Markeset[115] for high strength concrete are in agreement with the above description of compressive softening. One final remark should be made here. The

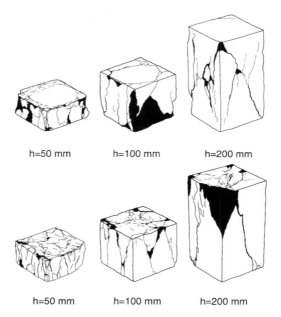

Figure 3.24 Broken specimens from uniaxial compression tests between rigid steel platens (top row), and from tests between loading platens with intermediate layers of teflon and grease (bottom row). From left to right the slenderness increases from 0.5, 1.0 to 2.0.[104]

25 mm high prism showed an increase of stress in the tail of the softening diagram (Figure 3.21b and d), indicating that at this stage the specimen has been crushed and the load is carried by the more stiff aggregates. This is also the reason that the fractured 25 mm high specimens were not included in Figure 3.24. In fact, the second rising branch at large axial deformations was postulated in a thought experiment by Armer and Grimer,[116] see Figure 3.25. In earthquake engineering, the second rising branch may be of importance as concrete may be crushed to extreme limits, and the remaining carrying capacity will depend on the strength of a "heap of aggregates".

3.2.5 EFFECT OF MATERIAL COMPOSITION

Throughout the years much information has been gathered on the stress-strain behaviour of concrete in compression. As we have tried to demonstrate in the foregoing sections, the stress-strain behaviour and, in particular, strength and post-peak deformation depend to a large extent on the conditions under which a test is performed. The effect of material composition can therefore only be considered for specimens loaded under identical circumstances. These circumstances are wider than one might suspect on the basis of the boundary and size effects debated in Section 3.2.4. In addition to these factors, strain- (or stress-) rate is also of importance. Loading/unloading cycles and/or fatigue, and other physical factors (e.g., humidity and temperature) may play a substan-

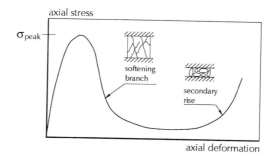

Figure 3.25 Secondary rising branch after compressive softening, after Armer and Grimer.[116]

tial role. Perhaps combinations of all these factors have an important effect on the mechanical behaviour as well. It is not the goal of this book to give an exhaustive overview of all these effects. Basically, the main interest is the derivation of (or at least a procedure to derive) parameters for fracture models of concrete, and to emphasize the close relationship between model and experiment. In this section, the information on the effect of material composition on stress-strain behaviour is kept relatively limited. This choice has been made because comparing data sets from different sources may be useless because of differences in experimental procedures. Moreover, exhaustive overviews can be found in literature already.[10,117] In this section we will discuss the effect of concrete quality on stress-strain behaviour, and the effect of lightweight aggregates. In the next section the effect of loading rate and load cycling will be presented.

As mentioned before, knowledge of the stress-strain behaviour of concrete is essential input for computer models. However, the analytical solutions require data which can serve as a basis for modelling. Basically, the same shapes of stress-strain curves are found for concretes with different compositions. Micro- and macrocracking seem to be the source for the observed non-linearities, and details in the fracture process may cause deviations, but the impression is that these are only marginal. In recent years there is a tendency to go to more refined concrete compositions with a higher compressive strength, see for example the proceedings of the high strength concrete symposium held in Lillehammer.[118] This is achieved mainly through the selection of a balanced aggregate grading. Some of these matters were discussed in Chapter 2. The uniaxial stress-strain behaviour for concrete for a variety of strength classes is shown in Figure 3.26. The results were obtained on 200 mm cubes, i.e., the German standard specimen for determining the compressive strength of concrete. Basically the shape of the stress-strain curves is the same: an initial, almost straight, ascending branch, a curvilinear part just before peak, the peak, and then a more or less shallow descending branch with a relatively long tail. Differences are of course observed in peak strength and a higher peak strain because the pre-peak behaviour tends to become more linear before peak. The

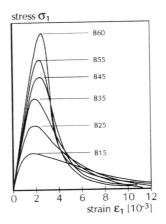

Figure 3.26 Stress-strain curves for different concrete qualities, after Wischers.[119] Note that also in the original figure, no stress scale was given. The difference in strength was expressed as is done here, where the terminology B80 refers to the characteristic cube compressive strength for that particular concrete.

post-peak behaviour seems more brittle for the higher strength concretes, but Wischers[119] claimed that this was caused by the amount of energy released in the elastic parts of the specimen. We will come back to energy balances in Chapter 5. Nevertheless, the behaviour seems more brittle for the high strength concrete. For even higher qualities as those shown here, specific experimental conditions should be met to obtain a stable post-peak response.[114,120,121] We will further discuss these matters in Chapter 4.

There are several formula to describe the compressive stress-strain behaviour of concrete. Usually they are fits of experimental data.[81,122,123] The formula proposed in the CEB model code[122] is as follows:

$$\sigma_c = -\frac{\dfrac{E_{ci}}{E_{c1}}\dfrac{\varepsilon_c}{\varepsilon_{c1}} - \left(\dfrac{\varepsilon_c}{\varepsilon_{c1}}\right)^2}{1 + \left(\dfrac{E_{ci}}{E_{c1}} - 2\right)\dfrac{\varepsilon_c}{\varepsilon_{c1}}} f_{cm} \quad for \quad |\varepsilon_c| < |\varepsilon_{c,lim}| \tag{3.4}$$

where E_{ci} is the tangent modulus, s_c is the compressive stress in MPa, e_c is the compressive strain, $e_{c1} = -0.0022$ and $E_{c1} = f_{cm}/0.0022$, where f_{cm} is the compressive peak stress. We will not delve into more detail here, but results as those presented in Figure 3.26 can be described very well using this formula. However, on basis of the discussion of Section 3.2.4 it will be obvious that, in particular the post-peak behaviour cannot be described uniquely in terms of stress and strain. There Equation 3.4 breaks down, and we have to search for new directions for structural design and modelling.

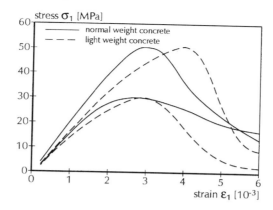

Figure 3.27 Comparison between lightweight concrete and normal weight concrete stress-strain behaviour for two different concrete qualities, after Wang et al.[81] (Repritned with permission of ACI, Detroit, MI.)

When lightweight concrete is loaded in compression, again the global shape of the stress-strain diagram is identical to the shape of the diagram for normal weight concrete. Wang et al.[81] investigated the stress-strain behaviour of lightweight concrete containing expanded shale aggregates, and compared it to the response of normal weight concrete measured on 75 mm diameter and 150 mm long cylinders. In Figure 3.27 a comparison is made between lightweight concrete and normal weight concrete stress-strain curves for two different concrete qualities, namely 30 MPa and 50 MPa. Obviously, the post-peak behaviour of the lightweight concrete is more brittle, displayed by the steep descending branch. There are also some differences in peak strains. The peak strains increase for lightweight concrete, but this is mainly caused by the reduced Young's modulus of the material, as was discussed in Section 3.2.1. Several authors report the development of cracks through the aggregate particles.[57, 82, 90] This is also found in numerical simulations, as will be shown in Section 7.2.1 for tensile loading. It seems that the mechanism described in Figure 3.11 is essentially correct. Moreover, the lightweight aggregates are usually rather porous, and a denser (and stronger) interfacial zone with a lower porosity may develop as the cement paste can enter some of the larger surface pores in the aggregates[55] (see also Section 2.2.2). Note however that some types of lightweight aggregates may have a rather dense shell, and the interfacial behaviour might change.

3.2.6 LOADING RATE AND LOAD CYCLING

The stress-strain behaviour, or rather stress-deformation response of concrete under uniaxial compression is also affected by the loading rate and (although not completely clear at the moment) on cyclic loading and fatigue. Again, the major tendencies will be shown, without going into too much detail.

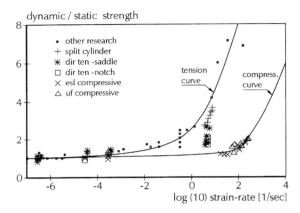

Figure 3.28 Effect of strain rate on dynamic tensile and compressive strength of concrete, after Ross and Kuennen.[129]

In this section only compressive loading is treated; some attention to tensile impact and fatigue will be given in Section 3.3.

3.2.6.1 Effect of loading rate

Loading rate effects may be important in several practical situations where danger of collisions exist. Most knowledge on loading rate effects concerns the strength of the material. In principle, we would like to describe the behaviour under various loading rates, based on detailed knowledge of microstructure and microcrack processes. Obviously, differential inertia effects due to differences between aggregates and matrix should be taken into account. There is an extensive literature on dynamic behaviour of concrete.[124-127] Most of it is related to tensile cracking and we will come back to this in Section 3.3. Of course, all the specimen-machine interactions described above will appear also in dynamic testing. Again, it seems sensible only to compare results derived with the same experimental technique.

An excellent overview of dynamic test and monitoring techniques was given by Reinhardt.[128] For very high loading rates extreme techniques are required such as drop weights, explosions in contact with the model under consideration, projectiles, loading in a gas pressure chamber or by blasting. In most cases only attention is given to strength. In Figure 3.28 some recent results of loading rate on compressive strength of concrete are given. A comparison is made with the impact behaviour in tension. Basically, the ratio of dynamic over static compressive strength remains constant for strain rates up to 10^{+2}. At that range, a slight increase of dynamic compressive strength seems to occur. The compressive results were obtained using a split Hopkinson bar.[129] As far as stress-strain behaviour is concerned, not much information is available concerning the complete curve. The instabilities that may appear in the post-peak range become very important under impact loading. Some in-

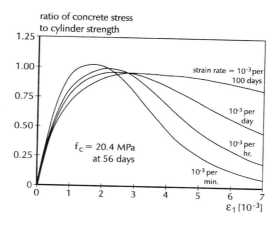

Figure 3.29 Uniaxial compressive stress-strain curves for concrete under various loading rates, after Rüsch.[93]

sight in compressive stress-strain behaviour at very moderate loading rates can be obtained from Figure 3.29.

Upon increase of the loading rates, the pre-peak behaviour tends to become more steep (and straight), the peak stress increases slightly as described above, and the softening behaviour is more brittle. These results are for a very limited loading rate regime (0.001 per 100 days to 0.001 per minute). The tendency for higher loading rates, however, is the same.[130] In these latter experiments hardened cement paste and mortar were tested for strain rates between 0.3×10^{-6} and 0.3. As mentioned, the same tendency was found as in Figure 3.29, except that the variation in initial Young's modulus was much larger. Note, however, that still some controversy exists concerning the initial Young's modulus which is related to the deformation measuring technique.[125,126] The various measurements do not give a unique answer. The scatter in Young's moduli is relatively large, and the increase of Young's modulus with increased strain rate which is mentioned in many text-books seems not valid. Probably, inertia effects may play some role. Moreover, moisture movement in the material may be responsible as well.[130,131] For the more homogeneous hardened cement paste, the brittleness increases enormously under dynamic loading.[130] This confirms what has been said in the previous section about the effect of material composition on softening behaviour. Recently, some results were published for high strength concrete,[132] specifically in the low loading rate regime. Basically, these results confirm the above.

3.2.6.2 Cyclic loading

Cyclic behaviour of concrete is of importance, especially for those practical circumstances where variable loadings exist. However, in the post-peak regime, unloading of various part of a structure may occur as soon as localization of cracking occurs in a neighbouring area. This can easily be demonstrated

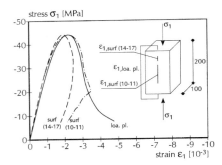
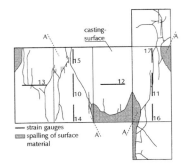

Figure 3.30 Surface unloading and overall softening measured on a 200 mm high prism loaded in uniaxial compression.[80]

from a detailed analysis of a prism compressive stress between brushes, as shown in Figure 3.30. At the surface of a 200 mm high prism, several strain gauges were glued, whereas the test was controlled using the average signal measured by several LVDTs mounted between the loading platens. The latter measurement gives the global response of the specimen, whereas locally larger differences may appear. The local effects are visible from the strain gauge measurements. As can be seen, strains measured with some strain gauges followed the global curve for some time, until probably one or two strain gauges broke. Other strain gauges immediately showed unloading as soon as the peak was reached. Close scrutiny of the cracked specimen after the test showed that these last strain gauges were placed on parts of the specimen that were completely split off. The increasing surface strains were measured with strain gauges 10 and 11, which were located exactly over the discrete cracks that led to failure.[80] The unloading behaviour measured with the strain gauges is representative for a small part of the specimen only. Not even the area and shape of the part that is split loose are known, and such results are not very useful, except that they elucidate the softening process. In order to determine the unloading, but also reloading behaviour of concrete, specimens must be loaded along such carefully prescribed paths. In the early sixties, many compressive tests were carried out under load-cycling.[133,134] In these early experiments no sophisticated electronics were available for measuring post-peak behaviour, and the unloading-reloading cycles were carried out in order to get some insight in post-peak behaviour. In Figure 3.31 an example of a stress-strain curve with a number of unloading/reloading cycles is shown. In the pre-peak regime three load-cycles are done, the next one is at peak, and a fifth unloading/reloading cycle is made in the post-peak regime. The first and second (pre-peak) cycles are hardly visible, and coincide almost with the ascending branch. The third cycle is more clearly visible. After an initial steep part in the unloading branch, the unloading curve approaches gently towards the x-axis of the diagram. The reloading curve increases more rapidly, but

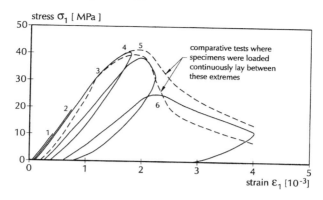

Figure 3.31 Cyclic compressive stress-strain curve for concrete, after Spooner and Dougill.[135]

clearly visible is that the slope has decreased to some extent in comparison with the initial ascending branch. At some moment, the reloading curve bends away, and the softening curve is followed again (point 5). The same is visible in the post-peak regime. The stiffness of the reloading curve decreases further, and upon reloading the diagram seems to reach the original post-peak path again. For comparison, the limits of monotonically loaded specimens are included in Figure 3.31, which suggests that the loading/reloading cycles hardly affect the shape of the post-peak diagram. Note that with each load-cycle, the amount of irreversible (in-elastic) strain increases. This is normally attributed to crack closure effects at the micro- and meso-scale.

The slope of the reloading curve is sometimes used as a measure for the amount of damage that has been done to the specimen. The progressively descending slope is associated with a damage variable, as is done, for example, in the plastic fracturing theory.[136] Spooner and Dougill[135] tried to define different energy dissipation sources on the basis of cyclic loading tests. Based on such tests, cumulative energy dissipation graphs were drawn, which served as a starting point for theories for describing the behaviour of *progressively fracturing solids*.[137,138] Spooner et al.[139] and Spooner and Dougill,[135] argued that concrete behaves similarly as an ideal material in which energy is dissipated by two processes. Due to the first mechanism (a), energy is dissipated during only the first loading over a given strain range. A second mechanism (b) provides for energy dissipation both during increase and decrease of strain. The energy dissipated due to mechanism (a) can be taken as a measure of the damage sustained to the material. The second mechanism (b) provides a damping effect similar to that observed in real materials under repeated loading. They argued that by subjecting a material sample to a series of unloading-reloading cycles the subsequent dissipating mechanisms can be distinguished. The cumulative dissipation energy curves have the same shape for uniaxial and multiaxial compression, as well as for uniaxial tension. For uniaxial compression, the total dissipated energy, or fracture energy, can increase up to a multiple of the tensile fracture energy.[80,89]

It is rather straightforward to relate the energy dissipation mechanisms proposed by Dougill and his co-workers to the micro-crack processes that were described in Section 3.2.2. Moreover, one might envision what would occur under fatigue loading, as the damping energy factor would increase. However, we will leave this point open to future debate. In literature, many different bounds have been proposed for cyclic tests, such as crossover points, inflection points, common points etc. [86,133,134] These points are all based on typical curvatures in the cyclic stress-deformation curves. They might be helpful for describing the phenomenology of stress-deformation curves. Here we take the point of view that the fracture process can be regarded as a progressive process, much in line with the ideas proposed by Dougill and his co-workers. However, by directly attacking the problem at the micro- and meso-scale, much of what is shown in this chapter can be explained in a simple and straightforward manner.

3.3 UNIAXIAL TENSION (MODE I)

Let us now continue the discussion on uniaxial tension. Many of the processes described above for uniaxial compression apply to uniaxial tension as well. When concrete was initially used as a structural material, no tension design was the rule, and researchers did not pay much attention to the behaviour of concrete under uniaxial tension. Even nowadays, large structures such as dams are built on the basis of no-tension design. The tensile capacity of the material is low in comparison to the compressive strength, i.e., $f_t < f_c/10$. The introduction of numerical methods has changed this completely. It may also be clear from Section 3.2, that at a smaller size scale, i.e., the micro- and meso-level, fracture of concrete in compression is essentially a tensile phenomenon. The introduction of the Fictitious Crack Model (FCM)[5] in 1976 by Hillerborg, Modéer and Petersson has been instrumental in the progress that has been made in understanding tensile fracture. Suddenly, concrete mechanics was drawn in the realm of fracture mechanics, although a few old fashioned continuum ideas survive to date which, as we will see, hamper a full-fledged application of the theory. Note that the overview in this chapter is not a strict historical overview. Some of the developments on tensile fracture preceded those in compressive fracture and vice versa. In this section we will break even more than in the previous one on compression with the classical notion of a continuum.

The Fictitious Crack Model is presented in somewhat more depth in Section 5.3.4. Here we will only mention the major parameters. The material parameters in the model are the tensile strength (f_t), the Young's modulus in tension (E_t), the shape of the descending branch, and the fracture energy (G_f), which is defined as the area under the stress-crack opening diagram. All parameters are summarised in Figure 3.32. Basically, the FCM assumes a linear stress-strain law for the pre-peak regime and a stress-crack opening behaviour for the

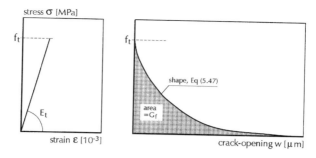

Figure 3.32 Overview of the parameters in the Fictitious Crack Model of Hillerborg et al.[5]

post-peak regime. A quantity that is derived from the above information is the characteristic length l_{ch}, which is defined as

$$l_{ch} = \frac{EG_f}{f_t^2} \tag{3.5}$$

The characteristic length is a measure for the brittleness of the material. For example, concrete with a maximum aggregate size of 16 mm will have a higher fracture energy and a lower tensile strength than a 2 mm mortar, whereas the Young's moduli will differ only slightly. The characteristic length is larger for the more heterogeneous concretes, which behave less brittle.

In the following paragraphs we will discuss the physical mechanism (microcrack processes) underlying tensile softening in concrete. As we will see, the physical processes deviate somewhat from the original assumptions made by Hillerborg and co-workers,[5] as well as from the assumptions made in the crack band model developed by Bažant and Oh.[140] The crack band model is almost identical to the FCM. The main difference is that in the FCM model a line-crack is used, whereas in the crack band model the crack opening is smeared over a certain crack band width. The aforementioned "deviations" are caused mainly from problems in defining the "real" crack tip in concrete. Next we will review the effect of material composition, age, humidity and curing conditions, load-cycling and fatigue on the tensile softening behaviour. In the discussion on material composition we will include some results for fibre concrete as this seems the most practical way to improve the tensile softening properties of cementitious composites. Finally, similar to the previous section on uniaxial compression, we will touch upon the effect of impact loading on tensile fracture.

3.3.1. MICROCRACKING AND MICROMECHANISMS IN TENSION

When a plain concrete specimen is subjected to uniaxial tension, a stress-deformation diagram as shown in Figure 3.33 is measured. Here we reproduce

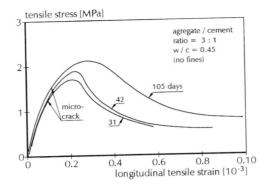

Figure 3.33 Stress-strain diagrams for concrete loaded in uniaxial tension, after Evans and Marathe.[141] (Reprinted from *Materials and Structures,* with kind permission of RILEM, Paris.)

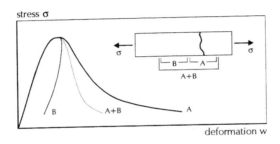

Figure 3.34 Tensile bar and stress-deformation diagrams for measuring devices A and B placed at different locations on the specimen.

the early results of Evans and Marathe.[141] These tests were among the earliest where post-peak behaviour in tension was actually measured in the laboratory. The principle of the test method used by Evans and Marathe is outlined in Section 4.1 (Figure 4.1b). The curves have essentially the same shape as the compressive stress-strain curves that were shown before. The important step that Hillerborg and co-workers made in 1976 was that they omitted the notion of strain in the post-peak regime, but rather included in their model that localization of deformations occurs in a narrow zone, as soon as the peak of the stress-deformation diagram was exceeded. Thus, as depicted in Figure 3.34, when a measuring device is placed over the localization zone, a softening curve is measured, whereas a measuring device placed outside the localization zone registers unloading.

In the original approach, the uniaxial tension test was proposed as the best means to determine the parameters needed in the model.[5,142] The micromechanisms were initially not very clear. Based on the similarity of the FCM to the Dugdale-Barenblatt plastic crack tip model, it was *assumed* that a zone of

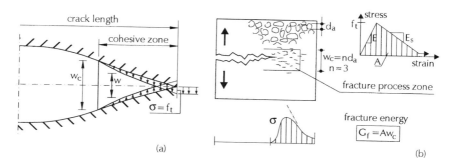

Figure 3.35 Definition of the ("fictitious") crack tip after Hillerborg et al.[5] (a) and definition of process zone after Bažant and Oh[140] (b). (Figure (a) is reprinted from *Cement & Concrete Research*, with kind permission from Elsevier Science, Ltd., Kidlington, U.K. Figure (b) is reproduced by permission of RILEM, Paris.)

distributed microcracking spread out in front of the tip of the macroscopic stress-free crack. In Figure 3.35a the FCM is shown. Over a certain length from the "fictitious" crack tip stress $\sigma(w)$ can be transferred. In the model by Bažant and Oh,[140] which is shown in Figure 3.35b, the crack tip is drawn to the edge of the inelastic zone containing microcracks. Because, as we will see in Chapter 5, nonlinear fracture models are based on energy considerations only, the exact fracture mechanism is assumed to be of less importance.

However, as stressed in the introductory chapter of this book, knowledge of physical mechanisms is considered essential information for deriving a fracture model with some "predictive capabilities". The confusion about the true mechanisms taking place during tensile softening was very clear from a report by Mindess.[143] From a thorough survey of literature he found that depending on the loading application and crack detection technique, process zone sizes between a few micrometers and half a meter were measured by various researchers. Such a situation is of course highly undesirable, and indicates only that the exact mechanism is not known. Some insight into the true behaviour in the softening regime can be derived on the basis of crack measurements in uniaxial tensile tests. We will try to summarize some of the observations here, but the true understanding will have to wait until Chapter 7 when experimental data are compared to the outcome of micromechanical computations.

Petersson[142] sprayed specimens with water while they were in the post-peak regime in tension, and could thus ascertain that a localised crack zone of a certain width was present. Moreover, he found that the width of the localised zone depended on the structure of the material. Coarser grained materials showed a wider zone than finer grained materials. As we will see in the next section, coarsely grained materials consume more energy. Again, as for the description of the compressive fracture process (Section 3.2), it is logical to assume that microcracks are present before any mechanical tensile load has been applied. These microcracks will extend under load, and much of the

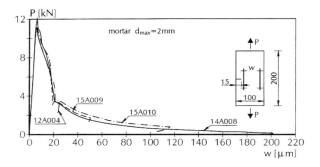

Figure 3.36 Examples of tensile load-average crack opening diagrams from the impregnation experiments.[144] In the inset the specimen geometry is shown.

details of the process will depend on the interaction of the microcracks with each other, but also on interactions between the microcracks and aggregate particles in the material. Moreover specimen geometry (size and shape) and boundary condition effects must be considered, but these matters will be discussed in other chapters (viz. Chapters 4 and 7).

3.3.1.1 Impregnation

The process can best be understood when the *assumed* mechanism of a "cloud of microcracks in front of a stress-free macrocrack" is omitted. In a series of vacuum impregnation experiments and optical microscopy tests much of the details of the fracture process have become clear.[57,144-147] The vacuum impregnation technique is explained in somewhat more detail in Section 4.4.3. Essentially it is possible to monitor internal crack growth in specimens. However, the technique is limited to detection of continuous macrocracks in contact with the outer surface of the specimen only. In the experiments, a single-edge-notched (SEN) prism of size $200 \times 100 \times 50$ mm was loaded in uniaxial tension to a prescribed axial deformation. The specimen is shown in the inset of Figure 3.36. Single-edge-notched specimens were selected because crack initiation would occur at a known location. Note that even in symmetric double-edge-notched specimens crack initiation will be non-symmetric due to the heterogeneous material structure[148] (Figure 4.63). Some examples of stress-average crack opening diagrams from SEN tensile tests on mortar are shown in Figure 3.36. Each test was halted at a different crack opening, viz. at $w = 10, 25, 50,$ 100 and 200 μm. After impregnation with a fluorescenting epoxy, the specimens were cut into six or seven slices and internal cracking could be studied. From the experiments, the extent of cracking in the specimens cross-section could be determined, but also the geometry of the crack in the direction of the tensile load. Impregnation experiments were carried out on four different concretes, namely a 2 mm mortar ($f_c = 49.8$ Mpa and $f_{spl} = 3.8$ MPa), 16 mm normal concrete ($f_c = 48.3$ MPa and $f_{spl} = 3.0$ MPa), 16 mm high strength

TABLE 3.1
Average "Direct" Tensile Strength, Splitting Tensile Strength and Compressive Strength for Four Different Concretes

Mixture	d_{max} [mm]	f_c [MPa]	f_{spl} [MPa]	f_t (n, st.dev) [MPa]([-],[MPa])	f_t/f_{spl} [-]	f_t/f_c [-]
Mortar	2	49.8	3.8	2.70 (8, 0.15)	0.71	0.053
Concrete	16	48.3	3.0	2.46 (5, 0.17)	0.82	0.051
HSC	16	87.8	4.5	2.92 (8, 0.35)	0.65	0.033
Lytag	12/4	56.5	3.8	2.90 (6, 0.09)	0.76	0.051

concrete ($f_c = 87.8$ MPa and $f_{spl} = 4.5$ MPa) and lytag lightweight concrete containing 12 mm lytag particles and sand up to 4 mm ($f_c = 56.5$ MPa and $f_{spl} = 3.8$ MPa). The strength values mentioned were all obtained after 28 days of casting using standard (Dutch standard) 150 mm cubes. The view that is given below was consistent over the complete test series, and in agreement with measurements carried out by others.[149-151] Note that the "direct tensile strength" measured in these displacement-controlled uniaxial tensile tests is much lower than the splitting tensile strengths mentioned above. In Table 3.1 the tensile strength values for these four concretes are summarised. From these results it can be concluded that there is no direct relation between tensile splitting strength f_{spl} and "direct" tensile strength f_t.

The "direct" tensile strength also depends on the boundary conditions of the experiment (Section 7.2.2) and, similarly to the compressive strength, it must be concluded that the tensile strength is a system characteristic and not a material parameter. Note that the "direct" tensile strength of the high strength concrete increases less as the splitting tensile strength. This can be seen by comparing the f_{spl} and f_t values for 16 mm normal concrete and 16 mm high strength concrete in Table 3.1. Obviously, increasing the compressive strength of the concrete by adding silica fume does not necessarily imply an increase of the tensile strength of the material. This is probably related to the interfacial behaviour between aggregate and matrix as discussed in Section 2.2.2.

But, let us continue with the fracture mechanisms. In Figure 3.37 an example of a crack map is shown. The tensile stress is applied in the vertical (y-) direction, and the photographs of Figure 3.37 show the crack growth at various locations x (Figure 3.38 shows a specimen after impregnation and sawn into slices). These cracks run over the thickness of the specimen, i.e., from $z = 0$ to $z = 50$ mm. By placing the photographs of the various slices next to each other as in Figure 3.37, we get an idea of the extent of the internal crack front. In the example of Figure 3.37, a 2 mm mortar specimen was loaded up to an average crack opening (measured at the four corners of the specimen) $w = 50$ μm. An intact core can clearly be seen. Cracks extend from the sides of the specimen into the interior, but do not connect. Note that each of the photographs also shows the geometry of the crack in the tensile direction. Now, what can we conclude from these impregnation experiments?

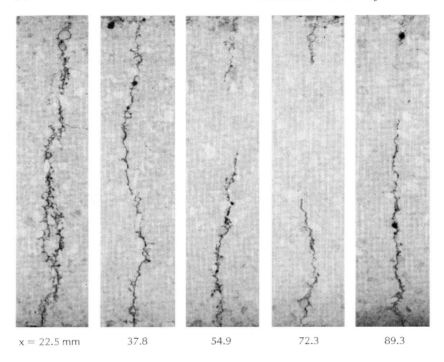

| x = 22.5 mm | 37.8 | 54.9 | 72.3 | 89.3 |

Figure 3.37 Example of a crack map. A 2 mm mortar specimen was loaded up to an average crack opening w_{unl} = 49.8 μm, F_{unl} = 1.65 kN.[146]

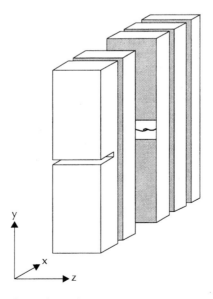

Figure 3.38 Slicing of a specimen after impregnation and definition of the coordinate system.[147]

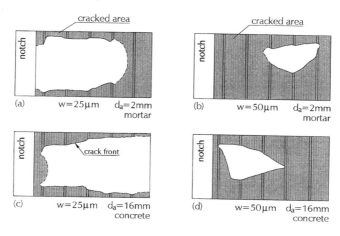

Figure 3.39 Crack fronts derived from the impregnation experiments: (a) in 2 mm mortar at w_{unl} = 25 μm, (b) in 2 mm mortar at w_{unl} = 50 μm, (c) in 16 mm concrete at w_{unl} = 25 μm, and (d) in 16 mm concrete at w_{unl} = 50 μm[145].

Let us consider the diagrams of Figure 3.39. In Figure 3.39a and b the crack fronts at an average axial crack opening of 25 and 50 μm are shown for 2 mm mortar. In Figure 3.39c and d, the crack fronts at 25 and 50 μm are shown for 16 mm normal concrete. Curiously, the (macro) crack seems to initiate at the "un"-notched side of the specimen for the 2 mm mortar. In contrast, for the 16 mm concrete (but also for the lytag concrete and the high strength concrete) crack growth started from the notch, as can be seen in Figure 3.39c. Probably the initial cracks are too small for the 2 mm mortar to intersect with the first saw cut, in which case the crack remains undetected, or alternatively, isolated microcracks are present, which cannot be filled with the impregnation technique either. Observations of detailed surface deformation measurements of the specimen of Figure 3.39a show that initially the largest deformations occur in the notch region, and only later at the side of the macroscopic crack, see Figure 3.40. The reason for the crack growth at the "un"-notched side of the specimen will be elucidated in Chapters 4, 5 and 7, and is related to the boundary conditions in the experiment. From the local deformation measurements of Figure 3.40, it might be concluded that cracking in the form of distributed microcracks occurs in the vicinity of the notch at load levels between 8 and 10 kN. This would be an indication that a microcrack zone exists before the macrocrack starts to propagate.

At w = 50 μm, the (macro) crack has extended well through the specimens cross-section, but for both materials an intact core seems to remain. Note, however, that this internal core may contain distributed microcracks, which cannot be detected with the impregnation technique. When the load-drop (i.e., the peak load minus the residual load in the post-peak regime) is plotted against the cracked area, Figure 3.41 is obtained. At w = 50 μm around 80% of the cross-sectional area is cracked, but the decrease in load is no more than 60 to

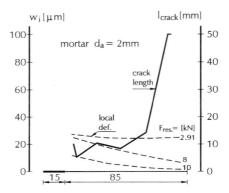

Figure 3.40 Indirect proof of distributed microcracking in mortar: local deformations measured at the specimen's surface at three stages of loading (post-peak), and crack length determined from the impregnation test result.[147]

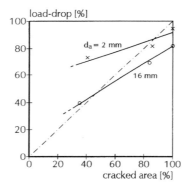

Figure 3.41 Load-drop (= peak load minus residual load) vs. cracked area for impregnation experiments on mortar and concrete.[147]

70%. Again, the difference might be explained from distributed microcracking, or alternatively the intact core might be pre-stressed by hygral stresses caused by non-uniform drying of the specimen before loading. Note that this simple interpretation is based on the assumption that a linear relation should exist between cracked area and load decrease. This is not necessarily correct, as stress-concentrations might have some effect. Essentially, the above reasoning corresponds to assumptions made in continuous damage models.[152,153] Such models "predict" that at final separation, the full cross-section of the specimen is cracked and no load can be transferred anymore. Thus, possible frictional effects or other mechanisms are excluded *a priori*.

Figure 3.41 indicates that the assumptions made in the continuous damage models are not in agreement with the impregnation experiments. At 100%

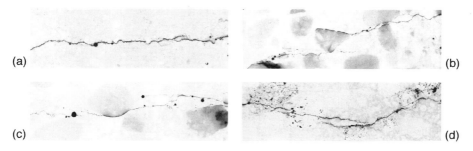

Figure 3.42 Examples of crack face bridging at $w = 100$ μm in 2 mm mortar (a), 16 mm normal concrete (b), high strength concrete (c) and lytag lightweight concrete (d).[57]

cracked area, the specimen can still carry load, even up to $0.2\sigma_{peak}$ for the 16 mm concrete. Obviously some other mechanism is active, which can be resolved by looking in more detail to the crack geometry at 100 μm crack opening. At this stage all specimens were fully cracked, but showed a non-zero residual stress. Four details in the (macro-) cracks in the specimen loaded to $w = 100$ μm are shown in Figure 3.42 (see also Plates 1 through 5). In Figure 3.42a, crack growth in 2 mm mortar is shown, and for 16 mm normal concrete, 16 mm high strength concrete and 12 mm lytag concrete in Figures 3.42b,c and d, respectively. Clearly the cracks are not continuous, but rather overlaps and branches exist. Note that these overlaps are not isolated events in individual specimens, but they have been detected in relatively large quantities.[146]

The type of crack overlap observed in concrete (Figure 3.42) is called *"crack interface bridging"*. Essentially the two crack faces are connected by the ligament between the overlapping crack tips, thus allowing for stress transfer between the crack faces. The mechanism is shown schematically in Figure 3.43a. Individual cracks approach one another, but instead of joining they seem to avoid each other.

This has been shown theoretically by Simha et al.[154] In Sections 5.3.2.4 and 7.2.1, these crack interactions will be addressed in somewhat more detail. At the time when these crack overlaps were observed for the first time, it was hypothesized that the overlap would fail if one of the crack tips would propagate and coalesce in the wake of the second crack as shown in Figure 3.43b. In optical microscopy experiments using a Questar long distance remote controlled optical microscope, we actually observed the failure of crack overlaps in concrete and sandstone specimens on a number of occasions.[57,155] The microscopy technique is explained in some detail in Section 4.4.3. Here, we will show a number of results.

3.3.1.2 Optical microscopy

Essentially the microscope can scan a larger area of a specimen's surface while the specimen is under load. In Figure 3.44, the load-average crack

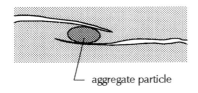 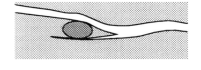

aggregate particle

Figure 3.43 Interacting overlapping cracks as fundamental load transfer mechanism in cracks in concrete (a) and failure mechanism(b).[146]

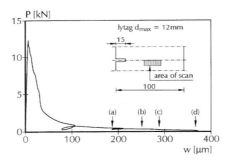

Figure 3.44 Load-average crack opening diagram of a lytag test. The area scanned with the long distance Questar remote measurement system is indicated in the inset, as well as the location of final rupture of a crack overlap that was monitored in real time (see also Figures 3.45 and 3.46).[57]

opening diagram of a uniaxial tensile experiment on a single-edge-notched lytag prism is shown. The specimen geometry and boundary conditions were completely identical to those in the impregnation experiments of the previous section. As can be seen the specimen was loaded to full separation, i.e., until no stress could be transferred anymore. In the inset of Figure 3.44 the area that was scanned with the microscope is indicated. In Figure 3.45 a mosaic of images has been made of the scan area. The crack pattern is shown at two stages of loading, at $w = 100$ μm and at 200 μm. Clearly two main crack branches can be observed, separated by an intact ligament. Note that the two crack branches contain smaller size crack overlaps as well, and also that discontinuous microcracks precede the main crack "tip". Obviously, defining a tip for such a crack system is not straightforward. Upon further loading, the upper crack was arrested (shielded by the lower crack branch), and at some stage even closed. The lower crack branch propagated and joined the upper crack. Details of the crack joining process are shown in Figure 3.46a-f. In Figure 3.46a, the tip of the lower crack branch A is just visible in the lower right corner of the figure. In Figure 3.46b the growth of the lower crack tip can be seen. Below every figure the load carried by the specimen and the average crack width measured at the four corners of the specimen are indicated. Remarkably, not the tip of the lower crack branch extends, but rather some

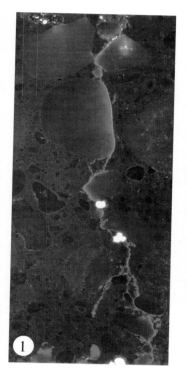

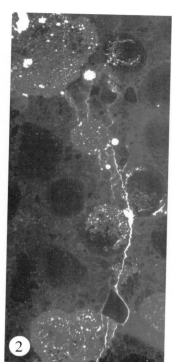

PLATES 1-5 Examples of crack face bridging and branching in high strength concrete (1), lightweight concrete (2,3), 16mm normal concrete (4) and 2mm cement mortar (5). The cracks have been revealed by means of fluorescenting epoxy impregnation (Section 3.3.1.1 and 4.4.3). The average crack opening in all figures is 100 μm. Photographs by the author.

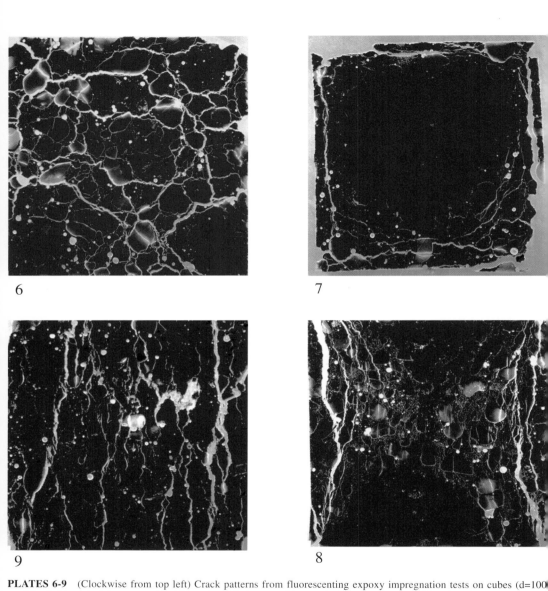

PLATES 6-9 (Clockwise from top left) Crack patterns from fluorescenting expoxy impregnation tests on cubes (d=100 mm) loaded in uniaxial compression. Plates 6 and 9 show cracking vertical to the direction of loading (9) and distributed cracks in the plane perpendicular to the applied load (6), for the case where loading is applied throught teflon platens. Plates 7 and 8 show the well known hour-glass failure mode when steel loading platens are used. See also Sections 3.2.4 and 7.4.2. Photographs reprinted by kind permission of R.A. Vonk.

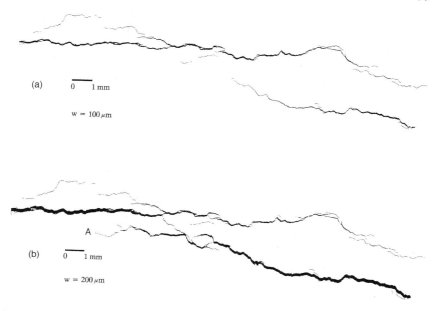

Figure 3.45 Crack growth at the surface of the lytag test of Figure 3.44. Crack-graphs are obtained from a mosaic of video images at a 62.5X magnification. In (a) the mosaic of the scan at $w = 100$ μm is shown; (b) shows the scan at $w = 200$ μm[57].

other imperfection in the wake of this tip leads to further growth above the original tip. This underscores the point made before, that the definition of the "true" crack tip in heterogeneous materials like concrete is very difficult indeed, if not impossible. In Figure 3.46c the tip of the lower crack branch has almost reached the upper crack, and in Figure 3.46d the failure of the ligament is observed. Failure is accompanied by the development of a flexural crack in the ligament. At this stage (Figure 3.46d), the LVDTs measuring the crack opening are out of range, and loading in this particular experiment was continued manually. Final rupture takes place in Figure 3.46e, and in Figure 3.46f the open (and stress-free) crack can be seen. Note that a small piece is torn loose during final rupture. Such debris may be responsible for inelastic effects frequently observed during load cycling, as discussed in Section 3.3.3. Clearly visible is the closure of the upper crack branch during growth of the lower crack branch and the subsequent crack coalescence.

The crack face bridging which has been shown here as a two-dimensional phenomenon is in reality a three-dimensional mechanism. A hypothesized 3D crack face bridge is shown in Figure 3.47. The crack face bridges are essentially "flap-like" structures that connect the two crack faces. Normally this type of fracture is visible when crack surfaces are studied in more detail. The 3D bridging mechanism was confirmed in tests on hardened cement paste.[156]

The crack overlap mechanism is not only found in concrete, but appears in many different brittle disordered materials over a large range of magnitudes.[57]

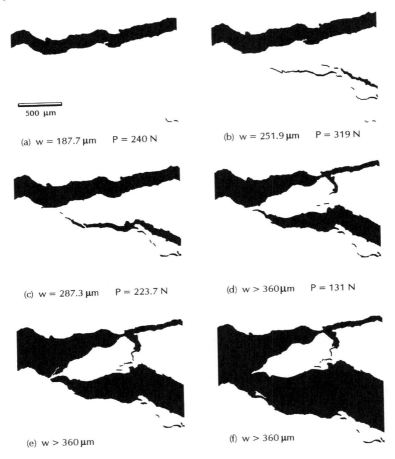

500 µm

(a) w = 187.7 µm   P = 240 N

(b) w = 251.9 µm   P = 319 N

(c) w = 287.3 µm   P = 223.7 N

(d) w > 360 µm   P = 131 N

(e) w > 360 µm

(f) w > 360 µm

Figure 3.46 Six stages in the final rupture of the crack overlap of Figure 3.45.[57]

Sempere and Macdonald[157] show crack overlaps at the kilometre scale at the bottom of the ocean. The mid-atlantic ridge is a spot where magma wells up from the earth's interior, and overlap type crack structures have been observed where the earth's crust separates. Other fine examples can be found in the East African Rift valley. At a scale that is of more interest here, crack overlaps have been found in sandstone in the millimetre size range,[155] and in non-transformable ceramics in the micrometre size range.[158,159] At the atomic scale, crack overlaps are found in numerical simulations.[160] More recently it was shown that the crack overlap geometry is fractal.[161] Later on, in Chapter 7, we will see that the overlap mechanism is a natural outcome of numerical micromechanics models where the heterogeneity of the material is included in the formulation.[162-164]

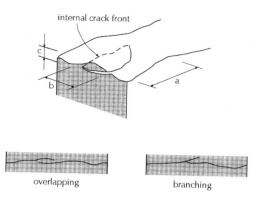

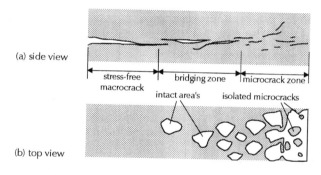

Figure 3.47 Three-dimensional "flap" on a crack surface. The flap is the remains of a 3D crack face bridge.[57]

Figure 3.48 Identification of subsequent mechanisms in mode I cracking of concrete.

Finally, in Figure 3.48 the mechanism of mode I crack growth in concrete is shown schematically. In Figure 3.48a the top view of a fracture zone is drawn, in Figure 3.48b a side view. The fracture shows the following features: distributed microcracking, bridging and branching zone, and final stress-free macrocrack. Seen from above, the crack face bridges appear as isolated islands in a stress-free crack. This schematic figure suggests that definition of a crack tip is extremely difficult. From all our fracture studies not much evidence for frictional stress transfer was found: in most cases, the (surface) crack faces seem to rotate away from one another. Frictional stress transfer was the mechanism hypothesized in other models for tensile fracture.[165] In such models uniformity of cracking was assumed over the specimens cross-section. How-ever, from the detailed analyses of Section 7.2, it will become clear that in a uniaxial tensile experiment a crack zone traverses the specimens cross-section during softening. This implies that in a uniaxial experiment all off the mechanisms shown in Figure 3.48 will be active in some part of the specimens cross-section at the same time. One might say that the assumption of uni-directional localization put forward by Hillerborg and co-workers is oversimplified. The process is truly three dimensional: *localisation occurs in three spatial directions.* Perhaps everything could be lumped in a 2D model, but then effects like non-uniform drying and the associated eigen-stresses should somehow be reflected in the model. The same is true for thermal effects due to heat of hydration in early-age concrete. Fracture roughness is also sometimes quoted as a factor controlling fracture toughness.[166] In normal concrete, more ductile behaviour is found when the crack faces are more tortuous. This factor seems related to the friction model proposed by Duda.[165] Most likely many of the factors mentioned here will contribute to fracture toughness in some way. Using numerical micromechanics, much of the fracture process can be elucidated as we will see in Sections 6.3.3 and 7.2.

3.3.2. EFFECT OF MATERIAL STRUCTURE ON TENSILE SOFTENING

As may have become clear from Section 3.2.2 on micromechanisms in compressive fracture and the above exposure on the tensile fracture mechanism, the softening response of the concrete will depend to a large extent on the composition of the material. From Figure 3.42 it could be seen that the size of the aggregates in the concrete has a profound effect on the width of the fracture zone. This has been demonstrated by others too, using different (more indirect) crack detection techniques.[142,167] Other factors are of importance too, like the w/c ratio of the matrix material as well as the age, which both directly affect the porosity of the interfacial zone between the aggregates and the matrix. The type of aggregate, i.e., the use of lightweight low-modulus and low-toughness aggregates has an important effect on the crack geometry. Addition of fibres may enhance bridging. Furthermore, the age at loading as well as some factors relating to the moisture content, and perhaps thermal gradients in young hardening concrete, all should have some effect on the softening behaviour of concrete. There are similarities in compressive and tensile behaviour, and some of the above matters were discussed in Section 3.2 as well.

3.3.2.1 Effect of particle size and particle type

From the impregnation experiments shown earlier in this chapter, the effect of the particle size on the tensile softening diagram can be retrieved. In Figure 3.49 tensile softening diagrams for 2 mm mortar, 12 mm lytag concrete and 16 mm normal concrete are shown. The size of the aggregates mainly affects the height of the tail of the softening diagram. Thus, the surface under the diagram associated with the fracture energy of the material increases. Note that the lightweight lytag concrete contains up to 4 mm sand particles which act as stiff aggregates. This can be seen in Figure 3.42d, where the cracks run around the (stiff) sand particles, but through the (less stiff) lytag particles.

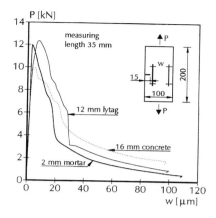

Figure 3.49 Stress-average crack opening diagrams for (a) mortar with $d_{max} = 2$ mm, (b) lytag concrete with 12 mm lytag particles and 4 mm sand, and (c) concrete with $d_{max} = 16$ mm.[57]

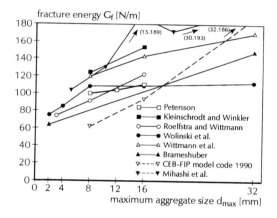

Figure 3.50 Dependency of fracture energy G_f on maximum aggregate size d_{max}.[84] (Reprinted with kind permission of Dr. Hordijk.)

The effect of particle size on fracture energy G_f was studied systematically by several researchers. In Figure 3.50 an overview is given of results obtained in these investigations. The relation that has been proposed in the CEB-FIP model code is included as well. Clearly, increasing the size of the (stiff) aggregates has an important effect on the fracture energy. Note that there are relatively large quantitative differences between the various investigations that are shown in Figure 3.50. However, the same trend is found in all cases. The physical explanation is that bridging and branching become more important in coarse grained materials. For lightweight concrete, cracks run through the aggregate particles, and a lower fracture energy should be expected. However, this is not always true because the w/c ratio can have an important influence as well. In high strength concrete, cracks can run through aggregates as well depending on the relative difference between aggregate and matrix toughness. Therefore, the fracture energy would probably decrease; if not, the tensile strength of the high strength concrete would increase and diminish the negative effect of the reduced crack face bridging. In general there is a tendency for high strength and more homogeneous materials to behave more brittle. However, similar as with compressive fracture, one should be careful in defining brittleness, as this is affected by specimen size and boundary conditions. The brittleness of the material in a given structural application can, for example in the case of the Fictitious Crack Model, be described through the characteristic length l_{ch}, Equation 3.5. The characteristic length is proportional to the fracture energy G_f and the Young's modulus E and inversely proportional to the square of the tensile strength (f_t^2). Thus, for a high strength concrete, G_f increases, but also f_t increases, whereas the Young's modulus will be comparable to the value for normal weight concrete. Depending on the relative increase of tensile strength to fracture energy, the fracture energy will increase or decrease. In general, a lower characteristic length is found for the more brittle materials like cement paste, mortar and lightweight concrete, whereas for coarser grained materials a higher characteristic length is found. For example, Petersson[142] reported l_{ch}-values ranging from 100 to 200 mm for mortar and 200 to 500 mm for normal concrete. More recently l_{ch}-values ranging from 335 mm for lytag lightweight concrete (with $f_c = 27.5$ MPa and $f_t = 1.96$ MPa) to 168 mm for high strength lytag concrete ($f_c = 61.9$ MPa and $f_t = 3.49$ MPa) were reported.[168] For ordinary gravel concrete l_{ch} varied between 474 and 599 mm, with the lower value for materials with a higher strength.[168] Note that the characteristic length defined by Hillerborg is not the only existing brittleness number. Others[169,170] proposed different definitions as well. We will not go into further detail here. In Section 8.4, a potential application of brittleness numbers will be presented.

For materials with extremely large particles, such as dam concrete with aggregates up to 120 mm, the fracture energy increases up to 175 to 235 N/m.[171] The large particles have a negative effect on the tensile strength. These experiments were carried out on three different concretes obtained from 30 year old dams.

Fibres added to the concrete may have a positive effect on bridging. Continuous fibres tend to spread the cracking over the entire specimen length,[172] similar to crack growth at regular spacing in reinforced concrete beams. Discontinuous fibres may have a somewhat similar effect, and Aveston et al.[173] were the first to demonstrate that cracking would take place in a wider zone when steel fibres are added to the concrete mix. Many types of fibres can be used, such as steel, polypropylene, aramide and glass fibres. Moreover, many different shapes have been designed to improve the bond between the fibre-reinforcement and the matrix material. Examples of different stress-crack opening diagrams for fibre-reinforced cements from uniaxial tension tests are shown in Figure 3.51. The behaviour can be quite diverse depending on the

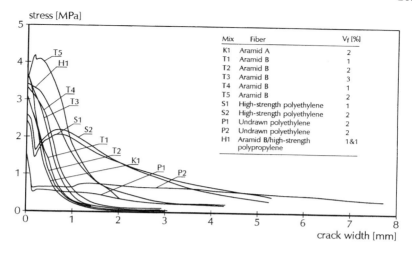

Figure 3.51 Stress-average crack opening diagrams for various fibre-reinforced cements.[174] (Reprinted with permission of ACI, Detroit, MI.)

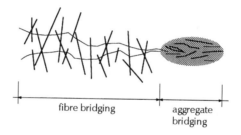

Figure 3.52 Fibre bridging mechanism in cracks.

properties of the fibres (length to diameter ratio, Young's modulus) and bond properties.

Basically, the fibres can act as an additional bridging mechanism as shown in Figure 3.52. When the crack face bridging by grains (Section 3.3.1) has failed, stress transfer over the crack is possible through fibres intersecting with the crack. The maximum bridging stress that can be reached depends then on the elasticity of the fibres, the number of fibres, the inclination of the fibres with respect to the crack faces, and the pull-out behaviour. Note that in Figure 3.51 the total crack width increases up to 4 to 6 mm, a factor 10 larger than the maximum crack width in plain concrete (i.e., the crack width at which the crack becomes stress-free). Several researchers attempt to model the behaviour of fibre-reinforced cements from a meso-level approach.[175,176] Essentially they try to optimise fibre-reinforced cements from fundamental knowledge of load transfer mechanisms in these materials. It carries too far in the context of this

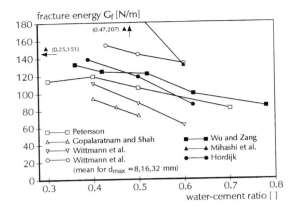

Figure 3.53 Effect of w/c ratio on fracture energy G_f.[84] (Reprinted with kind permission of Dr. Hordijk.)

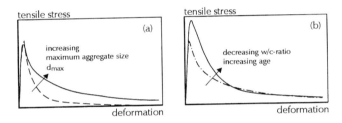

Figure 3.54 Schematic representation of the influence of maximum aggregate size (a) and w/c ratio and age (b) on the tensile stress-crack opening diagram.

book to discuss these mechanisms in more detail. In a recent book these matters were treated exhaustively.[177]

3.3.2.2 Effect of water-cement ratio and porosity

As mentioned before, the w/c ratio has an important effect on the porosity of the matrix material and the interfacial zone between aggregate and matrix. The porosity has an important influence on the fracture behaviour, as will be discussed in Section 6.3.3. As the tensile strength increases with decreasing porosity, as was shown in Section 2.1.3, the fracture energy should change with decreasing w/c ratio as well. In Figure 3.53 fracture energies for concrete mixtures made with different w/c ratios are plotted. With decreasing w/c ratio, the porosity decreases and an increase of fracture energy is observed. Because the w/c ratio has an important effect on the tensile strength of the material, the increase in fracture energy at low w/c ratio can be explained largely from the increase of the tensile peak in the softening diagram. The main effects of particle size and w/c ratio on tensile softening can be summarised as in Figure 3.54. Note that an increase in age results in a decrease of porosity (at least when

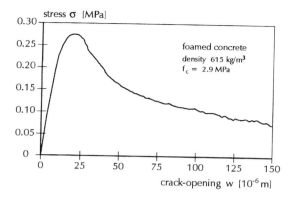

Figure 3.55 Tensile stress-crack opening diagram for foamed concrete.[83]

sufficient water is available for the hydration of the cement to continue). Thus, the effect of age is comparable to that of a decreasing w/c ratio.

A material with an exceptional porosity is foamed concrete. In this material, foam is added to the cement, and a high volume fraction of the material comprises pores. The density of the material is generally between 400 and 1200 kg/m³, i.e., much lower than normal gravel concrete.[178] The Young's modulus of the material is very low, i.e., 1500 MPa, whereas the tensile and compressive strength is 0.28 MPa and 2.9 MPa, respectively. Surprisingly, the fracture energy of the material is relatively high, i.e., G_f = 25 MPa.[83] A stress-crack opening diagram of foamed concrete is shown in Figure 3.55. The salient characteristics are the low tensile strength and the relatively high ductility, which is manifested in a rather high descending branch. No physical explanation was given for this peculiar behaviour,[83,178] but it seems that the pores act as crack arresters and deflectors, as will be discussed in Section 6.3.3.

3.3.2.3 Effect of curing conditions

Hygral and thermal gradients in a specimen may cause eigen-stresses. In Figure 3.56 the effect is shown schematically. When a specimen is allowed to dry, i.e., when the ambient atmosphere has a lower relative humidity than the specimen itself, moisture gradients will develop in the material. As we have seen in Chapter 2, cement paste is susceptible to moisture changes. The distance between the CSH particles will decrease when the moisture content decreases. In other words, the material will shrink. Because only those areas of the specimen where the moisture content decreases will shrink, internal stresses will build up. Depending on the moisture gradient that will develop during drying, these eigen-stresses might exceed the tensile strength of the material, thereby causing crack nucleation and growth. Because equilibrium must exist, the tensile stresses at the surface must be balanced by compressive stresses in the interior.

Similar effects may happen when thermal gradients are caused by the hydration of the cement and the subsequent cooling of the material. The result

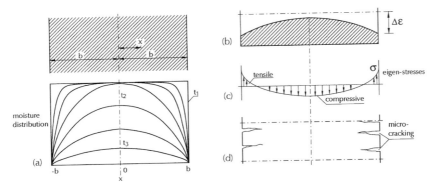

Figure 3.56 Effect of drying on hygral gradients and eigen-stresses in concrete.

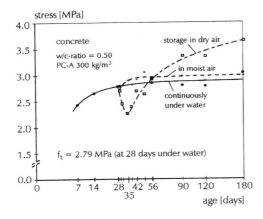

Figure 3.57 Effect of moisture condition on tensile strength of concrete.[179]

is in general a deviation from the peak load for a homogeneous (in terms of temperature and moisture content) specimen. Several researchers have studied the effect of curing conditions on tensile strength and softening.[84,179-182] As far as tensile strength is concerned, the specimen strength will decrease when tensile eigen-stresses are present. For example, for concrete drying in air (laboratory conditions), a decrease of the tensile strength of 20% may occur after three to five days of drying, whereafter the relative tensile strength will increase again as the moisture gradients slowly disappear. Note however that diffusion in concrete is an extremely slow process,[183] and it may take years in thick concrete elements before equilibrium is reached. When a specimen is constantly kept under water, or when it is stored in a fog-room, no decrease of apparent tensile strength is observed. This can be seen in Figure 3.57.

When specimens are dried in an oven at 105°C, the apparent tensile strength of the specimen increases.[84,182] There are probably no moisture gradients present, and all the physically absorbed water has been removed from the concrete, forcing the CSH particles to closer distances, see Section 2.1.3. The bonding between the particles will improve, and the macroscopic strength will increase.

Hordijk[84] and Brameshuber[181] measured the change in fracture energy for different drying methods. For specimens left to dry in the laboratory at 60% Relative Humidity, Hordijk observed an increase of fracture energy from 80 to 100 N/m for normal weight concrete with an age varying from 16 to 21 days, whereafter a more or less equilibrium state developed of around 100 N/m. Brameshuber measured the change in G_f for concrete at very early ages, e.g., from 6 to 72 hours. He first observed a decrease of G_f both for sealed concrete and concrete stored under water. Later, the fracture energy would increase. Thus, the curing condition has an important effect on crack propagation as well.

A clear example of the effect of curing and the associated shrinkage cracking on (mechanical) crack propagation in hardened cement paste (w/c ratio = 0.5) is shown in Figure 3.58. Specimens were cured in water either for 14 or 28 days. After this period, the specimens were kept in the laboratory environment at a RH of 50%. The age at testing varied between 35 and 40 days. The specimens that were kept 14 days under water showed a large quantity of shrinkage cracks. When mechanical (uniaxial tensile) load was applied, the crack propagated along these pre-existing shrinkage cracks, and the material seemed to behave more ductile. When the material was kept under water for an extended period of time, no surface cracking occurred, and a smooth crack developed in the specimen under uniaxial tension. The fracture surfaces are compared in Figure 3.58b. Clearly visible is the difference in smoothness. Surprisingly, the descending branch of hardened cement paste shows an increase of stress during a short interval in the descending branch, see Figure 3.58a. This is a strong indication that linear elastic fracture mechanics is applicable to this rather homogeneous material as will be discussed in more depth in Section 5.3.2.3. For concrete such significant differences will be difficult to visualise as aggregates and pores may further affect the crack tortuosity.

It will be clear that the exact fracture energies depend on the porosity and permeability of the concrete and the conditions in which the material is stored. Moreover, the specimen dimensions will have an effect on the hygral (but also on the thermal) gradients. Therefore, the situation should be studied from case to case, and it is difficult to give quantitative results for the many different situations that may be of interest in practice. This will not be attempted in this book. In the ideal situation, the numerical simulation programs (see Chapters 6 and 7) should include subroutines to calculate the effect of hygral and thermal

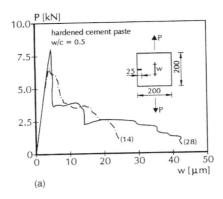

(a)

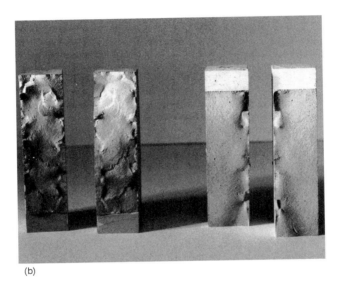

(b)

Figure 3.58 Load-crack opening diagrams for hardened cement paste (a). The number between brackets indicate the number of days of underwater curing. Testing was done after 35 to 40 days of hardening. (b) shows the effect of shrinkage cracking on the final crack path after 14 and 28 days of underwater curing.[184]

gradients on the properties of the material. Attempts to model these phenomena are well under way.[185,186]

3.3.3 EFFECT OF LOAD-CYCLING, FATIGUE AND IMPACT

The final topic that we would like to include in this section on mode I cracking is load-histories. In uniaxial tension load-histories are limited to loading/unloading and reloading cycles much in line with what has been discussed in Section 3.2.6 for compressive loading. Moreover, the loading rate and/or the number of cycles can be varied. This leads to fatigue and impact. Some pioneering work on the effect of load-cycling on the stress-crack opening

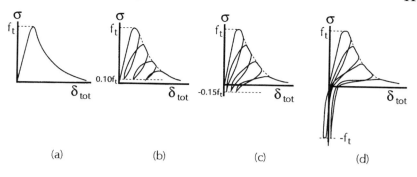

Figure 3.59 Monotonic and cyclic load-paths.[82]

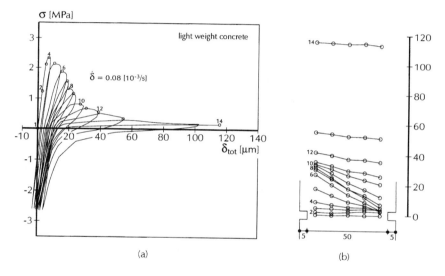

Figure 3.60 Cyclic stress-crack opening diagram for lightweight concrete (a), and distribution over the width of the specimen (b).[82] The scan numbers in (b) correspond to the stages of loading indicated in (a).

diagram in uniaxial tension is from Reinhardt and co-workers.[82,84,187-190] Among others, Gylltoft[191] and Hordijk[84] have attempted to model the cyclic behaviour in tension. Load-cycling, like in compression, can be done between a variety of stress limits. In Figure 3.59, examples are shown of cyclic load-paths in uniaxial tensile tests carried out by Cornelissen et al.[82] Load cycles were carried out between a lower stress-level (viz. $0.10f_t$, $-0.15f_t$ and $-f_t$) whereas the upper stress level varied and corresponded to the tensile stress-crack opening envelope curve. Note that in the last two load-paths a compressive lower bound stress was selected.

An example of a cyclic test type (d) is shown in Figure 3.60. The specimens were double-edge-notched prisms (with a notch depth of 5 mm) of size 250 ×

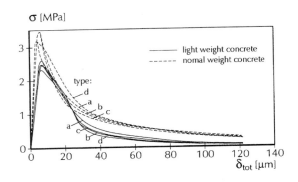

Figure 3.61 Envelope curves for load-paths (a) through (d) of Figure 3.59.[82]

60×50 mm³. The effective area between the notches was 50×50 mm². In Figure 3.60b the distribution of axial deformations are shown over the specimen width. Note that in this last diagram, each data point reflects the average deformation measured at the front and back face of the specimen. The Fictitious Crack Model (see Figure 3.32 and Section 5.3.4) *assumes* that opening in the fracture zone is uniform. However, Figure 3.60b shows clearly that non-uniform deformations occur during scans 4 through 10. This indicates that crack growth is extremely non-uniform, and that an important assumption of the Fictitious Crack Model is violated. For a long time, the non-uniform opening was neglected by Reinhardt and co-workers. The belief in the theory was simply too strong. In Section 4.1.3.2 we will come back to these matters in more detail.

The axial deformation in Figure 3.60a is the average deformation measured at 10 locations at the front and back surface of the specimen. The deformation includes the elastic deformation in the measuring length of the extensometers (see also Figure 4.4). Thus, the pure crack opening is not shown. The diagram very much resembles the cyclic compression curves that were presented in Section 3.2.6.2. The same question can be raised here: does the cyclic envelope curve correspond to the monotonic softening curve? In Figure 3.61 we see the result of the cyclic tests of Cornelissen et al.[82] The average envelope curve (always of three tests) are compared for the load-paths of Figure 3.59. Results are shown both for Normal-Weight Concrete (NC) and for Lightweight Concrete (LC). No significant effect of load-path on softening behaviour was found, and it can be concluded that the softening curve is not affected by the earlier load history, at least for the four load-paths investigated. A more detailed investigation of the cyclic behaviour of concrete was carried out by Hordijk,[84] who performed tests with load-cycles within a larger loop. Unloading/reloading curves are important when crack closure effects occur in structures. In Figures 3.45 and 3.46 we already saw an example of crack closure in a tensile experiment where two overlapping cracks developed. In that example, some loose debris was formed, which may cause irreversible deformation.

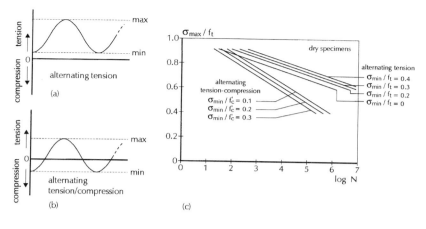

Figure 3.62 Average *S-N* curves for constant amplitude tests for various lower stress limits, for dry concrete subjected to cyclic tension or to alternating tension-compression.[188]

A large number of tensile fatigue experiments have been carried out by Cornelissen.[188] In general load cycles are carried out with a given frequency between a constant upper and lower stress level. Of main interest is the number of cycles that can be performed until failure occurs. In Wöhler diagrams (or *S-N* curves), the upper stress level (σ_{max}/f_t) is plotted against the logarithm of the number *N* of cycles to failure, as for example shown in Figure 3.62. Average *S-N* curves (at a frequency of 6 Hz) are shown for different lower stress limits (σ_{min}/f_t), for dry concrete subjected to cyclic tension or alternating tension-compression. For concrete, no fatigue limit as found for steel has been detected until now. The combined effect of the lower and upper stress levels on the number of cycles to failure is generally shown in a so-called Goodman diagram. Some results can be found elsewhere.[192] More recently, in a small number of tensile fatigue experiments, Hordijk tried to find out whether the softening curve can be used as fatigue limit.[84] However, the number of experiments was too small to draw final conclusions.

Tensile impact experiments have been carried out by a number of authors.[129,193-195] Figure 3.63 shows the effect of loading rate on the relative impact tensile strength. The results show a large increase of relative strength for very high loading rates, i.e., strain rates larger than 1/s. Recently, Rossi et al.[196] argued that wet concrete was substantially more rate-dependent than dry concrete. It should be noted, however, that in this last publication only a very limited range of loading rates was investigated, i.e., up to stress rates of approximately 50 MPa/ms, i.e., in the lower branch of Figure 3.63. Ross and Kuennen[129] performed numerous uniaxial tensile and splitting tensile impact tests in a Hopkinson bar. Their results, which were included in Figure 3.28, are in agreement with those presented in Figure 3.63. Measuring deformations under impact tensile loading (and perhaps also in compressive impact) is not a straightforward task. Weerheijm[126] carefully tried to improve these methods

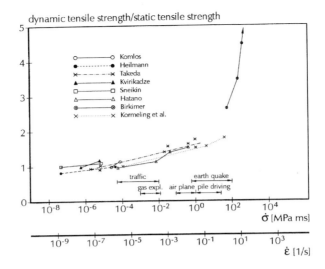

Figure 3.63 Effect of loading rate on the relative tensile strength of concrete.[195]

for Hopkinson bar experiments. However, some controversy remains, mainly because separate load-time and deformation-time diagrams must be combined to retrieve a stress-deformation diagram. We will not delve into too much detail here; static loading remains the main topic of the book. Finally, when concrete fractures under tensile impact, sometimes multiple crack planes develop.[197] The development of multiple cracks is sometimes used as an explanation for the apparently increased energy consumption under impact loading.

3.4 SHEAR (MODE II, III AND MIXED MODES)

Up till now we have discussed the behaviour of concrete under "simple" uniaxial compression or uniaxial tension. In structures, shear may occur as well, which has been subjected to heated debates in the past decades. In principle, shear can be considered as a biaxial tensile-compression state of stress. This can be shown under a simple rotation of coordinate axes. From this point of view, it would be sufficient to consider the multiaxial behaviour of concrete only: the tensile-compressive regimes of the failure surface for concrete would then include all shear stress states. However, the situation is not as simple as that. After concrete has fractured, or when the material becomes very anisotropic, for example through the addition of many short steel fibres as in SIFCON (Slurry Infiltrated Fibre CONcrete), shear failure may occur in weak planes in the material. Shear in existing cracks has led to vivid discussions at conferences in the past decade. Several researchers have faith in the existence of shear fracture energy of concrete, in the spirit of the Fictitious Crack Model. Moreover, the extension has been attempted to mode III (out-of-plane shear) following classical assumptions from linear elastic fracture mechanics. A short deviation to the crack modes distinguished in linear elastic fracture mechanics seems in place here. The theory is presented to some extent in Chapter 5: in Figure 5.20 the three basic fracture modes are shown. They are, tensile opening (mode I), in-plane shear (mode II) and out-of-plane shear (mode III). Moreover, combined or mixed-modes can occur. The question is now can we simply use these three crack modes from linear elastic fracture mechanics, apply the philosophy behind the Fictitious Crack Model, and determine the various fracture energies, i.e., for mode II, III and the mixed-modes in a simple test as we tried to do for mode I loading in Section 3.3? For mode I we already mentioned shortly the problem of non-uniform loading in a fracture zone, and moreover, the seemingly difference of the fracture mechanisms from the originally *assumed* "cloud of distributed microcracks in front of a stress-free macrocrack". The heterogeneity of the material causes deviations from such assumptions, although, on the other hand, they also seem to be partially true. Undoubtedly, the heterogeneity of the material might cause similar problems when mode II, III and mixed-mode fracture energies are determined. It is not at all certain whether the material will simply follow our assumptions. Unbiased close observation of the material behaviour in experiments is essential for progress in the field. At the same moment we must consider theoretical predictions, and compare them to the experiments. In this section we will present the debate on shear (mode II) fracture of concrete. The contradicting views will be compared, and some results of experiments will be presented. First we will discuss several approaches originating from different fracture mechanics theories, to mode II, III and mixed mode fracture. Then, the debate on whether mode II fracture really exists will follow, and finally, we will present some attempts to measure mode III fracture in concrete.

3.4.1 DIFFERENT APPROACHES TO SHEAR FRACTURE

The level of observation is of extreme importance in discussing shear and mixed-mode fracture of concrete. Let us first assume that the continuum representation is applicable. Two schools can be distinguished in shear fracture of concrete. In the first school, definitions originating from classical linear elastic fracture mechanics are used. Thus, the experiments are designed conform assumptions made in LEFM: under pure mode II a crack should initiate and propagate under uniform shear stress.

For example, Bažant and Pfeiffer[198] define a shear crack as shown in Figure 3.64a. This particular definition is an extension of the LEFM definition, although it is assumed that as a precursor to the development of the final shear crack, an array of inclined tensile (mode I) microcracks develops in the material. At failure, the inclined cracks are *supposed* to join, in order to form a single shear plane as indicated. A second approach would be an extension of the non-linear Hillerborg, Fictitious Crack Model for shear. In this approach, a process zone — originally defined as a region of distributed microcracking

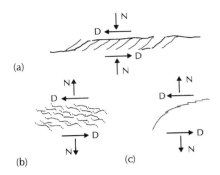

Figure 3.64 Mode II crack growth through joining of inclined mode I microcracks (a), process-zone subjected to shear (b), and curved-crack propagation under mixed-mode I and II loading (c).

in front of a stress-free macrocrack (see Figure 3.35) — is subjected to in-plane shear. Such a process zone under shear is shown — hypothetically — in Figure 3.64b. We might refer to this situation as a continuum approach to shear fracture. A third situation occurs when the shear zone is not sufficiently confined, i.e., when a mixed-mode state of stress is allowed to develop. In Figure 3.64c this possibility is sketched. The crack will deviate from linearity and will propagate along a curved path. Thus, the main question remains the definition of a mode II crack or a mixed-mode I and II crack. In the past few years, many attempts have been undertaken to measure the mode II fracture energy of concrete,[198-201] but also the mixed-mode I and II fracture energy,[202,203] whereas others have attempted to extend the Fictitious Crack Model to mixed-mode I and II situations.[204-206] When measuring mode II crack propagation, ideas from linear elastic fracture mechanics are used. In mode II, crack growth should be confined to a plane. *This assumption is essential!* The same applies to the extension of the Fictitious Crack Model to include shear: the process zone should be confined to a small region, and no crack extension outside this area along a curved path or secondary cracking in inclined directions is accepted. Now, let us first review tests for measuring mode II and mixed-mode I and II fracture energy. Subsequently, we will present the extensions of the Fictitious Crack Model under shear. Later, in Chapters 4 and 7, we will analyze shear fracture using a simple micromechanics model.

3.4.1.1 Linear fracture mechanics approach

Bažant and Pfeiffer[198] have most strongly argued in favour of a mode II fracture energy for concrete. The mechanism underlying their claim was shown before in Figure 3.64a. The idea is that crack propagation is confined to a narrow region, preferably a line. The most common test specimens to reach this goal are the Iosipescu beam geometry[257] with two notches, or a symmetric punch-through-specimen. Both the beam geometry, used by Bažant and Pfeiffer, and the punch-through-cubes by Davies[199,200] are shown in Figure 3.65 as representatives for this class of experiments.

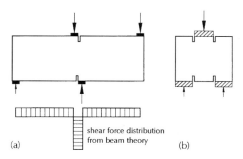

Figure 3.65 Four-point-shear beam and shear load distribution from beam theory (a), and punch-through-cube (b).

In the four-point-shear beam a narrow zone exists where, following beam theory, a highly uniform shear load develops. This is shown in Figure 3.65a. In the same region, a linear varying bending moment is present. According to Bažant and Pfeifer, a shear crack would develop in the narrow shear zone in the concrete beam. The actual distribution of external loads acting on the beam depends on the specimen dimensions and the relative distance of the supports. In the four-point-shear beam of Bažant and Pfeiffer, the ratio between the loads in the inner and outer supports was 1:15. Similar four-point-shear beams (some with axial pre-stressing along the axis of the beam) were tested by Swartz and Taha.[208] Recently, the experiments of Bažant and Pfeiffer were repeated, using exactly the same specimen dimensions and a concrete mixture which resembled that of Bažant and Pfeiffer as closely as possible.[209] In the punch-through specimens of Davies (Figure 3.65b), zones of pure (in-plane) shear are also *assumed* to develop between sets of opposite saw-cuts in cubical specimens. The cubes are easy to handle in the laboratory (in fact, more simple than the four-point-shear beams) and can be loaded in a standard compression machine. As we will see, some additional measures are essential for loading the four-point-shear beam under unrestrained loading (see Section 4.1.2.2). Bažant and Pfeiffer claimed that the beams failed in pure mode II. However, in their experiments a stroke-controlled compression machine was available, i.e., the displacement of the piston in the hydraulic actuator of the compression machine was used as control variable. As we will see in Chapter 4, selecting the proper control variable in four-point-shear experiments is not a straightforward task. The piston displacement undoubtedly will have caused an explosive failure around peak. This makes the test not very suitable for fracture studies and, moreover, incorrect peak loads may have been measured as instabilities cause premature failure in the pre-peak regime(!), see also Figure 4.11. More seriously, it can also be argued whether beam theory applies to the beam of Figure 3.65a. The height-to-span ratio is 1:8, but then we consider the complete beam and neglect the short span between the middle support and middle loading point. In fact, Ingraffea and Panthaki[210] raised serious questions against the pure mode II fracture claimed by Bažant and Pfeiffer. According to Ingraffea

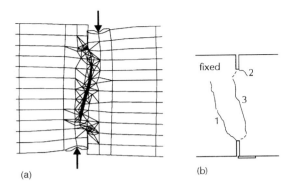

(a)   (b)

Figure 3.66 Curved-crack growth from the notches in a four-point-shear beam, and growth of an inclined, tensile splitting-crack (a),[210] and growth of curved cracks from the notches in a stable, displacement-controlled, four-point-shear test (b).[152,209] (Figure (a) is reprinted by permission of the publisher [ASCE].)

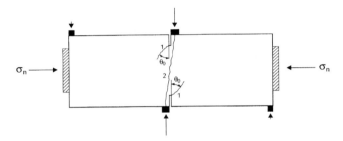

Figure 3.67 Effect of axial confinement on the spacing between the curved cracks in four-point-shear tests.[208] (Reprinted from *Engineering Fracture Mechanics,* with kind permission from Elsevier Science, Ltd., Kidlington, U.K.)

and Panthaki, two curved cracks would first develop from the upper and lower notches of the specimen as shown in Figure 3.66. In Figure 3.66b the outcome of a recent experiment is shown, confirming the development of curved cracks in carefully controlled shear experiments.[209] Similar observations were made by Swartz and Taha.[208] In their experiments, the curved-crack branches were more closely spaced because confinement in the direction of the beam axis was applied, see Figure 3.67.

On the basis of numerical analyses, Ingraffea and Panthaki[210] concluded that as soon as the curved cracks were fully developed, an inclined, tensile splitting-crack would lead to complete rupture as shown in Figure 3.66a. The situation resembles a Brazilian splitting test. Thus, the conclusion might be that curved mode I (or perhaps mixed mode I and II) cracks develop from the notches, and axial splitting failure (in mode I) finally leads to complete rupture of the beam. This is far from the supposedly mode II fracture. The result suggests that

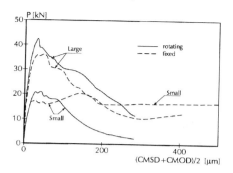
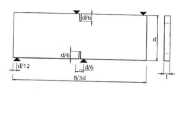

Figure 3.68 Effect of boundary restraint in four-point-shear beam tests on beams of two different sizes, $d = 150$ mm and 300 mm (a). The beam is shown in (b).[156, 209]

extreme care should be taken in the interpretation of fracture experiments. Again, similar as in the mode I cracking of concrete, many assumptions are made, often based on a vague similarity in behaviour observed in other materials. The four-point-shear test will be analyzed critically in Sections 4.1.2.2 and 7.3.1.

One last result is shown here, namely the effect of boundary restraint on load-displacement behaviour. Friction in the supports may cause confinement in the narrow shear zone in the four-point-shear test, as will be discussed in detail in Section 4.1.2.2. The difference between experiments between freely rotating supports and fixed supports is shown in Figure 3.68. Basically, restraining the supports leads to a higher tail in the post-peak curve. This was observed both for small and large beams of concrete and sandstone.[156,209] The crack pattern shown in Figure 3.66b was found in a test between fixed supports.

In the punch-through experiments of Davies, similar behaviour may have occurred. In that particular experiment, two shear cracks should develop simultaneously from the two sets of notches. In concrete, this demand is in strong conflict to the heterogeneous nature of the material. We will not develop further the discussion on the punch-through specimen in this book.

3.4.1.2 Mixed-mode I and II fracture energy?

The second issue mentioned is mixed-mode I and II fracture energy. RILEM committee 89-FMT developed a Round Robin test for measuring the mixed-mode fracture energy of concrete.[211] The proposed test is a variant of the four-point-shear test mentioned in the previous section, but then with a single notch as shown in Figure 3.69. In the beam, a crack will grow along a curved trajectory under presumed mixed-mode I and II loading. Many experiments were carried out by Carpinteri et al.[202,203] Their main conclusion was that the mixed-mode I and II fracture energy would increase by about 30% with respect to the mode I fracture energy. Later, it was shown that the increase of fracture energy could be attributed to frictional restraint in the supports.[212] It was shown

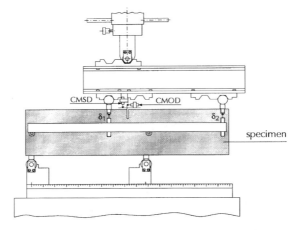

Figure 3.69 Four-point-shear test on single-edge-notched beams as proposed for the RILEM 89-FMT Round Robin test.[211] (Reprinted by kind permission of Chapman & Hall, Andover, U.K.)

that only opening perpendicular to the crack occurred. Simply, a curved mode I crack propagates through the beam. When experiments are carried out using the set-up of Figure 3.69, frictional restraint in the supports and loading points will confine the growth of the crack. The loads on the inner and outer supports and load-points should be in a fixed ratio during the complete test. This can be achieved only when the beam is loaded between freely rotating pendulum bars in a tensile machine as discussed in Chapter 4. We will not discuss these matters here in detail; more information can be found elsewhere.[156,212]

3.4.1.3 Extension of the Fictitious Crack Model for Shear

In Figure 3.64b we showed a process zone subjected to shear. In view of the three basic crack modes defined in linear elastic fracture mechanics, extension of the Fictitious Crack Model to shear seems a straightforward development step. In the late 1980s, experimental set-ups were developed at three European Universities, namely by Hassanzadeh et al.[213] in Lund (Sweden), Reinhardt et al.[206] and Nooru-Mohamed[214] at the Stevin Laboratory of Delft University of Technology, and Keuser and Walraven[215] at the University in Darmstadt (Germany). The idea behind these experiments was to extract information of shear sliding at narrow crack openings for so-called aggregate interlock theories. Aggregate interlock models describe the frictional restraint in cracks, sometimes including the effects caused by confinement normal to the crack plane, and/or dowel action in the sliding crack caused by steel reinforcement.[216-219] A reinforced crack under shear is shown schematically in Figure 3.70. The reinforcement, which may be either a steel rebar of larger diameter (typically 6 to 50 mm) or short steel fibres (0.5 to 1.0 mm diameter), or any other type of fibre made of different materials, adds to the shear transfer in the crack. In

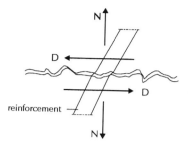

Figure 3.70 Aggregate interlock and dowel action in (fibre) reinforced concrete under shear.

plain concrete and mortar we have to rely on the interlock between the two rough crack faces. In Figure 3.70, the crack width is larger than w_c (see also Figure 3.32), i.e., the crack has completely "softened" and normal stresses cannot be carried by the concrete. In rough cracks sliding is accompanied by an uplift (or dilatancy), which may of course be hindered by steel reinforcement normal to the plane of the crack.

Shear transfer in unreinforced cracks is caused by sliding friction along contacts between the two crack faces. Most likely, loose debris in the crack (see Figure 3.46f), may have some effect on sliding as well. For narrow crack openings, but beyond w_c, the areas of contact are relatively large and the shear capacity increases. The model developed by Walraven explains these effects from the crack roughness caused by the structure of the concrete quite well. The aggregate interlock models developed in the 1970s and 1980s could be verified only against evidence from experiments at relatively large crack openings. For example, in Walraven's experiments crack openings between 10 and 400 µm were studied. However, in view of the method of producing the initial cracks (by splitting between steel rods in a load-controlled set-up), it is rather doubtful whether the crack width has not exceeded these prescribed values, at least during some interval of the loading sequence. The match between theory and experiments usually is less good for narrow cracks, perhaps because of the above problems in creating narrow cracks, but most likely also because the softening of the concrete is omitted completely. The aggregate interlock models were derived primarily for use in (smeared) finite element models for simulating the behaviour of reinforced concrete structures. The aggregate interlock component was expected to explain — at least part of — the shear capacity of cracked-reinforced concrete structures. We will not delve further into aggregate interlock theories here; a short introduction can be found in Chapter 5. Instead we will focus the discussion on mechanisms that might become important when a crack develops under combined tension and shear.

As mentioned, several experimental efforts in the late 1980s and early 1990s have aimed at a better understanding of shear in cracks at narrow openings, or at crack growth under a combined tensile and shear load. The experiments that

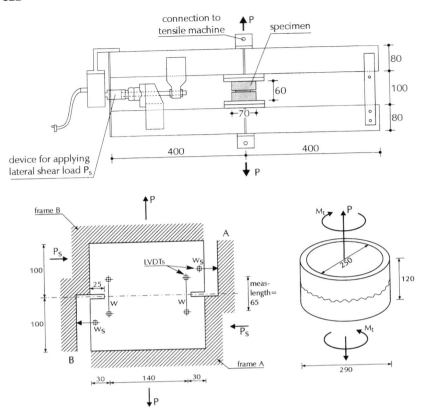

Figure 3.71 Mixed-mode set-up after (a) Hassanzadeh,[205] (b) Nooru-Mohamed[214] and (c) Keuser and Walraven.[215] (Figures (a) and (b) are reprinted by kind permission of the authors.)

were devised are shown in Figure 3.71. Hassanzadeh in Lund, tested relatively small specimens under combined tension and shear in the set-up of Figure 3.71a. The specimen size was small (70 mm cubes with a circumferential saw-cut of 15 mm deep at half height) in order to overcome instability problems during tensile cracking. Also, he attempted to create a uniform process zone in the specimen without having the nonuniform fracture problems that were touched upon in Section 3.3.3, and which will be further elucidated in Chapters 4, 5 and 7. The specimen was glued between two parallel steel beams. The complete set-up was placed in a tensile machine, and in displacement control (see Section 4.1), a narrow crack opening could be achieved, i.e., small enough to remain in the softening regime. Subsequently, a shear load could be applied by means of a manual device. The philosophy underlying the experiment was that in real concrete structures, cracks would propagate such that always a linear or parabolic relationship was maintained between the crack-opening displacement and the crack-sliding displacement. Therefore, only linear paths

$w = (tan \ \alpha).w_s$ and parabolic paths $w = \beta\sqrt{w_s}$ were studied (where w is the normal crack opening and w_s is the sliding displacement).

In Delft, no initial guess was made about the load-paths that should be expected in structures, but instead seven different load-paths were investigated in order to elucidate the material behaviour under biaxial tension/shear.[214] The load-paths varied from simple displacement controlled tension, to shear followed by tension, or alternatively to shear followed by tension. The set-up used in these experiments was a complete biaxial load-frame fitted with independent servo-controls in the two loading directions. The specimens were double-edge-notched plates of different size. In Figure 3.71b the loading on a plate is shown, complete with control LVDTs. Later on, in Section 4.3, we will describe the set-up in somewhat more detail. The third device, i.e., the apparatus developed in Darmstadt, is shown schematically in Figure 3.71c. Notched hollow cylinders were pulled in axial tension, and shear was applied as torsion on the pre-cracked cylinder.

The tests in Lund suffered least from nonuniform opening, but crack growth could not be monitored because of the circumferential notch. In contrast, rotational instabilities were largest in the Darmstadt apparatus because of the relatively large specimen size, see also Section 4.3.4. The largest freedom in load-paths was available in Delft, where the specimens were designed to allow for crack growth monitoring during testing as well.

The first tests carried out in Delft focused on studying the decrease of shear stiffness at very narrow crack openings, i.e., up to 100 μm.[206] A problem was that the shear modulus for uncracked concrete was too low, most likely because the state of stress was far from ideal in the punch-off type specimen that was selected in these early experiments. Because of the expected problems with nonuniform opening, it was decided to use double-edge-notched (DEN) specimens instead, which allowed for monitoring crack initiation and growth along all sides. In Figure 3.72, the shear capacity of pre-cracked concrete plates (200 × 200 mm² and 50 mm thickness) is shown. Specimens were first pre-cracked in displacement control, unloaded as soon as the required crack-opening was reached, and subsequently sheared-off at constant (zero) normal load. At the time that these shear tests were carried out, the servo-control in the lateral shear direction was not available, and explosive failure occurred as soon as the maximum shear load was reached. The diagrams of Figure 3.72 should be read as follows. First we follow a given P-w curve, for example the one marked 65 μm. The unloading curve can be seen and as soon as $P = 0$, the shear curve P_s-w is followed. These latter diagrams show the crack-opening displacement w under shear, but do not reveal the shear displacements in the plane of the crack. Note the sharp increase of w just before $P_{s,max}$ is reached. The crack patterns in the experiments varied as shown in the inset of Figure 3.72. For crack openings below 250 μm, inclined secondary cracks always developed as soon as the shear load reached its maximum. On the other hand, sliding was observed when the opening of the initial crack was larger than 250 μm.

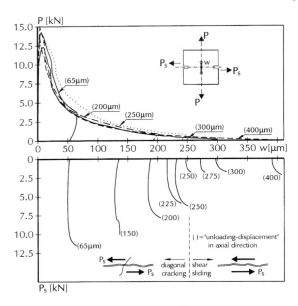

Figure 3.72 Shear resistance for pre-cracked mortar (d_{max} = 2 mm).[220] Note the change of failure mode at w = 250 μm.

These values were obtained for 2 mm mortar, and perhaps this change-over point in failure modes changes when a coarser grained material is used. These experiments are quite important, as they indicate that the earlier aggregate interlock theories are applicable only in situations where larger crack openings appear, viz. > 250 μm. The inclined, secondary cracking is normally not included in the aggregate interlock models, and might explain relatively large differences between different models at small crack openings. The growth of the inclined cracks will vary from experiment to experiment, depending on the size of the specimen and the exact boundary conditions. In the experiments of Figure 3.72, the shear load was "tensile", i.e., the direction of applying shear was found to be important because irregularities in the cracks would cause different crack growth under shear. Examples of so-called compressive shear (the shear load P_s was applied in the direction shown in Figure 3.71b) after pre-cracking are shown in Figure 3.73. In Figure 3.73a-c, examples are shown of specimens loaded up to 50, 100 and 150 μm crack opening, followed by compressive shear while $P = 0$ = constant, whereas in the examples of Figure 3.73d-f, a constant confinement $P = -1$ kN was applied normal to the crack during shearing. Typically, a larger ductility was measured when confinement was applied normal to the crack (Figure 3.73d,e). The failure modes changed from sliding at zero confinement to strut splitting when confinement was applied. The strut develops as shown in Figure 3.74. It can only develop when confinement is applied, otherwise shearing-off the two crack faces becomes

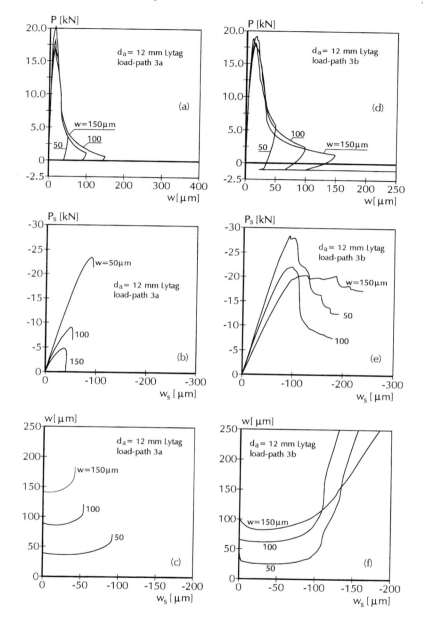

Figure 3.73 Shear strength of lytag lightweight concrete under zero normal stress (a-c) and under small normal confinement of -1 kN (d-f).[214] (Reprinted by kind permission of Dr. Nooru-Mohamed.)

easy as uplift can occur unrestrained. Finally, it should be mentioned that the inclined tensile cracks found in the tests of Figure 3.72 cannot develop in these compressive shear experiments of Figure 3.73.

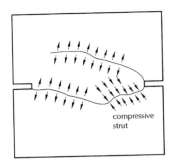

Figure 3.74 Compressive strut mechanism.[214] (Reprinted by kind permission of Dr. Nooru-Mohamed.)

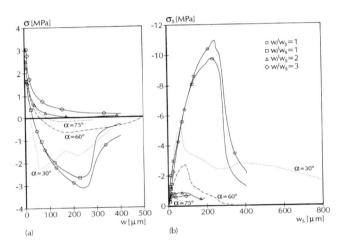

Figure 3.75 Comparison of linear displacement paths of Hassanzadeh[205] and Nooru-Mohamed.[214] The σ-w diagrams are compared in (a) and the σ_s-w_s diagrams in compared in (b). (Reprinted by kind permission of Dr. Nooru-Mohamed.)

As mentioned, linear and parabolic displacement paths $w = (tan\ \alpha).w_s$ were investigated by Hassanzadeh.[205] Some of the linear displacement ratio experiments were carried out by Nooru-Mohamed[214] as well, namely $w/w_s = 1.0$, 2.0 and 3.0. These ratios can be translated to $\alpha = 45°$, 63.4° and 71.6°, respectively. Tests were carried out on specimens of three different sizes, namely square DEN plates of size 50×50 mm^2, 100×100 mm^2 and 200×200 mm^2, with a constant thickness of 50 mm. Thus, we can compare the results of the small size DEN specimens with the experiments carried out by Hassanzadeh, but only in a qualitative sense. In Figure 3.75 we show normal stress-crack-opening diagrams (σ-w) and shear stress-sliding diagrams (σ_s-w_s) for $\alpha = 30°$, 60° and 75° obtained by Hassanzadeh, and make a comparison with the linear load-path experiments of Nooru-Mohamed.

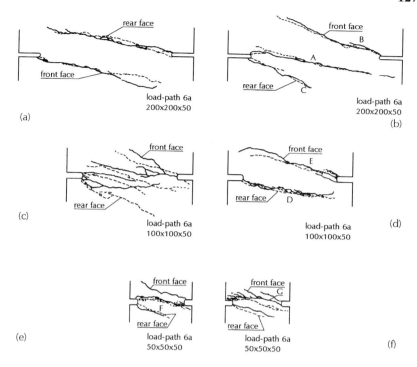

Figure 3.76 Crack patterns of concrete under a constant displacement ratio path (w/w_s = constant), for specimens of three different sizes, after Nooru-Mohamed.[214] The specimens were DEN plates (see Figure 3.71b) of 200×200 mm² (a,b), 100×100 mm² (c,d) and 50×50 mm² (e,f) with a constant thickness of 50 mm. (Reprinted by kind permission of Dr. Nooru-Mohamed.)

The tendencies are quite similar. A cross-over from tensile stress to compressive stress with increasing crack opening w is observed in the σ-w diagrams (Figure 3.75a). A compressive peak is present. The results suggest that at some stage compressive confinement normal to the crack plane is needed in order to stabilize crack growth. The shear stress-sliding diagrams of Figure 3.75b, have qualitatively the same shape, but large differences are found in quantitative sense. The 50 mm specimens tested by Nooru-Mohamed consistently showed a higher shear capacity, which may of course be caused by the higher strength of the concrete, whereas differences in test-technique might be of importance as well. Moreover, differences in fracture modes may have occurred, but most unfortunately the circumferential notch in the experiments of Hassanzadeh does not allow for comparison. In these experiments crack growth occurs in the specimens interior, and is most likely confined. The crack patterns observed by Nooru-Mohamed for three different specimen sizes are shown in Figure 3.76.

In these linear displacement ratio experiments (w/w_s = 1.0) the failure modes changed from inclined splitting to distributed inclined cracking in the small

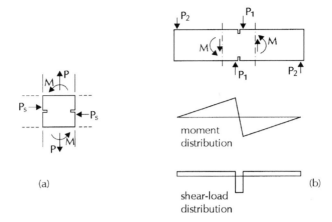

Figure 3.77 Comparison of DEN plates loaded in the Delft biaxial frame (a) and the loading situation in a DEN confined four-point-shear beam (b).

specimens. This suggests that in the specimens of small size, the crack growth is controlled by the material structure, at least to some extent. At present, it is not clear why not all specimens failed in the same manner, but perhaps inaccuracies in the specimen size or in the handling prior to loading should be improved. The fracture patterns of Figure 3.76b-f can be compared to those observed in the DEN four-point-shear beams mentioned before (Figure 3.66, fixed supports). This similarity might be explained from the comparison of the forces and moments prevailing in these respective experiments. If the square DEN plates are rotated by 90°, the situation sketched in Figure 3.77a arises.

Shear loads act at the top and bottom of the plate, normal load is applied as sketched, and for maintaining equilibrium bending moments must develop as well. Note that bending moments are generated due to nonuniform crack growth in a uniaxial tensile test (see Sections 5.3.2.3, 4.1.3.2 and 7.2.2). If a square element is considered in the DEN four-point-shear beam, we find the same loading, in particular when a comparison is made to the four-point-shear beams tested by Swartz and Taha,[208] where confinement is applied in the direction of the beam axis (Figure 3.77b). Based on this similarity, we should expect similarity in failure as well. The situation will be further analyzed in Chapter 7.

3.4.2 DOES MODE II FRACTURE IN CONCRETE EXIST?

The results presented thus far are not very promising. It seems that in mode II fracture experiments the cracks tend to nucleate and propagate in mode I. Local stress disturbances around notches may be one reason; however, it is felt that an erroneous interpretation of four-point-shear experiments on the basis of simple beam theory and the heterogeneity of the material under consideration are the main problems. The confined DEN four-point-shear beams tested by

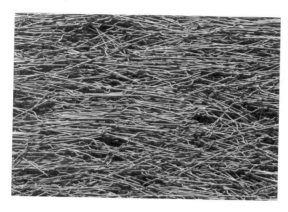

Figure 3.78 Fibre structure of SIFCON before impregnating with slurry. In the example 30 mm long Dramix fibres were sprinkled in a narrow (50 mm) plexiglass mould forcing the fibres to align.[229]

Swartz and Taha[208] suggest that the size of the zone where cracks grow decreases, i.e., the curved cracks are closely spaced. One might envision situations where the confinement has increased to the extent that a single shear plane develops. In fact, shear banding is observed in concrete under triaxial confinement as will be shown in Section 3.5 (Figure 3.98). In triaxial or multiaxial experiments, active confinement is applied at the specimen boundaries.

Second, confinement may also be applied from within the material, viz. by adding fibres to the concrete. Fibres act as an additional bridging mechanism as shown before for mode I fracture and, more importantly, they seem to distribute the fractures in many small cracks. Yin et al.[221] used this internal confinement concept as an explanation for the increased strength of fibre-reinforced concrete under biaxial loading. The ultimate fibre concrete is SIFCON, Slurry Infiltrated Fibre Concrete, which contains fibre volumes as high as 21%. The material is produced by first sprinkling (steel) fibres in a mould and subsequently impregnating the fibre mass with a low-viscosity cement slurry. Normal practice is to add silica fume and/or superplasticizer to increase the workability of the slurry. Lankard and Newell[222] were the first to propose the material for repair of concrete bridge decks. Since then, interest in the material has increased.[223,224] Structural applications of the material have been proposed recently.[225-228] Naaman proposed to use it as plastic hinges in reinforced concrete frames and Van Mier is working on a SIFCON truss system. The material is highly anisotropic because the fibres are oriented in layers.[229] An example of the fibre structure of SIFCON before impregnating with slurry is shown in Figure 3.78.

The layered structure is clearly visible and when saw-cuts are made parallel or perpendicular to the fibre direction, sections as shown in Figure 3.79 are

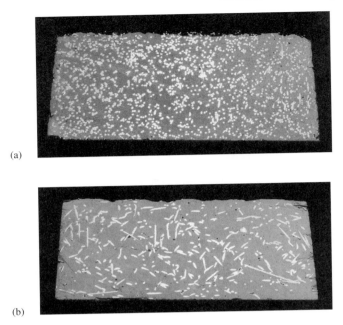

(a)

(b)

Figure 3.79 Anisotropy in SIFCON: saw-cut perpendicular to the fibre direction (a), and parallel to the fibre direction (b).[230]

found. This demonstrates the enormous anisotropy. It has been found that the direction of loading with respect to the direction of the fibres has a major effect on the shear behaviour of the material. Especially when the shear load is applied parallel to the fibre direction, a very low shear strength is measured, whereas a much higher shear load is found when the fibres are oriented normal to the shear direction. We will not delve into all the details of shear in SIFCON here, but most interesting and relevant to the foregoing discussion is that shear fractures, as defined by Bažant and Pfeiffer[198] (i.e., Figure 3.64a), have actually been observed in SIFCON. In Figure 3.80 two stages of shear cracking in a DEN 100×100 mm² SIFCON specimen (similar geometry as shown before in Figure 3.71b) are shown. The fibre direction is horizontal, as was the shear load. Clearly widening of inclined cracks can be observed when the globally applied shear displacement increases. The crack patterns are mosaics of only part of the total crack patterns. The mosaic photographs were taken with a Questar long-distance optical microscope, which is described in Section 4.4.3.

These SIFCON results actually suggest that shear fracture in cement composites is possible. However, the material outside the shear zone should be confined sufficiently to circumvent crack extension outside the (narrow) shear zone. In the SIFCON, confinement comes from the fibres in the material itself. Always some fibres will have inclined orientations; the alignment is of course never perfect as can be seen in Figure 3.78. As mentioned, in Section 3.5 it will

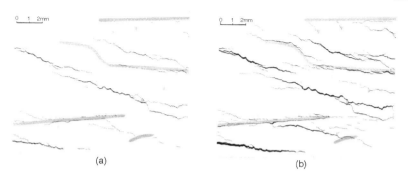

Figure 3.80 Two stages of cracking in a shear experiment on a SIFCON DEN plate (100 × 100 mm²), with the fibre direction parallel to the direction of the shear load.[229]

be shown that shear bands can develop under external (active) confinement in multiaxial experiments as well.

Finally, a third situation where shear fractures may arise is perhaps under dynamic loading. Thus far, shear fractures in plain concrete have been reported by Bažant and Pfeiffer[198] and Davies,[199] but in all their experiments load-control, or control of the piston in the hydraulic actuator, was used and explosive collapse of the specimens must have occurred (see Section 4.1). One might argue that cracks propagate very rapidly under explosive collapse, and it would be no surprise if the crack would select the shortest path through the material. It is quite likely that the rapid loading does not allow for the growth of curved overlapping cracks as shown in Figure 3.66. This remains a point of future investigation. The experiments by Bažant and Pfeiffer have been instrumental for obtaining a better understanding of shear fracture in plain concrete. However, their view was perhaps too simple, but the discussions and controversy that emerged after their proposition of shear fracture energy has been a prerequisite for progress in the field. We will present results of a number of meso-scale numerical analyses of the four-point-shear test in Section 7.3.1.

3.4.3 SHORT NOTE ON MODE III FRACTURE

Information on mode III fracture, i.e., out-of-plane shear or tear is very limited for concrete. Bažant and Pratt,[231] Yacoub-Tokatly and Barr[232] and Xu Daoyuan and Reinhardt[233] performed experiments on notched cylinders subjected to torsion. Bažant et al.[234] applied torsion to (un-notched) beams with a square cross-section. The difference of testing notched or un-notched specimens is shown in Figure 3.81. In un-notched specimens, a crack develops perpendicular to the major (tensile) principal stress in the beam, whereas a deep notch might force the crack to grow in a plane. A discussion similar to that in the previous section on mode II can be held here. Is it really mode III, or is tensile stress the controlling factor at the meso-level?

Most authors agree that a deep notch is needed in order to confine the concrete in the middle section. For example Xu Daoyuan and Reinhardt[233]

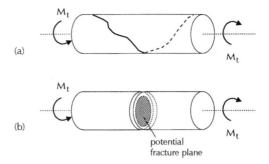

Figure 3.81 Mode III fracture tests on (a) un-notched and (b) notched cylinders.

proposed to perform tests on cylinders with a notch-depth-to-diameter ratio of at least 1/10. They present results in terms of G_f^{III}, which can be computed from complete torque-rotation curves, but find that the mode III fracture energy depends on the size of the cylinder. Bažant and Prat[231] reported conical failure modes for mode III fracture, which would support the idea that tensile mechanisms control the process at a lower level.

3.5 MULTIAXIAL STATES OF STRESS

Thus far only results have been shown for concrete subjected to (supposedly) simple uniaxial, tensile or compressive stress, and a limited number of results on shear have been discussed. In all these experiments, the stress-state is generally far from uniform. A triaxial state of stress is present, at least in parts of the concrete specimen under consideration. In Figure 3.18 we already showed that regions of triaxial compression can develop in the concrete specimen when no steps are taken to reduce frictional boundary restraint between specimen and loading platens. This implies that there is interest in obtaining information on triaxial behaviour of concrete. Moreover, in finite element packages for simulating the behaviour of reinforced concrete structures a fully three-dimensional constitutive law should be available. Thus far we only discussed the behaviour of concrete in simple uniaxial compression, uniaxial tension and two basic fracture modes originating from linear elastic fracture mechanics. Here we will make the step towards general bi- and triaxial behaviour, and focus on micromechanisms of fracture (as an extension of the mechanisms mentioned earlier for uniaxial compression). Next we will discuss experimental measurements of failure surfaces for concrete, stress-strain behaviour under generalised stress, and fracture modes of confined concrete. Later on in the book, in Chapters 5 and 7 we will return to limit surfaces for concrete, in particular to mathematical formulations and numerical simulations.

In multiaxial testing, three stresses (σ_1, σ_2, σ_3) and three strains (ε_1, ε_2, ε_3) must be measured. Conforming to the literature in this field of research, we will

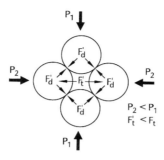

Figure 3.82 Particle interactions under biaxial compression.

assume that σ_1 is the major principal stress, σ_2 the intermediate principal stress and σ_3 the minor principal stress. Again, compressive stresses are negative and tensile stresses are positive. This means that, for example, in a triaxial compressive state of stress, σ_1 is the major principal stress. The numerical value of σ_1 is the largest absolute number, but on a normal scale it is the smallest (i.e., most negative) number. In the appendix principal stresses and invariants are presented in some detail.

3.5.1 MICROMECHANISMS

Following the discussion in Section 3.2.2 for uniaxial compression, we can use the same particle stack models to explain the behaviour of concrete under multiaxial compression. In Figure 3.82 the interaction between four rigid aggregate particles embedded in a soft matrix are shown. Basically the same mechanism develops as discussed with Figure 3.7b, except that now an external compressive stress σ_2 is applied perpendicular to the vertical stress σ_1. The external stress σ_2 counteracts the wedge splitting forces caused by the vertical stress, i.e., the tensile splitting force between the particles in the concrete is reduced. When another confining (compressive) stress σ_3 is applied in the out-of-plane direction, the same mechanism develops in that direction as well. As a result of the confinement, the local tensile splitting forces are reduced, which implies that a higher external stress σ_1 must be applied to fail the composite. Note that in a biaxial compressive state of stress, splitting is still quite easy in the free (un-stressed) direction, and that the failure load will increase only marginally. On the other hand, when a tensile load is applied in the lateral direction(s), the effective local splitting stress between the particles increases, and failure occurs at relatively low levels of external load. Thus, by considering the particle structure of the concrete, and by using the local splitting tensile stress as a criterion for failure, qualitatively the failure of concrete under three-dimensional states of stress can be explained. As will be shown in Section 7.5, heterogeneity in combination with a simple tensile/shear criterion for fracture between particles is sufficient information for computing the biaxial failure envelope. Other mechanisms might be active as well, such as the interaction

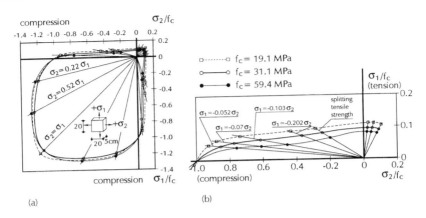

Figure 3.83 Biaxial failure contours for three different concretes, after Kupfer.[236] In (a) the complete contours are shown (normalised with respect to the uniaxial compressive prism strength) and (b) shows the tensile-compressive regime enlarged.

between wing-cracks. We will return to this in Section 3.5.4. Also, under multiaxial stress, the behaviour can change substantially when the material is saturated and pore-pressure builds up, see Visser.[235] We will not go into these matters in this book.

3.5.2 FAILURE CONTOURS

The strength of concrete under triaxial stress is defined by means of a failure contour or limit surface. The limit surface contains all possible failure points under any triaxial state of stress. The shape of the failure contour for concrete in 3D stress-space is rather complicated. Let us therefore first consider the biaxial envelope. After that we will show failure contours measured in standard triaxial tests on cylinders, and finally we will show an example of results measured in a true triaxial machine in which cubical specimens are tested. The results presented here are believed to be representative for a large bulk of data that have been obtained over the past few decades. This implies that in the experiments sufficient attention is given to reducing boundary shear and inter-actions between adjoining loading platens. These important experimental is-sues will be addressed in depth in Chapter 4, more specifically in Sections 4.1.3.1 and 4.1.3.3.

3.5.2.1 Biaxial failure contours

Biaxial compression and compression-tension experiments have been car-ried out by a great number of researchers. The results of Kupfer[236] are among the most widely quoted. In Figure 3.83 the biaxial failure envelopes are shown for three different types of concrete with uniaxial compressive strength of 19.1, 31.1 and 59.4 MPa, respectively. Specimens of size $200 \times 200 \times 50$ mm^3 were loaded between so-called brushes bearing platens, a reliable method to reduce friction at the specimen boundaries, as will be shown in Section 4.1.3.1. Using the same equipment the range of materials tested was extended to lightweight concrete, foamed concrete, hardened cement paste and gypsum by Linse and Stegbauer.[237] The shape of the biaxial failure contour is more or less the same for all these different materials. Here we will only show the results of Kupfer for normal weight gravel concrete. In Figure 3.83a, the full biaxial envelope is shown and in Figure 3.83b the tensile-compressive regime is shown enlarged. It can be seen that the strength under equi-biaxial stress (i.e., $\sigma_1 = \sigma_2$) increases by about 15 to 20% compared to the uniaxial compressive strength. The largest increase of strength in the biaxial compressive regime is measured when the ratio between the two principle stresses is 0.5. Also when the biaxial compressive regime is considered only, we see that the concrete quality has not much effect on the shape and the relative size (i.e., all graphs are proportional to the uniaxial compressive prism strength) of the biaxial failure contour.

In the tension-compression regimes the following observations can be made. When a small lateral tensile component is applied, the strength decreases significantly in the compressive direction. This can easily be explained from the micro-mechanism of the previous section. From an experimental point of view, the biaxial tension-compression experiments are most difficult to perform. Usually an increased scatter is observed in the tensile/compressive regimes, and moreover, test-data are extremely scarce especially in the triaxial tension-compression regimes.

The effect of concrete quality is most pronounced in the biaxial tension-compression regimes. The biaxial tensile/compressive strength decreases most when the concrete compressive strength increases. As we will see in Section 3.5.4, a transition of failure modes from shear to tensile cracking occurs in the tensile/compressive regime.

Many researchers have measured the biaxial failure contours for concrete. In an international comparative research[238] it was shown that the use of unrestrained loading systems is essential. Figure 3.84 shows qualitatively the effect of boundary restraint on the biaxial failure envelope. Unrestrained testing leads to a lower bound for the concrete compressive strength. We already discussed this issue in Section 3.2.4, but in Chapter 4.1.3 we will come back to these matters in more detail.

An important issue is the path dependency of the biaxial failure contour. It seems that irrespective of the load-path followed, the same biaxial failure contour is found.[239] However, when specimens are pre-cracked under stable displacement-control into the post-peak regime, the failure envelope seems affected,[80] but these measurements concern triaxially stressed concrete. Probably the effect of load-path on failure contour will depend on the amount of damage sustained to the specimen in previous loadings. Noteworthy is the effect of initial anisotropy of the concrete on the shape of the biaxial failure envelope.[79] The effect of initial anisotropy on strength was discussed earlier for uniaxial compressive states of stress (Figures 3.9 and 3.10).

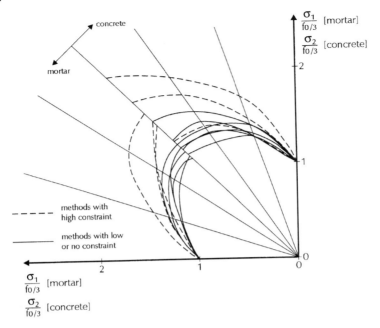

Figure 3.84 Effect of boundary restraint on the biaxial compressive failure contour.[238] Restrained boundaries means that tests are conducted between rigid steel platens; unrestrained testing implies that friction reducing measures have been taken. (Reprinted with permission of ACI, Detroit, MI.)

3.5.2.2 Triaxial failure contours

Visualizing the failure envelope in triaxial stress-space is not easy. The failure contour for concrete is a complicated three-dimensional cone-like surface with the top of the cone in the triaxial tensile octant. In the triaxial compression regime the cone seems open-ended, although this was never confirmed, since experiments have never been carried out above a hydrostatic pressure of 150 MPa.[240] The only exception are a number of elastically confined tests (uniaxial strain tests that were confined in a steel block) on hardened cement paste.[241] In those tests axial stresses up to 2068 MPa were reached. Figure 3.85 shows the three-dimensional surface. In the early days of triaxial testing, cylindrical specimens were always used, where fluid pressure was applied to the circumference of the cylinder in a pressure cell, and axial loading was applied in a standard compression machine. Among the earliest and best-known cylinder tests are those of Richardt et al.[242] Cylinder tests set a limitation to stress states that can be investigated. Always two stresses are equal, namely $\sigma_2 = \sigma_3$ due to the circumferential fluid pressure. This implies that only a single cross-section in triaxial stress space can be investigated. This plane is called the Rendulic plane, and is shown in the appendix in Figure A6. Two meridians are distinguished, namely the compressive meridian where the vertical stress σ_1 is smaller than the radial stresses: $\sigma_1 < \sigma_2 = \sigma_3$ (stresses are assumed to be

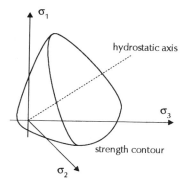

Figure 3.85 Limit surface (qualitative) for concrete in triaxial stress space. The top of the cone is in the triaxial tensile regime; the cone is open ended in the triaxial compression regime.

negative in compression), and the tensile meridian where the radial confinement is smaller than the axial stress ($\sigma_1 > \sigma_2 = \sigma_3$). One should be careful because in many textbooks different notation is used and often compression is defined positive. In view of all discussions on tensile fracture, we have adopted the normal sign convention throughout the book (i.e., tension is positive and compression is negative).

The full triaxial stress space can be explored using a truly triaxial device only (in Section 4.1.2.2 the necessary test equipment is discussed). In that case cubical specimens must be used, which have the disadvantage that stress-concentrations appear in the corners. Special measures must be taken to reduce frictional end-restraint as will be discussed in Section 4.1.3.3. The cylinders have the advantage that load application is more easy. The length of the cylinders is normally large enough to reduce end-effects such that a pure stress-state is present in the middle part of the cylinder. The application of fluid pressure along the circumference of the cylinder is almost unrestrained. In fact, load application through fluid pressure seems the best available method to date.

A good impression of the shape of the failure contour can be obtained from cylinder tests. Many cylinder experiments have been carried out in the past.[242-249] In Figure 3.86 the compressive and tensile meridians are shown for four different concretes. These tests were carried out by Newman.[244] Note that the compressive meridian is cut off at $\sigma_2 = \sigma_3 = 0$, i.e., no experiments can be carried out where circumferential tension must be applied. The tensile meridian is extended in the regime where σ_1 becomes tensile (i.e., positive, implying that the ratio σ_1/f_c becomes negative). Note that the axes have been normalised with respect to the uniaxial compressive cylinder strength f_c for each of the materials. The compressive strength increase in the axial direction is quite impressive. Richardt et al.[242] reported that the confined concrete compressive strength (σ_1) was equal to the uniaxial strength f_c plus approximately four times the confining stress. Newman[244] derived a slightly more advanced relation,

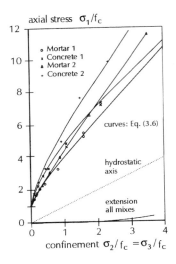

Figure 3.86 Compressive meridians and tensile meridians for four different concretes and mortars tested by Newman.[244] The tensile meridians are almost identical for the four materials tested and are situated near the $\sigma_2/f_c = \sigma_3/f_c$ axis.

$$\frac{\sigma_1}{f_c} = \sqrt{A \cdot \left[\frac{\sigma_2}{f_c}\right]^2 + B \cdot \frac{\sigma_2}{f_c} + 1} \qquad (3.6)$$

where A and B are constants. A increases from 4.87 for mortar 1 to 8.22 for concrete 2 (Figure 3.86). The parameter B is equal to 10 for both mortars and 20 for both concretes. σ_1 is the axial stress and $\sigma_2 = \sigma_3$ is the confining stress.

Triaxial extension tests are scarce, especially when we are interested not only in strength but also in the post-peak behaviour. In Figure 3.86, the results obtained by Newman are plotted (see the lower curve near the $\sigma_2/f_c = \sigma_3/f_c$ axis). Almost the same behaviour was observed for the four materials investigated. Note that for low confining pressure $\sigma_2 = \sigma_3$, tensile stress must be applied in the axial direction to fail the specimen. The equi-biaxial strength is defined by $\sigma_1 = 0$. In triaxial extension tests, the state of stress can be separated in a hydrostatic part ($\sigma_1 = \sigma_2 = \sigma_3 = c$) and an axial component $\sigma_1 - c$. The axial component yields an effective tensile stress (Figure 4.15). As a consequence, the cylinders will fail through the formation of an axial cleavage crack. This was clearly demonstrated by Newman[244] in his experiments, and more recently by Visser and Van Mier[249] in stable displacement controlled extension tests. In the latter tests a complication was added because the pressure fluid could enter the crack after initiation, see Chapter 4.

Figure 3.86 shows all possible states of stress that can be investigated using a conventional triaxial cell and cylindrical specimens. The results of such tests

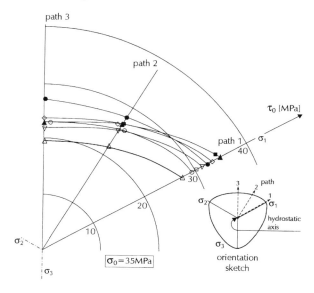

Figure 3.87 Deviatoric section of the triaxial failure surface. Data points after Gerstle et al.[238] (Reprinted with permission of ACI, Detroit, MI.)

can be very conveniently translated into a hydrostatic component and a deviatoric stress as explained in the Appendix. Truly triaxial tests can only be carried out using cubical specimens. In that case, three independent loading frames are needed. Thus, generally the experimental set-up will be more complicated as will be discussed in Section 4.1.2.2. The international cooperative research of Gerstle et al.,[238] which was mentioned in the previous section, was aimed at the measurement of triaxial failure envelopes as well. Their results were presented in terms of octahedral normal stress, octahedral shear stress and the orientation of the latter stress component in the deviatoric plane (see Appendix). A section of the failure contour normal to the hydrostatic axis (i.e., a deviatoric section) from this international cooperative research is shown in Figure 3.87. Only one sixth of the complete deviatoric section is shown. This was done because of the assumed isotropy. Several contours are shown for results obtained from different experimental set-ups. Results are shown that were obtained from experiments with boundary restraint (rigid steel platens) and with friction reducing measures. Note that the data points along the σ_1-axis and the (negative) σ_3-axis can be measured in a standard triaxial cell as shown in Figure 3.86. These data points are located on the triaxial compressive meridian and triaxial extensile meridian, respectively. The intermediate points ($\theta = 30°$) can only be measured in a truly triaxial device. Figure 3.87 shows a notable effect of boundary restraint on triaxial strength, except that it seems less significant as compared to the biaxial results presented in Figure 3.84.

A different view of the triaxial envelope is given in Figure 3.88. Here we see the results obtained in the true triaxial machine of Eindhoven University of

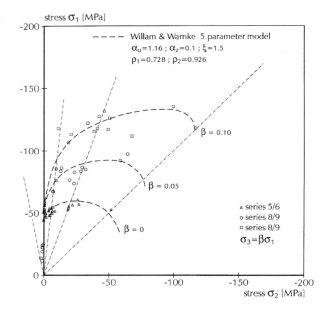

Figure 3.88 Comparison of triaxial data obtained with the Eindhoven true triaxial machine[80] and the Willam and Warnke[250] failure surface.

Technology.[80] Contours are shown in the σ_1-σ_2 plane for different values of σ_3 = $\beta\sigma_1$. In Figure 3.88, a comparison is shown with the Willam and Warnke[250] failure contour that is described in detail in Section 5.3.1. Note that the contour was fitted to earlier true triaxial tests of Schickert and Winkler.[251] The fit of the Willam and Warnke criterion to the more recent Eindhoven data is remarkable.

Some models require as input a series of yield surfaces. This is especially important in plasticity models. The subsequent yield surfaces are normally defined based on a certain amount of plastic strain needed to reach a given surface. The plasticity can be translated into microcracking for concrete. Kotsovos and Newman[245] and Kotsovos[246] define, for example "yield surfaces" which are referred to as OSFP, OUFP and ultimate. OSFP is the stress where stable crack propagation starts; it can be identified from the first deviation from linearity of the stress-strain curves under various states of stress. OUFP is the onset of unstable crack propagation; it was associated with the fatigue limit of concrete. Ultimate was of course the true failure surface containing all combinations of triaxial stress where the concrete fails. We will not delve into these matters deeper. In this book, the crack process is regarded as a continuous process and no distinct boundaries are distinguished.

It would be interesting to study the multiaxial behaviour for concretes of different composition such as w/c ratio, aggregate grading and so on. However, only limited and scattered information is available.

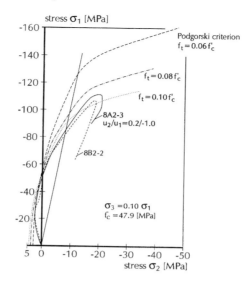

Figure 3.89 Stress-paths σ_1-σ_2 for tensile/compressive constant displacement-ratio paths $u_2/u_1 =$ 0.2/-1.0, with $\sigma_3 = \beta\sigma_1$.[80]

3.5.2.3 Bounding surfaces from constant displacement-ratio tests

A particularly interesting load-path in triaxial space is a stress-path generated by keeping a constant ratio between two or three displacements or strains. Such experiments, when done in stable displacement control, are interesting because some of the stress-paths can be compared to the triaxial failure contours mentioned before.[80] Most interesting are experiments where two principal displacements are held in a constant ratio ($u_1/u_2 = -1.0/0.2$, i.e., the displacement u_1 is compressive and u_2 is tensile), and a small confining stress $\sigma_3 = \beta.\sigma_1$ is applied in the third direction ($\beta = 0.05$ or 0.10). Applying a constant displacement ratio u_1/u_2 implies a non-linear relation between σ_1 and σ_2. In Figure 3.89 stress-paths measured in constant displacement-ratio experiments following the above mentioned path are shown. Basically, the ratio σ_1/σ_2 is initially negative, i.e., the path starts off in the tension/compression regime. In fact, we start in the t/c/c regime because the third principal stress is always compressive. The curves reach a tensile peak (σ_2), after which the second principal stress decreases and eventually becomes compressive. Thus, the state of stress changes from t/c/c to c/c/c. The reason for this is that the lateral deformations caused by the two principal stresses σ_1 and σ_3 become so large that the displacement u_2 would increase faster than prescribed by the path $u_2/u_1 = 0.2/-1.0$. The servo-control (see Chapter 4) will counteract this tendency by supplying more pressure in the 2-direction. Finally, a second peak stress is reached, however now in the major principal direction σ_1. After this point is exceeded, the specimen will fail with shear bands developing in the plane through directions 1 and 3 (Section 3.5.4).

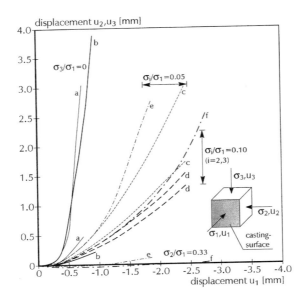

Figure 3.90 Non-linear displacement paths from constant stress-ratio experiments $\sigma_2 = \alpha\sigma_1$, σ_3 = $\beta\sigma_1$.[253]

When a constant stress-ratio is maintained, the relationship between the various displacements becomes highly non-linear. This is demonstrated in Figure 3.90. Results are shown of six constant stress-ratio experiments, σ_2/σ_1 = α and σ_3/σ_1 = β, with α = 0.10 and 0.33, and β = 0, 0.05 and 0.10, respectively. In this graph the major compressive displacement u_1 is plotted along the horizontal axis, and u_2 and u_3 (which are generally positive) are plotted along the vertical axis. It should be mentioned that for each triaxial experiment, two curves u_1-u_2 and u_1-u_3 have been plotted in Figure 3.90. Depending on the amount of stress applied in the second and third direction, the deformations increase slower or faster as can be clearly seen. From Figure 3.90 (and identical graphs for other triaxial load-paths), we can construct displacement contours containing all displacements (u_1,u_2,u_3) at peak stress. Figure 3.91 shows examples of critical displacement contours.[253] The contours are plotted for concrete loaded with the major principal stress either perpendicular or parallel to the direction of casting, and for different levels of σ_3/σ_1 = β (namely β = 0.05 and 0.10). The contours in Figure 3.91 are the counter part of the stress contours in Figure 3.88. When these graphs are placed side by side, we can clearly see the purpose of a constitutive model, namely to find a mapping procedure between stress and strain (or displacement).

Figure 3.91 shows only a small part of displacement space. Much more testing has to be done to make the picture complete. However, one additional point should be raised here. If we return to the stress-paths of Figure 3.89, it

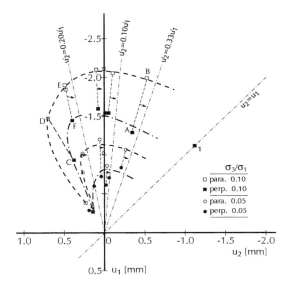

Figure 3.91 Displacement contours from triaxial experiments. The contours connect the displacements measured in experiments when the maximum stress combination is reached.[253]

can be seen that the paths follow to some degree the failure envelope according to the Podgorski.[252] (Note: the Podgorski criterion compares quite well to the Willam and Warnke criterion shown in Figure 3.88; it is also a five-parameter model.) Moreover, when the stress-paths are compared to results from constant stress-ratio experiments, it can be seen that these latter failure points are very close to the stress-paths. This would suggest that in a displacement-ratio experiment, specifically when one of the displacements is forced to remain tensile, the stress-path follows part of the failure contour. The above figures suggest that such a trend exists; it is of course no proof. However, if a part of the (strength) failure contour could really be measured using displacement-ratio experiments, the number of triaxial experiments could be reduced substantially. For the stress-paths of Figure 3.89 it would mean that no separate tests have to be carried out between the first and second peak of each stress-path.

3.5.3 STRESS-STRAIN BEHAVIOUR

In the previous section we showed displacement contours but did not discuss the relation between stress and strain under general triaxial stress. The simplest and most systematic manner first shows stress-strain curves from biaxial stress-ratio experiments and from standard triaxial cylinder experiments, and then moves to true triaxial experiments on cubical specimens where the effect of intermediate and minor principal stress can be measured independently. Also of importance is cyclic loading and effects due to initial anisotropy

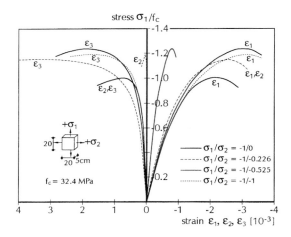

Figure 3.92 Biaxial stress-strain curves for concrete (biaxial compressive states of stress only).[236]

and load-path effects. However, we will limit the information to six graphs showing the major trends.

In Figure 3.92 stress-strain response measured from biaxial constant stress-ratio experiments are shown. The experiments were all carried out in the biaxial compression regime with σ_2/σ_1 = 0/-1, -0.225/-1, -0.52/-1 and -1/-1 (with $\sigma_3 = 0$ for all tests). The shape of the various curves is always the same, viz. similar to the uniaxial stress-strain curves that were treated exhaustively in Section 3.2. The largest difference is found between a uniaxial and an equi-biaxial compressive test. In the uniaxial experiment, the two lateral strains ε_2 and ε_3 are both tensile (positive), whereas in the equi-biaxial test both ε_1 and ε_2 are compressive strains, and only ε_3 is tensile. This already gives a clue for the failure modes, namely that the uniaxial compressive test will fail with cracks developing in the two free directions, whereas in the equi-biaxial experiment the specimen will fail through out-of-plane splitting (i.e., normal to the plane where the stresses σ_1 and σ_2 are applied). We will return to failure modes in the next section. Kupfer[236] showed that similar stress-strain behaviour is found for different concretes, i.e., the trends shown above remain the same. Gerstle et al.[238] showed that the scatter in biaxial stress-strain behaviour decreases substantially when friction reducing loading systems are used. Most biaxial experiments are limited to the pre-peak regime. Kupfer[236] managed to round the peak, but almost immediately after peak the specimens would fail explosively. Nelissen[254] carried out biaxial experiments in a fully displacement-controlled testing machine, but he published pre-peak stress-strain diagrams only. An example of a biaxial cyclic experiment loaded into the softening regime is shown in Figure 3.93. It concerns a biaxial constant stress-ratio experiment σ_2/σ_1 = -0.05/-1.0 with load-cycles to the envelope curve (compare to the uniaxial cyclic tests discussed in Figure 3.31). Only strains in

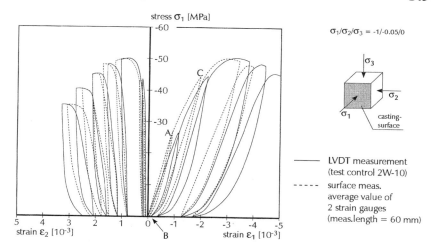

Figure 3.93 Example of a biaxial cyclic stress-strain curve.[80]

the two loaded directions are shown. The strain ε_1 in the major stressed direction is compressive, and ε_2 is tensile in spite of the small compressive stress in the σ_2-direction. Strains were measured both between the loading platens (in this particular example brushes were used, see Section 4.1.3.1), and on the surface of the specimen by means of 60 mm long strain gauges. These results identify a problem in deformation measurement in multiaxial experiments, namely that no strain gauges can be used when a specimen is fully covered by loading platens. Moreover, there seems to exist a difference between surface strain measurements using strain gauges and measurement of strains from LVDTs mounted between two opposite loading platens. Mainly lack of full contact between the loading platen and specimen leads to the difference between overall strains and surface strains. Basically the experiment shows that the shape of load-cycles under biaxial stress (and also triaxial stress[80,253]) is similar to the shape of cycles under uniaxial stress (either tensile or compressive, see Sections 3.2 and 3.3).

Under triaxial stress some changes appear, but these can be explained quite easily using the particle mechanism described before (i.e., Figure 3.82). In Figure 3.94 nominal stress-strain curves from confined cylinder tests are shown.[248] The term nominal is used here to indicate that the stress has been computed relative to the net area of the specimen (cylinder of diameter $d = 110$ mm and aspect ratio 2). Using nominal stress as a basis for comparison is of particular importance for triaxial experiments where strains can increase beyond 10%. The area of the specimen increases substantially. With increasing confining stress, a gradually increasing failure stress is measured, but the initial slopes of the curves remain almost constant. For confining pressures $\sigma_2 = \sigma_3$ up to 25 MPa brittle behaviour is observed, for higher levels of confining stress

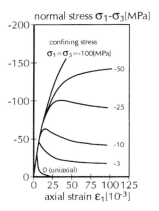

Figure 3.94 Stress-strain curves from confined cylinder experiments ($\sigma_2 = \sigma_3$).[248]

the behaviour becomes more ductile. The low confinement tests failed through the development of a clear discrete shear band (Section 3.5.4). For higher confinement levels (i.e., $\sigma_2 = \sigma_3 < -50$ MPa) no visible damage could be detected at the specimen's surface. The brittle-ductile transition (i.e., the transition from softening to plastic response) obviously occurred between confining stresses of -25 MPa and -50 MPa, or, stated differently, the brittle-ductile transition occurs when $\sigma_3/\sigma_1 \approx 0.20$-$0.25$. In similar experiments carried out by Willam et al.,[255] utilising small cylindrical NX size specimens (size approximately Ø 54 mm and length 108 mm; the designation NX originates from rock mechanics), the brittle to ductile transition occurred at the relatively low level of $\sigma_3/\sigma_1 \approx 0.14$. From rock testing it is known that there may exist some variability in the brittle to ductile transition depending on the type of rock, see for example in Paterson.[256] The transition tends to be higher when the rocks are more fine-grained. The experiments carried out by Jamet et al.[248] and Willam et al.[255] point to the same tendency for concretes. For the tests of Jamet et al., the ratio $d/d_{max} = 110/5 = 22$, whereas for the coarser concrete used by Willam et al. in combination with a smaller specimen size, we find $d/d_{max} = 5.4$.

Using cylinder tests, extensile states of stress can also be explored. Newman[244] carried out extension tests on four different concretes and mortars (Figure 3.86). The cylinders were first loaded hydrostatically to the pre-defined level of confining stress, whereafter the axial stress was gradually decreased. In all experiments carried out, both where failure occurred under axial compression or under axial tension, a single discrete fracture plane normal to the axis of the cylinder was found. The maximum confining stress applied in the extension tests did not exceed twice the uniaxial cylinder compressive strength, which is above the level where a brittle to ductile transition is measured along the compressive meridian. Indeed, from rock mechanics literature it is known that the brittle to ductile transition is much delayed under triaxial extension. For

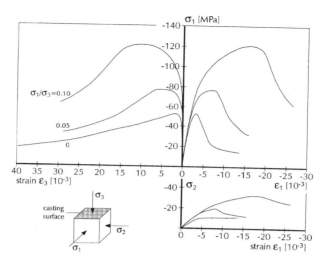

Figure 3.95 Effect of minor principal stress $\sigma_3 = \beta\sigma_1$ on the stress-strain curves for plane-strain tests.[80]

concrete the transition has never been found, but then, experiments in the extensile regime are very limited to date. Recently, Visser[235] started to explore extensile failure, but for the low pressure regime where axial tensile stress must be applied to fail the specimen. We will not go into more detail here.

Let us now show some stress-strain diagrams measured in a true triaxial device under stable displacement control. The first stable displacement controlled true triaxial tests on concrete were carried out by Van Mier.[80] For rocks such tests were also scarce up to that date, e.g., Michelis and Demiris[257] and Michelis[258,259] and more recently Kamp and Cochran[260] carried out displacement controlled true triaxial tests on marble. The type of confined compressive tests on cylinders as presented earlier in this section is more common for rocks, mainly because rock samples are obtained usually as cores. The stress-strain behaviour from true triaxial tests can be separated in the effect of the minor principal stress (designated σ_3), and the effect of the intermediate principal stress. In Figure 3.95 the effect of the minor principal stress is shown, for "plane-strain" tests ($u_2'/u_1' = 0/-1.0$) on a 16 mm concrete. "Plane strain" is placed between quotation marks; it means that one of the control displacements (u_2') was kept constant at zero. However, u_2' is not the pure specimen deformation u_2, but part of the loading platen deformation is included as well.[80] Again, contact effects between loading platen and specimen are responsible. At that time, no sophisticated computer was available to correct the control signal for the loading platen deformations in real time. It should be mentioned, however, that the error is relatively small, i.e., between 2 and 4%. The plane-strain tests fail through a discrete shear band in the plane through the major and minor compressive stress, i.e., large positive deformations are measured in the 3-

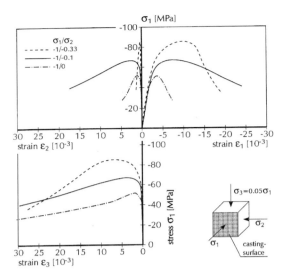

Figure 3.96 Effect of intermediate principal stress in triaxial constant stress-ratio tests with $\sigma_3/\sigma_1 = 0.05$ and $\sigma_2/\sigma_1 = \alpha$, where $\alpha = 0$, 0.10 and 0.33, respectively.[80]

direction and relatively large compressive strains are measured in the 1-direction. The stress σ_2 increases up to peak, after which it gradually falls back as can be seen in the lower part of Figure 3.95. With increasing confinement in the 3-direction ($\sigma_3 = \beta\sigma_1$, with $\beta = 0$, 0.05 and 0.10), the peak stress increases substantially. The pre-peak stress-strain curve (σ_1-ε_1) shows with increasing confinement a gradually longer plateau before the peak stress is reached. In contrast, the post-peak curve seems not affected by the level of confining stress. Basically, these plane-strain tests can be compared to the extension tests mentioned before. Failure occurs in a single direction. Therefore, a much higher transition from brittle to ductile behaviour is expected, as argued earlier in this section.

The effect of the intermediate principal stress is shown in Figure 3.96. Three constant stress-ratio experiments with the same $\sigma_3/\sigma_1 = 0.05$ are shown for $\sigma_2/\sigma_1 = \alpha$, where $\alpha = 0$, 0.10 and 0.33, respectively. The experiment with $\alpha = 0.10$ lies close to the compressive meridian in stress space and it can be seen that the specimen behaves more ductile as compared to the other two tests. The results indicate that a larger ductility is measured in a region around the compressive meridian, but also that the brittle behaviour generally observed in extension experiments is dominant in a larger part of stress space.

Finally, the influence of initial anisotropy on triaxial stress-strain behaviour will be discussed. In Figures 3.94 and 3.95 we have indicated the casting surface in the cubical specimens that were drawn beside the graphs. This is done because initial anisotropy plays a relatively large role in multiaxial

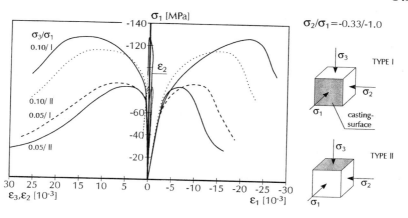

Figure 3.97 Effect of initial anisotropy on the stress-strain behaviour in multiaxial compression. Constant stress-ratio tests -1.0/-0.33/ß, with ß = 0.05 and 0.10.[80]

testing. In Figure 3.9 we showed that the direction of loading with respect to the direction of casting has a significant effect on the stress-strain behaviour measured in uniaxial compression. As reason for the increased strain at peak and the larger ductility for specimens loaded parallel to the direction of casting, the presence of porous areas near large aggregates was mentioned (Figure 3.10).

For multiaxial compression the effect has been observed as well, but now the strains increase even more under parallel loading as can be seen from Figure 3.97. Constant stress ratio tests -1.0/-0.33/β are shown for β = 0.05 and 0.10. We see that for type I experiments where the direction of casting is parallel to the direction of the major principal stress, peak strains that are approximately 20 to 25% larger than in the type II experiments are present. Again, an explanation can be found in the mechanism of Figure 3.10. The results shown here and those for uniaxial compression are all on the same concrete quality (a medium strength concrete with a uniaxial compressive strength between 45 and 50 MPa). For low strength concretes, bleeding can be substantially larger than in the concrete used here, and the effects from initial anisotropy can even be much larger. On the other hand, following the same line of reasoning, the number and size of voids near large aggregate particles decreases in high strength concrete where exceptionally low w/c ratios are used and the effects of initial anisotropy could be substantially less. From Figure 3.97 it can be seen that for the particular load-path followed, $\varepsilon_2 \approx 0$, which means that it is almost a plane-strain test. Indeed, comparing this result with the plane-strain tests from Figure 3.95 shows that in these latter experiments $\sigma_2/\sigma_1 \approx 0.3$. The advantage of such experiments is that failure occurs in the plane formed through the 1 and 3-directions. The (numerical) simulation of such load-paths is quite accessible.

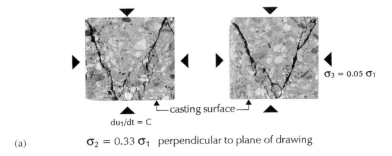

(a) $\sigma_2 = 0.33\,\sigma_1$ perpendicular to plane of drawing

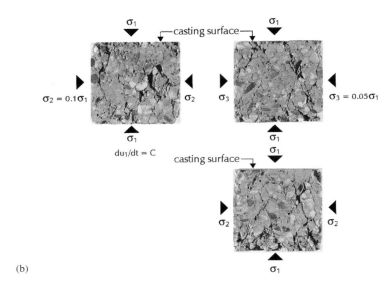

(b)

Figure 3.98 Fracture modes of concrete under triaxial compression.[80] (a) Uni-directional failure under stress ratio path -1/-0.33/-0.05 and (b) two-directional failure under constant stress ratio path -1/-0.10/-0.05.

3.5.4 FAILURE UNDER MULTIAXIAL STRESS

In the previous section, reference was made to failure modes in almost every example. Much of the modes of failure under generalised multiaxial stress can be deduced from the stress-strain curves already. Tensile strains usually imply that cracks develop normal to that direction, at least for moderate levels of confining stress (i.e., below the brittle-ductile transition). Depending on the actual state of stress, the strains can be positive in one, two or three directions. In the triaxial compression regime, positive strains can develop when the confining stresses in the second and/or third direction are substantially less than in the major compressive direction. In Figure 3.98 we show two typical different failure modes that can be observed under multiaxial compression. The first mode resembles uniaxial compression, and failure can occur in two directions. The second mode resembles equi-biaxial compression (extensile failure), where failure occurs in a single direction only (i.e., the least loaded direction). The failure modes illustrated in Figure 3.98 concern a triaxial compression test following a constant stress-ratio path -1/-0.10/-0.05. The stress-strain curves for this particular experiment were shown in Figure 3.96, and as can be seen both ε_2 and ε_3 are both tensile (positive). Moreover, the values of these strains are relatively large. The failed specimen shows an abundant amount of inclined shear fractures on all side surfaces. The failure mode resembles that of a uniaxially loaded cube, except that surface splitting cannot occur now because all sides of the specimen are covered by loading platens. The other type of failure is shown in Figure 3.98b. Here a uni-directional failure mode is shown for a constant stress-ratio test -1/-0.33/-0.05, for which the stress-strain curves were shown in Figure 3.96 as well. For this particular experiment very large tensile strains are measured in the minor compressive direction, and small tensile strains in the intermediate compressive direction. The failed specimen shows clear, discrete shear bands, which have been found in many types of rock as well.[98,261,262] These failure modes clearly suggest that for triaxially loaded concrete, below the brittle-ductile transition, localization of deformations occurs and the notion of a global strain becomes rather debatable. The movement of the more or less intact blocks between the shear bands is affected by the allowable displacements and rotations at the specimens boundaries. Such behaviour can be simulated using a simple interface fracture model.[89,163,263] The shear bands seem to propagate through the specimen's cross-section, much in line with observations under uniaxial tension as mentioned in Section 3.3 (see also Section 4.4 where we will show an interesting photo-elastic experiment).

The growth of a shear band can be deduced from Figure 3.99 where the fracture patterns of two constant displacement ratio experiments ($u_1/u_1 = $ -1.0/-0.33) are shown. The tests were carried out with different confining stress in the third direction, namely $\sigma_3/\sigma_1 = 0.05$ and 0.10 for the tests of Figure 3.99a and b, respectively. Both samples were loaded to the maximum deformation that could be sustained by the brushes that were used to load the concrete cubes, namely up to $\varepsilon_3 = 3\%$ (which implies $u_3 = 3$ mm). In the test with low confinement, the shear band was fully developed, whereas in the other sample only a partial shear band is visible. In the low confined specimen, the steep part of the σ_1-ε_1 curve is almost at the residual level, whereas the σ_1-ε_1 curve just rounded the peak. Other experiments at the higher confinement level showed fully developed shear bands as well, so it is to be expected that when the residual stress level is reached in the higher confined specimen, the shear band will be fully developed.

In rocks, the growth of a shear band in the post-peak regime was monitored in great detail by Hallbauer et al.[262] Again, the differences in behaviour between concrete and rock are only marginal. It should be mentioned here that

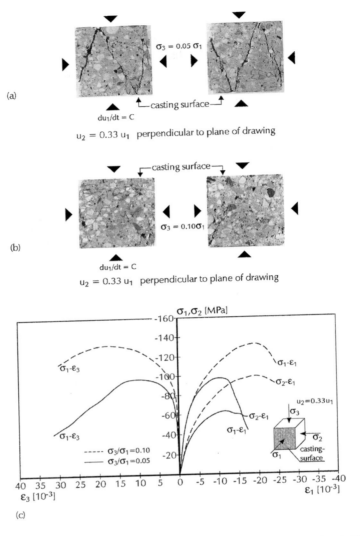

Figure 3.99 Shear band propagation observed in constant displacement ratio experiments.[253] Fully developed shear bands under $u_1/u_2 = -1/-0.33$ and $\sigma_1/\sigma_3 = -1/-0.05$ (a) and partially developed shear band under $u_1/u_2 = -1/-0.33$ and $\sigma_1/\sigma_3 = -1/-0.10$ (b), and stress-strain curves (c).

the concrete material structure seems to affect the width of the shear band. For example, the sample shown in Figure 3.98a has a rather wide shear band at the bottom where a large aggregate particle seems to have a widening effect on the shear band. Much about the shear band — material structure — boundary condition dependency still has to be learned. Future experiments will have to show what effects are caused by the material itself, and which are the results

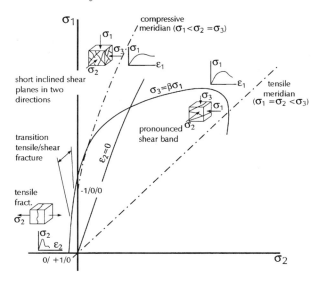

Figure 3.100 General overview of failure modes under multiaxial stress.

from the loading (test) configuration. For tension much more is known at present, as will be discussed in Chapters 4 and 5.

To conclude, in Figure 3.100 we show an overview of failure modes observed under multiaxial states of stress. Basically we can distinguish between uni-directional modes of failure and two-directional modes of failure. The uni-directional failure mode is found under triaxial extension (the best example is equi-biaxial compression), and under states of stress with a dominant tensile component. The angle of the shear bands under extensile states of stress is between 20° and 25° (angle between σ_1 and the shear band)[80] for tests between brushes. Under biaxial compression between brushes, values between 21° and 27° have been reported.[236] Similar values are known from rock and soil experiments. The two-directional mode of failure is observed for triaxial compressive states of stress near the compressive meridian. The best known example is the uniaxial compression experiment. As indicated in Figures 3.19 and 3.24, the shear band direction seems to depend on the type of loading platen used in the experiments. The fractures are increasingly more inclined when the frictional restraint between loading platens and specimen increases. The same might occur under triaxial states of stress, but much experimentation is needed before this issue can be completely solved.

The growth of the localized shear bands is also not fully understood. Again, for uniaxial compression, we discussed the growth and development of shear cones on top of rigid aggregate particles in the material. At the particle level, miniature shear bands develop, and the mechanism hypothesized in Figure 3.7 might be valid for triaxial compression as well. Much debate has been going on in rock, soil and ice mechanics,[264] but in concrete mechanics much of the research on the behaviour has halted after the decline of activity in the area of nuclear containment vessels. From the fundamental point of view, however, many issues are open to debate and resolving these matters would be highly desirable. For example, Horii and Nemat-Nasser[265] developed a micromechanical model which predicted the development of splitting failure under low confining stress from relatively larger pre-existing tensile cracks, to the growth of wider shear zones from an array of many small cracks under high confinement. Their model was based on model experiments using two-dimensional wing cracks. More recently, Dyskin et al.[266] reported on the growth and interaction of three-dimensional wing cracks in PMMA and bore-silica glass, but the phenomena can be observed quite easily also in ice.[267] The three-dimensional wing cracks seem to avoid each other, and complex interactions have been observed, which are hardly understood to date.

3.6 FINAL REMARKS

The mechanical behaviour of concrete has been discussed in depth in this chapter. After an exposure of failure under uniaxial compression, attention is given to uniaxial tension (mode I in fracture mechanics language), shear (mode II and III) and multiaxial states of stress. Traditionally, concrete text books contain many graphs, and all kinds of phenomenological information is given. We have tried to mimic this style in this chapter, although at some places it cannot be resisted to give a (partial) explanation. Facts are abundant, but does it really lead to true insight in the mechanical behaviour? The answer must be a blunt *no*. Therefore, the exposure of endless rows of facts will stop here. In the next part of the book, experimental and modelling tools will be described, which can be helpful in interpreting the observations presented in the first part (viz. Chapters 2 and 3).

PART 2
EXPERIMENTAL AND MODELLING TOOLS

Chapter 4

EXPERIMENTAL TOOLS

Even good tools can be misused,
and like any other workers we will do
better if we understand how our tools work.

Daniel C. Dennett

In the previous chapter, macroscopic and mesoscopic observations of fracturing of concrete — and sandstone — were described. Since the 1960s it has been first recognised that microcracking forms the basis for the observed non-linearities in the stress-deformation diagrams of these brittle, disordered materials under a variety of loading cases. It has been mentioned several times already that the observed behaviour depends to a large extent on the tools that are used. More specifically, post-peak response and stable crack growth depend very much on the adopted testing techniques, as well as on the boundary conditions in the experiment. In the post-peak regime large cracks are observed to develop, and the interactions of these fractures with the environment are of crucial importance. Understanding the results presented in the previous chapter is only possible through a thorough knowledge of the tools that were used to do these observations.

In this chapter, the experimental tools are described. Loading techniques, test-control systems and data-acquisition equipment are explained, as well as specimen selection and manufacturing criteria. This latter aspect is quite important because the material structure in the sample may depend to a large extent on the method of manufacturing, as well as on the shape and size of the specimen. Size/scale effects are one of the salient features of fracture of brittle, disordered materials. In spite of this, we will touch on this topic only shortly. Of importance will be to understand the physical mechanisms underlying size/scale effects. Several size effect laws have been proposed to date, in which a sound physical mechanism is lacking. In view of the phenomenological nature of such "laws", expanding them to larger sizes does not seem possible. A detailed account is given of available crack detection techniques, which are crucial for the determination of fracture mechanisms. A single paragraph has been included in which the theory of statistical designs is explained. Statistical design is a useful tool for designing test series from which unbiased conclusions can be drawn. Finally, but certainly not least important, the need of standard tests will be addressed. Both for the determination of tensile softening and compressive softening curves, standard test methods have recently been proposed, and it is expected that, very soon now, these recommendations will become effective.

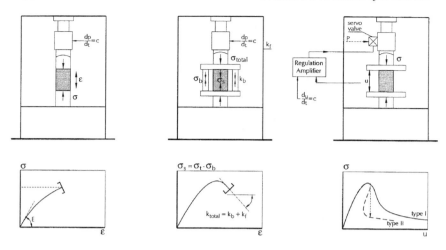

Figure 4.1 Response measured in a conventional load-controlled experiment (a), a load-controlled experiment with parallel-bar arrangement (b), and possible response observed in a displacement-controlled experiment (c).

4.1 LOADING EQUIPMENT

4.1.1 LOAD- VERSUS DISPLACEMENT-CONTROLLED TESTING

In the early days of materials testing, load-controlled experiments were the only possible means to get insight in the mechanical behaviour. When it was recognised that some materials — like concrete, rock and ceramics — have some remaining carrying capacity after the maximum peak stress was reached, new test methods from which the post-peak behaviour could be measured started to emerge. Figure 4.1 shows the three steps in equipment development. In conventional strength tests, load-control is applied, which means that a specimen will fail in an uncontrolled (explosive) manner as soon as the peak load is reached. In Figure 4.1a the principle is shown for a uniaxial compression test: pressure is gradually increased in the actuator until failure of the sample occurs. Only the pre-peak stress-strain curve can be measured. Traditionally, such experiments gave sufficient information for design of concrete structures, namely the initial Young's modulus of the material and the maximum strength. Note however that the strength determined in such a test depends not only on the specimen shape and size, but also on the boundary conditions used in the experiment, as discussed in Section 4.1.3 for uniaxial and multiaxial compression tests.

The first attempts to measure post-peak response of concrete date back to the 1960s and 1970s. In those days, fast electronics were not available and by placing stiff bars parallel to the specimen, at least part of the softening curve could be measured. For example, by using this technique, tensile softening curves were obtained by Evans and Marathe,[141] Petersson[142] and Schorn and

Berger-Böcker,[268] and for uniaxial compression by Grimer and Hewitt.[269] The principle of such experiments is that the stiff bars (either steel or aluminium) parallel to the concrete specimen take over the excess load that cannot be carried by the specimen when it starts to soften. This method is limited because the total stiffness (K_M) of the loading frame (K_F) plus parallel bar arrangement (K_B) should always be larger than the negative stiffness of the softening curve. Thus, for brittle materials with an extreme steep softening curve or, alternatively, when the stiffness of the parallel bar system is poorly dimensioned, the specimen will fail at the particular location in the descending branch where the specimen (tangent) stiffness matches the machine stiffness.

Nowadays, a more advanced method is available, which can be used under a wider range of conditions. The development of fast electronic devices has been pivotal in the possibilities that exist today. In a closed-loop servo-controlled experiment it is, in principle, possible also to measure the post-peak softening curve. The control variable in the experiment is not the load, but rather the axial displacement measured over the total specimen length, or in some experiments a displacement measured at a critical location at the surface of a specimen. In Figure 4.1c two situations are sketched that might happen in a displacement-controlled experiment. In a stable test, the control displacement increases monotonically in time. There is a balance between the elastic energy released by the specimen and the loading frame during the softening process. When this balance is lost, and when more energy is released by the machine and the specimen, unstable situations may arise. In the latter case snap-back behaviour is measured. In rock mechanics, these situations in compressive testing are normally referred to as a type I experiment (stable softening behaviour) and a type II experiment (unstable softening behaviour).[270] In a normal displacement-controlled experiment, the type II curve cannot be measured by using the axial deformation as control variable, and a straight drop of load is observed at the point where the snap-back occurs as indicated in Figure 4.1c. However, by selecting another control variable type II behaviour can in most cases still be measured. A few examples will be given in the next section. For this book, where we are mainly interested in fracture properties of the material, displacement-controlled testing is the most essential tool, and most of the discussion in this chapter will deal with this specific type of test control.

The choice of a specific test-control system is however only part of the entire experimental set-up. A global overview of a displacement-controlled uniaxial tensile (or compressive) experiment is given in Figure 4.2. The specimen is generally loaded in a stiff loading frame through a hydraulic or pneumatic actuator. The demand for a stiff loading frame still exists but, as we will see, by selecting a proper control variable much can be achieved, even in a so-called "soft" or low-stiffness machine. In Figure 4.2 a double-working hydraulic actuator is shown. A servo-valve is mounted on the hydraulic actuator. It is used to control the flow of oil to either of the two chambers in the actuator. The actuator is connected to a high-pressure oil pump. Of course the capacity of the

TENSION/COMPRESSION SET-UP   FLEXURAL TESTING

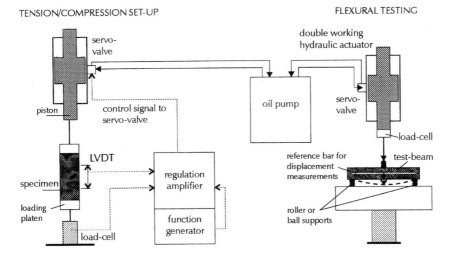

Figure 4.2 Overview of a displacement-controlled, uniaxial tensile or compressive experiment. At the right part of the figure a conventional four-point-bend experiment is shown, where the test beam is placed on roller bearings.

pump (normally expressed in litres of oil per minute) should correspond to the capacity of the servo-valve. The load is measured by means of a load cell. The principle of a load cell is generally based on measuring a strain on a calibrated steel cylinder with very accurate dimensions, see Figure 4.3. The cross-section of the calibrated steel cylinder is shown. The strains are, for example, measured in a full Wheatstone bridge configuration of strain gauges that must be glued to the calibrated steel cylinder with the utmost accuracy. To avoid mechanical damage to the strain-gauges, a protective cover has been placed around the instrument as shown. It is essential to make the cylinder as thin as possible to have the largest possible deformation in the measuring device, thereby increasing the accuracy. The signal is amplified by a standard measuring amplifier.

The demand for a thin steel cylinder often makes the load cell the least stiff part of the entire experimental set-up. The load cell should be placed as close as possible to the specimen. For example, in early designs of a tensile machine,[84] frictional restraint in a part of the test frame hindered an accurate measurement of the specimen load. The loading frame had to be adjusted, after which reliable and highly accurate results were obtained. Sometimes such errors remain undetected. We will return to the issue of tensile testing in various parts of the book, viz. Section 4.1.3, 5.3.4 and 7.2.2.

The other parts of the experimental set-up are shown in Figure 4.2 as well. The deformation of the specimen can be measured using a Linear Variable Displacement Transducer (LVDT), which can be mounted at any location of the specimen. In a uniaxial tensile or compressive experiment, the LVDT is usually placed in the same direction as the load is applied. Additional measure-

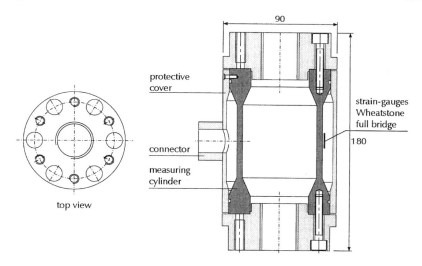

Figure 4.3 Principle of a load cell: vertical cross-section and top view.

ments can of course be made at the specimens surface, or internally, but this
will be discussed in Section 4.3. The deformation measured on the specimen
is used as a control variable in the displacement controlled experiment. Through
a conventional measuring amplifier, the signal of the LVDT is amplified and
fed into a regulation amplifier as indicated in Figure 4.2. Also, the servo-valve
is connected to the regulation amplifier as shown.

The principle of the closed-loop testing system is as follows. A signal from
a function generator (ramp selector) is fed into the regulation amplifier, which
sends another signal to the servo-valve to open. Oil pressure builds up in one
of the two chambers of the hydraulic actuator, the specimen is loaded and a
certain deformation will be measured at the specimen's surface. Subsequently
the response of the specimen, through the LVDT signal, is compared to the
input signal from the function generator. These two signals should match,
otherwise a correction is sent to the servo-valve and more or less oil is let into
the relevant pressure chamber of the hydraulic actuator. The oil pump gives a
constant pressure throughout the experiment. However, one should be aware
that the capacity in litres per minute is sufficient to warrant that no pressure
drops occur which might impair the experiment.

In the system of Figure 4.2, a number of components is essential to ensure
a fast response time. First of all, the electronics should react rapidly to changes
detected in the regulation amplifier. Furthermore, one of the most important
parameters is the capacity of the servo-valve, whereas the pump capacity and
the stiffness of the oil leads, as well as frictional losses (hysteresis) in the
actuator have larger or smaller effects on the response time. In addition to this,
the stiffness of the loading frame, including the stiffness of the specimen itself,

has an effect on the response time, and partly determines the slope of the softening branch that can be measured in a stable manner. The response time of the servo-valve can usually be adjusted through a so-called PID (Proportional-Integral-Derivative) feed-back algorithm of the regulation amplifier. Manufacturers of displacement-controlled test facilities specify the necessary procedures to derive an optimal setting of the regulation system for a given experiment. Note that the response times, and the performance of the loading system should be set for every new experiment as the stiffness and response of the specimen itself plays an active role in the closed-loop regulation system.

For load-controlled testing, the same equipment can be used. However, in that case the load measured by the load-cell rather than the deformation of the specimen plays a role in the control system. The principle remains the same, but one should be aware that explosive failure will occur as soon as the peak strength of the specimen is reached. Normally the electronics can be used to safeguard the test set-up from possible damage under explosive failure. Most common is to give an upper bound for the load that can be reached.

A final remark is in place. The (loading) capacity of an experimental system depends on the maximum pressure generated by the oil pump, the area of the piston in the hydraulic actuator, and, of course, the strength of the loading frame. In general, the loading frame is designed to have sufficient stiffness and the strength demand is not important. In that case the capacity of the actuator determines the capacity of the experimental set-up.

4.1.2 SELECTING THE CONTROL VARIABLE
4.1.2.1 General set-up of a stable fracture experiment

It may be clear from the previous section that the choice for a given control variable depends on the experiment, and the information that should be retrieved. In fracture experiments, the demand is usually most severe. A stable displacement controlled experiment is needed for measuring fracture energies and the shape of the softening curve. The stability of a fracture test depends on the balance between the energy released from elastic parts of the specimen and the loading frame during crack propagation. The response time of the servo-controlled system plays an important role in this. In some materials, crack propagation proceeds rather rapidly, and extremely high crack velocities may occur. For example, in ordinary hardened cement paste or in high strength concrete rather straight crack paths are measured, which lead to an extremely brittle response. A very fast servo-system is needed, and the elastic energy stored in the loading frame and the specimen should be reduced as much as possible. One of the first steps normally taken is to reduce the size of the specimen, or at least of the control length of the LVDT used in the closed loop system. For example, in the standard tensile test in the Stevin Laboratory of Delft University, a control length of 35 mm has proven to give stable results for most of the materials tested to date. Of course this length is based on all the equipment available in that particular laboratory.

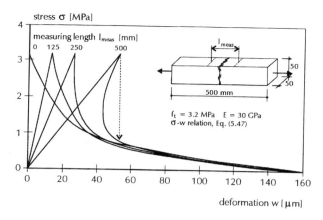

Figure 4.4 Effect of measuring length on the stress-displacement diagram in uniaxial tension.[84] (Reprinted with kind permission of Dr. Hordijk.)

In Figure 4.4 the effect of using shorter or longer measuring length in a uniaxial tensile experiment is shown. This figure is obtained from a theoretical analysis using Equation 5.47.[84] During softening, the elastic uncracked parts (see also Figure 3.32) unload and energy is released. By adding the energy release in these elastic unloading parts and the energy consumption in the fracture zone (which is in this case assumed to be a line crack), the relative importance of the elastic zones can be demonstrated. When the measuring length is equal to zero the fracture zone is covered exactly, and the softening diagram corresponds to Equation 5.47. Note that the pre-peak strain is zero when the measuring length decreases to zero. In that case only crack opening is measured. The elastic deformation increases when the measuring length increases. This follows directly from the diagram. When the measuring length has increased to 500 mm, the elastic unloading becomes so large that snap-back behaviour is observed. This may also happen in experiments, and the use of a sufficiently short control length is imperative.

For tensile tests, or other experiments where a single crack develops (or where a dominant crack governs the response of the entire structure), a short control length can be selected, at least when the location of crack growth is known. For uniaxial tensile experiments on prismatic or cylindrical specimens this means that special measures must be taken because in principle the fatal crack may appear at any location along the specimen length. The general solution to the problem is to make a notch in the specimen at the location where the crack should initiate.[187,271] It has been argued, however, that the notch affects the crack initiation and growth processes, and several authors suggested use of a method where un-notched specimens are used.[142,268,272] Nowadays it is possible to design a self-adjusting control system that selects automatically the critical control LVDT, as for example sketched in Figure 4.5. The electronics constantly monitor and compare the readings of several LVDTs mounted at the

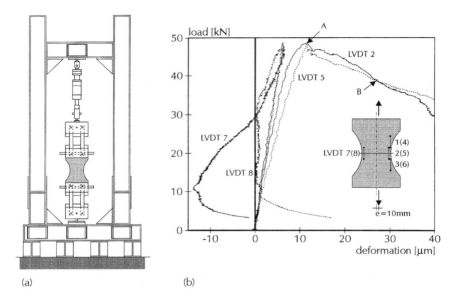

(a)   (b)

Figure 4.5 Intelligent control system for tensile testing of concrete. The electronics select the critical feed-back signal for the servo-controlled system. At point A, the control switches from LVDT 5 to LVDT 2, and at point B back from LVDT 2 to LVDT 5.[310]

surface of the specimen, and select the largest displacement as feed-back signal for the closed-loop servo system. As we have seen in Chapter 3, the shape of the softening diagram depends on the material structure, specifically on the aggregate size, the w/c ratio and the age at loading. The brittleness may vary depending on the material composition. This implies that the effect of elastic (unloading) parts may be more or less important depending on the concrete composition. When in Chapter 5 the issue of non-uniform opening is analyzed using non-linear (Hillerborg type) fracture models, we will see the importance of brittleness.

For compression the same problems may be encountered. However, in that case an additional factor arises because cracks will be abundant throughout the entire specimen volume. Thus, in principle, the complete specimen length must always be selected as control length. In particular for high strength, brittle concretes displaying type II behaviour (Figure 4.2), problems may be encountered. Note however that a fundamentally ill-posed assumption is made in the compressive testing of high strength concrete as will be discussed in Section 4.2. If, for whatever reason, type II behaviour is expected to occur, it is possible to control an experiment over a different displacement which should increase continuously in time. When, for example, cylinders are used in compression experiments, the circumferential strain can be used as a feedback signal. The lateral deformations in a compression experiment increase more gradually, in

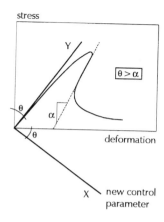

Figure 4.6 Test-control using a combination of force and displacement as feed-back signal.

particular in the post-peak regime as shown earlier in Figure 3.12. Some manufacturers of displacement-controlled testing systems have specially designed extensometers for lateral displacement control in type II uniaxial compression experiments. For prisms or cubes this type of test control is more hazardous, and the average lateral displacement with contributions from each of the free directions in the specimen should be used. Averaging becomes more important in that case.

The above discussion points towards two important steps in the selection of a control parameter. First of all, it should be attempted to reduce the elastic energy release (or elastic unloading) in the measuring length as much as possible. Second, one should try to find a displacement somewhere in the complete specimen-machine system which increases (or decreases) continuously in time. In other words, the rate $\partial w / \partial t$ should have the same sign during the entire experiment, where w is the response variable measured with the control device. This response variable is not necessarily a displacement, but can be a combination of force and displacement.

In Figure 4.6 the principle is outlined. In terms of stress and displacement, a type II softening diagram is measured in, for example, a uniaxial compression experiment on high strength concrete. The coordinate system should be rotated such that the x'-axis in the rotated system can be used as control variable. Obviously, following a simple coordinate transformation, we can see that we are now controlling the experiment over a combination of force and displacement. Such a system has been proposed and used by several authors.[120,121] Thus, when type II behaviour occurs in one particular coordinate system, one should look for another coordinate system where such behaviour cannot occur. As we discussed, this can be done by selecting another displacement in the specimen-machine system, or alternatively by rotating the entire coordinate system and using a combined parameter as feed-back signal.

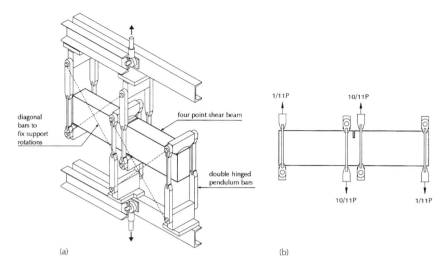

(a)   (b)

Figure 4.7 Four-point-shear experiment on a single-edge-notched beam loaded between freely rotating pendulum bars. To fasten the pendulum bars the dashed diagonals are mounted as shown. The inset shows the theoretical distribution of loads and reactions on the specimen.[156,274]

4.1.2.2 Examples of advanced fracture experiments

Let us now give some examples of advanced fracture experiments and also address some of the possible pitfalls in the design of a test set-up. In the previous section only some general remarks were given. It is difficult to give a simple recipe for the design of a stable fracture experiment. Therefore, a few examples (from personal experience) will probably clarify the matter much better than a lengthy theoretical chapter. Subsequently, we will discuss the choice for a control variable in a four-point shear experiment, in a hydraulic fracture experiment, and in a true triaxial experiment. In Section 4.1.3.2 we will discuss in more detail the uniaxial tensile experiment, which is central to fracture mechanics studies of concrete.

a. Four-point-shear experiments

In Section 3.4 we discussed shear fracture in concrete. A particularly interesting experiment that has attracted the attention of many researchers in the past decades is the four-point-shear test. The experiment has not only been used for concretes and rocks, but finds its origin in metal testing.[207] In Figure 4.7 the loading frame as it was used in the Stevin laboratory is shown; in the inset the loading scheme on a single-edge-notched beam is included.

In Figure 4.8 a photograph of the experimental set-up is shown. Double-edge-notched beams can be tested in the same set-up as well. Basically, a region of high shear develops in the area between the middle loads, at least when we assume that beam theory applies to this geometry. These matters are discussed in Section 3.4, and some analyses are shown in Section 7.3.1. The

Figure 4.8 Photograph of the experimental set-up for four-point-shear testing between rotating supports.

common approach to the experiment is to load the beam in a standard compression machine.[198,202,208] The beam is supported by roller bearings, but this will have an effect on the fracture process as demonstrated by Gjørv et al.[273] for a simple three-point bend test. The reason is that the roller bearings are never truly frictionless (see also Section 4.1.3). In the four-point-shear test of Figure 4.7, the specimen is held between freely rotating pendulum bars and the applied load is tensile rather than compressive. Depending on the length of the pendulum bars, the beam is more or less free to move in the horizontal direction, and crack growth in the central part between the notches is almost unrestrained. An additional advantage of testing between pendulum bars is that the reactions and loads can easily be measured by instrumenting the individual pendulum bars

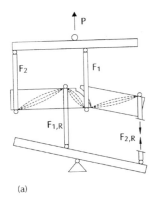

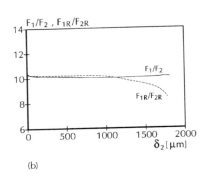

(a)

(b)

Figure 4.9 Distribution of loads and reactions in a four-point-shear test on a single-edge-notched beam loaded between freely rotating supports.[274]

with strain gauges. The loading situation in that case is exactly known during the experiment.

The initial design of the frame involved an asymmetric set-up, in which the rotational freedom of the lower pendulum bars was not the same as in the upper pendulum bars. After the machine was converted into a completely symmetric loading case, which is shown in Figure 4.7, a perfect distribution of loads and reactions could be maintained throughout the experiment. This is illustrated in Figure 4.9. The ratios between the loads and reactions is almost constant throughout the experiment. Only for large deflections, some deviations start to appear. Note that the ratios between loads deviates slightly from the theoretical distribution shown in Figure 4.7. This can be explained from the fact that the hinges at the ends of the pendulum bars have a finite dimension, and for practical reasons the length of the pendulums must be limited as well.

Because still some friction can appear in the hinges that are fixed at the end of the pendulum bars, the set-up was designed to allow for testing between rigid supports as well. Testing between fixed (i.e., non-rotating) supports is achieved by mounting diagonal bars between the pendulums as shown in Figure 4.7 (dashed lines). The fracturing of the beam is highly sensitive to the fixation of the supports as shown in Sections 3.4 and 7.3.1.

The beam was both notched at the top and bottom by making a saw cut with a rotating diamond saw. In the experiment it was decided to measure both the notch opening (Crack Mouth Opening Displacement or CMOD) and the sliding displacement in the notch (Crack Mouth Sliding Displacement or CMSD). In all, four LVDTs were used for measuring the CMOD and four LVDTs for the CMSD. They were mounted as shown in Figure 4.10, viz. at the front and back faces of the beam, both near the top and bottom notch.

The problem was to select a reliable control parameter in the experiment. Using the lattice model, which is presented in Chapter 5, a numerical simulation of the experiment was carried out. The average CMOD and CMSD were

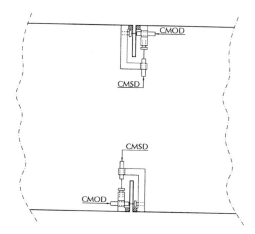

Figure 4.10 Measuring CMOD and CMSD on a double-edge-notched, four-point-shear beam.[156,274]

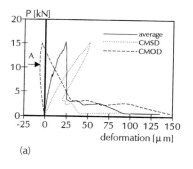

(a)

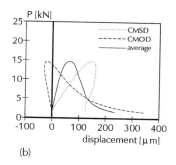

(b)

Figure 4.11 Selecting a test-control parameter in the double-edge-notched four-point-shear experiment. The numerical simulation with the lattice model suggested that the average of CMOD and CMSD would be the best parameter (a). This was confirmed in the experiments. Here we show a test on Felser sandstone (b).[274]

plotted separately, but also the average CMOD and CMSD were calculated. In Figure 4.11a the outcome of the simulation is presented, and the results suggest the following. When the loads are applied to the beam, initially closure of the notches occurs as can be seen from the CMOD measurement; they are negative in the pre-peak regime. Later on in the simulation, the CMOD changes sign, and opening of the notches is detected. The change of polarity makes the CMOD not suitable for test-control. The beam would fail in a brittle manner as soon as notch closure would reverse to notch opening, i.e., at point A in Figure 4.11a. In this particular example not even the maximum load was reached. If the CMSD would have been selected as control variable a similar problem would have been encountered. Namely, as soon as the steep part of the descending branch is reached, the simulation predicts type II response. In other

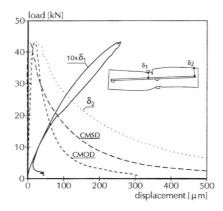

Figure 4.12 Displacements δ_1 and δ_2 measured in a four-point-shear test between rotating supports on a SEN beam. Load-displacement curves showing that δ_2 is a better control parameter than δ_1, but that the CMOD and CMSD could have been used equally well.

words, a snap-back would occur, leading — at least for a small part of the experiment — to unstable crack propagation. The solution to this problem was found by selecting the average of CMOD and CMSD as control variable. In that case, a continuously increasing displacement in time is found and no instabilities can occur. Stable tests could indeed be performed by selecting the average CMOD and CMSD as control parameter in the experiment, as may be clear from the sandstone test of Figure 4.11b. The experiment eventually led to a deeper insight in shear fracture behaviour of concrete and rock. The outcome of the experiments, and the simulations of crack growth with the lattice model are presented in different parts of this book (Sections 3.4 and 7.3.1).

As mentioned earlier, the four-point-shear set-up was also suitable for testing single-edge-notched (SEN) beams. One particular interesting case was the contribution to a Round Robin experiment carried out by RILEM technical committee 89 FMT ("Fracture Mechanics Test Methods for Concrete"). A test designed by Carpinteri et al.[202,203,211] was suggested for measuring the mixed-mode fracture energy of concrete. Similar experiments were carried out earlier by Ingraffea and Panthaki.[210] The situation is shown in Figure 4.12. In the Round Robin experiment, use of the far-end displacement δ_2 instead of the displacement δ_1 under the load point was suggested. Figure 4.12b shows why. Beyond peak δ_1 simply unloads, whereas δ_2 keeps increasing. The experiment carried out in the Stevin laboratory with rotating supports (Figure 4.7) showed that the CMSD and CMOD could be used as feed-back parameter equally well. However, as soon as the supports were fixed, a second crack would develop in the specimen that would make the entire experiment unstable. In that case it was essential to use δ_2 as feed-back signal. Note that Carpinteri carried out experiments in a compression machine in which the same situation always

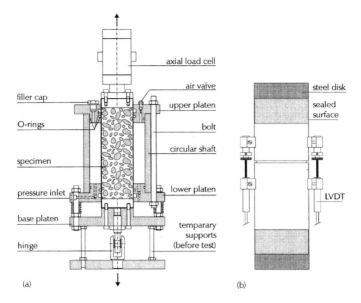

Figure 4.13 Experimental set-up for hydraulic fracture experiments of concrete and sandstone. Vertical cross-section of the pressure cell (a), specimen and LVDTs for test control (b).[275]

occurred as in the pendulum set-up with fixed supports. It must be concluded that care should be taken in defining the correct boundary conditions in the experiment.

b. Hydraulic fracture tests

The second example is the construction of a triaxial cell for testing concrete and rock under hydraulic pressure. In the oil industry for stimulating the production of oil wells, but also in off-shore platforms and irrigation dams, cracks may exist that are filled with a pressurised fluid. In order to investigate the crack tip processes under such conditions, a special hydraulic cell was designed in which a cylindrical specimen was loaded by fluid pressure along the circumference. The idea was that the pressurised fluid would enter the crack during propagation, whereas fluid should also be allowed to leak-off in the porous concrete or sandstone that was used in the experiment. Furthermore, it was decided that the specimen could be loaded in axial direction as well, either in tension or in compression. All these demands led to the design of a triaxial cell, where a 300 mm long, 100 mm diameter concrete cylinder could be tested. A cross-section of the cell, plus an instrumented specimen are shown in Figure 4.13. A photograph of the open cell containing an instrumented specimen is shown in Figure 4.14.

In conventional triaxial cells, no contact exists between the pressurised fluid and the specimen, and in general a membrane is used to separate the two media.

Figure 4.14 Photograph of the open cell showing an instrumented specimen.

However, in the present set-up the membrane was omitted. Instead, the specimen surface was impregnated using a low viscosity epoxy, except for a small region in the middle section. Furthermore, in conventional cells, the specimen is completely contained in the cell, but here it was decided to use longer specimens such that the epoxy connections between the specimen and the steel platens needed for the axial loading of the cylinder were located outside the pressurised fluid reservoir. This meant an extremely careful specimen preparation because the concrete or rock would be in contact with the seals (O-rings) in the top and bottom plate of the pressure cell. For preventing leak-off along these seals, or even through the porous material, surface impregnation of the outer ends of the specimen using a low viscosity epoxy as mentioned before was found essential. Having solved all leakage problems and gluing procedures, the next problem was test-control as we wanted to investigate the effect of crack tip pressure under stable propagation. The pressure fluid was water, which implied that a water/oil separator should be used in order to allow for the use of our standard hydraulic pressure pumps. The selection of the control parameter in the closed loop servo-system was more straightforward here than in the four-point-shear experiment of the previous section.

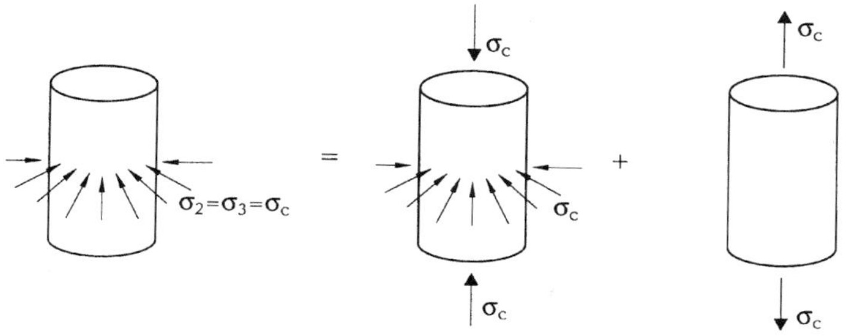

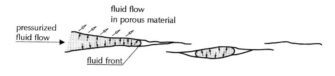

Figure 4.15 Breakdown of the biaxial compressive stress in a hydrostatic component and an axial tensile component.

fluid flow
in porous material

pressurized
fluid flow

fluid front

Figure 4.16 Fluid pressure in a crack.

By applying circumferential compression, the cylinder will fail by axial splitting. In principle a biaxial state of stress is generated, and out-of-plane fracture will occur. This can be visualised by separating the biaxial load into a hydrostatic component and an axial tensile component as shown in Figure 4.15.

So, basically, a lateral compressive load leads to an axial cleavage crack, and the most straightforward control parameter is therefore the axial deformation of the cylinder. During crack propagation the pressurised fluid enters the crack and gives an additional driving force for crack propagation as shown in Figure 4.16. Therefore, an extremely fast servo-hydraulic system was needed.[235,275] In order to have a short control length, a circumferential notch was sawn in the cylinder. At this location the splitting crack would develop. An example of a stable, displacement-controlled test under circumferential compression (zero axial load) is shown in Figure 4.17. With this machine essentially the tensile meridian of the failure contour of concrete and rock can be tested, an area of research that is completely neglected because of the inherent difficulties of the test method. Note, however, that in a present application, the fluid enters the crack and fluid leak-off in the porous material has a more-or less important effect on the failure contour.[249]

c. Multiaxial compression experiments

A third example concerns the development of a true triaxial test machine at Eindhoven University of Technology.[80] This truly triaxial machine allows for

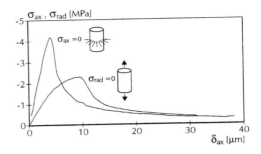

Figure 4.17 Examples of stress-axial deformation curves under biaxial circumferential stress and zero axial load and axial tension only.[249]

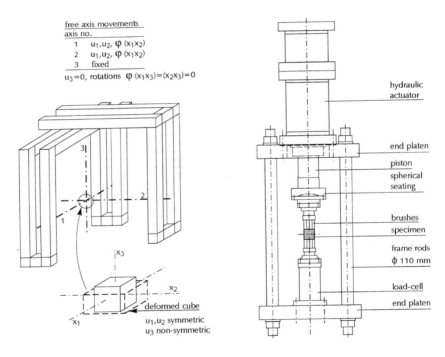

Figure 4.18 Truly triaxial machine at Eindhoven University of Technology. (a) Basic scheme and (b) construction of a single loading axis.[80]

testing 100 mm cubes. A threefold servo-control allows for almost unrestricted multiaxial testing of concrete. The apparatus was the first in the world that allowed for strain-softening testing under arbitrary multiaxial loading paths. In Figure 4.18 an overview of the test apparatus is shown. Essentially, three "simple" uniaxial compression machines are interwoven to form a truly triaxial

Figure 4.19 Photograph of the Eindhoven multiaxial machine.[80]

machine for testing cubical specimens. The three axes are hung in cables, which allow for free movement of all axes in the horizontal plane. The frames were not allowed to move in the vertical direction because the cables were fixed, but currently the machine is upgraded and contra-weights are installed such that all three axes can move freely in the vertical direction as well. The advantage of free moving axes is that the centre of the specimen always remains in the centre of the machine, and symmetric deformations of the specimen are allowed. If all three axes are connected to one another, the specimen is pushed in a corner of the machine and, especially under extreme large deformations, this may cause problems in the loading platen. The truly triaxial machine developed at the Technical University of Munich is a so-called one-part machine (Linse[276]), whereas the multiaxial apparatus at the "Bundesanstalt für Materialprüfung" (BAM) in Berlin is a three-part machine in which the specimen can deform symmetrically.[251,277] In the initial design of the Eindhoven multiaxial machine the situation was therefore not ideal (the allowable deformation of a specimen are sketched in the inset of Figure 4.18a), but for experiments with a small deformation in the vertical axis no appreciable problems arise. Each of the axes of the Eindhoven machine consists of a single hydraulic actuator of 2000 kN capacity in compression and 1400 kN in tension. In each loading frame, a load cell, a spherical seating (which is fastened during an experiment), and the loading platen are identified. A single loading axis is sketched in Figure 4.18b. A photograph of the set up is given in Figure 4.19. Each loading axis is provided with a servo-valve, and three independent regulation circuits are available. In principle, this allows for testing a large

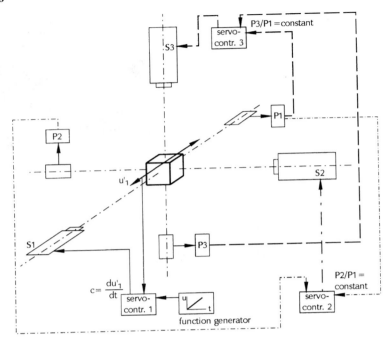

Figure 4.20 Circuit of regulation for a constant stress ratio experiment (σ_1/σ_2 = constant and σ_1/σ_3 = constant), with displacement control in the direction of the major principal stress (du_1/dt = constant).[80]

variety of load-paths. In Figure 4.20, the circuit of regulation for a constant stress-ratio experiment (σ_1/σ_2 = constant and σ_1/σ_3 = constant), with displacement control in the direction of the major principal stress (du_1/dt = constant) is shown. At the time that the machine was built, the regulation circuit had to be realised through hard wiring the various regulation amplifiers. Nowadays computer-control is easily available, which allows for unrestricted multiaxial testing as well.[214]

Using the Eindhoven multiaxial machine, also cyclic load-paths and constant displacement-ratio experiments have been carried out.[80] A major problem in multiaxial experiments is the selection of a platen through which unrestrained load transmittal to the specimen is possible. This is particularly important as neighbouring platens may interact, and may cause errors in the determination of the failure stress. This point will be further elucidated in Section 4.1.3. In the Eindhoven triaxial machine brush platens were used as load transmitting medium. At the University of Colorado in Boulder a truly multiaxial machine has been developed where load is applied to the specimen through fluid cushions,[278-280] Figure 4.21. Fluid cushions give, in principle, a constant stress on a specimen, as elucidated in the next section. This implies that the deformation over the specimen is not constant, and displacement

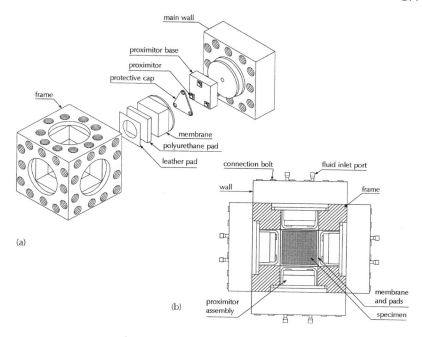

Figure 4.21 Multiaxial cell at the University of Colorado in Boulder.[278] (Copyright ASTM, reprinted with permission.)

control is becoming rather tedious. In fact, displacement-controlled testing can be done only when a reliable averaging procedure for the deformation measurements is developed first. Therefore, the Colorado machine operates in load-control only, thereby limiting the range of application. A possible control parameter could be the volume of the pressure fluid in the fluid cushion. But then, of course, care should be taken that the fluid cushion is strong enough to prevent the cushion to enter into cracks in the softening regime. There is enough room for speculation in this particular case. Note that the fluid cushions cannot be used for applying a tensile stress to the specimen. If combined tensile/compressive states of stress are investigated, the Colorado machine is slightly modified, and in the tensile direction aluminium brushes are used.[278]

4.1.3 THE IMPORTANCE OF BOUNDARY CONDITIONS

In fracture experiments boundary conditions play an important role. Often, these effects are simply ignored, which may lead to an erroneous interpretation of test results. There are two major boundary effects, namely frictional influences and rotational influences, which both may affect the experimental result considerably. The effects were mentioned before in specific paragraphs, viz. in Sections 3.2.4, 4.1.2. Later on in this book some of the effects will be analyzed in greater detail. The boundary effects seem to be of major importance in

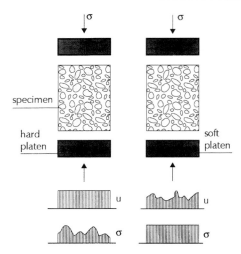

Figure 4.22 Effect of hard and soft loading platens on the stress and deformation distribution in a concrete specimen subjected to uniaxial compression.

understanding the fracture of brittle, disordered materials like concrete and rock. For the specific case of multiaxial testing interactions between adjoining loading platens are also of importance. The effect is somewhat related to frictional boundary restraint in uniaxial compression experiments. In this section we will focus in detail on frictional boundary restraint, in particular on the effect of boundary restraint in compression tests and three-point-bend tests, as well as on the rotational stiffness of the entire specimen-machine system.

4.1.3.1 Frictional restraint

Frictional restraint can have a significant effect on the outcome of a compressive experiment, but also in three-point-bend (or other) tests on beams. In Section 3.2.4 we discussed the effect of boundary restraint on softening. Let us go in somewhat more detail here and discuss different solutions to overcome the problem as much as possible.

a. Uniaxial compression experiments

In Figure 3.18 we showed the effect of rigid and soft platens on the specimen response in compression. The most common type of rigid loading platen is a steel platen which is in direct contact with the specimen (sometimes also referred to as dry platen). In such a set-up, the boundary of the specimen is forced to remain straight during an experiment. As a consequence of the heterogenous nature of the concrete, the stress-distribution becomes highly non-uniform as sketched in Figure 4.22. In addition to this non-uniform state of stress, which has its origin in the material itself, non-uniformities arise due to the friction between the loading platen and the specimen. The friction is caused by the difference in Poisson's ratio between the loading platens and the specimen. In Figure 4.22 we show the effect of material heterogeneity only. When, on the other hand, soft loading platens are used, the effect of frictional restraint reduces considerably. In this case the deformations of the plane of the specimen that is in contact with the loading platen are non-uniform. The stresses are now uniformly distributed. The fluid cushions used in the Colorado multiaxial cell (Figure 4.21) can be considered as soft platens. The only restraint comes from the membrane of the fluid cushion, and from leather pads that are inserted between the fluid cushion and the specimen. These leather pads are needed in this multiaxial cell in order to avoid contact between the fluid cushions in the other loading directions during an experiment; more about that in Section 4.1.3.3. Thus, the axial stiffness of a loading platen has an effect on the axial stress and deformation distribution in the specimen. The Poisson's ratio of a loading platen, and the possible difference with the value for the concrete (or rock) specimen can lead to frictional restraint, and consequently to a non-uniform state of stress in the specimen as well.

Over the years, researchers have tried to develop special loading platens in order to overcome the frictional boundary restraint. In uniaxial compression experiments, one of the measures that evidently can be taken is to load long prismatic or cylindrical specimens. For multiaxial experiments this is, however, impossible, and if one is interested in the post-peak behaviour, frictional restraint should be avoided throughout the experiment. This is a very severe demand because the Poisson's ratio of concrete changes with load level, whereas the Poisson's ratio of the loading platen is generally constant.

In Figure 4.23 several loading platen that have been proposed in the past are summarised. The different loading systems shown here were compared in an International Cooperative Research on the multiaxial behaviour of concrete.[238] Next to the dry platen and the standard triaxial test, in which the axial load is applied through rigid steel platens and lateral confinement through fluid pressure (separated by a membrane), there are four so-called friction-reducing platens. The systems were classified to friction-reducing ability and uniformity of the generated boundary stress distribution as shown in Figure 4.24. The dry platen system has the largest axial rigidity, but also the largest frictional restraint. The fluid pressure system used in the conventional triaxial cell and the fluid cushions as used in the Colorado cubical cell are at the other end of the spectrum: their axial rigidity is zero and the frictional restraint is minimal. The other systems fall somewhere in between of these two extremes.

Lubricated platens are almost similar to dry platens, except that a friction-reducing pad or some grease is placed between the platen and the specimen. The frictional restraint can be reduced considerably, but this depends on the type of anti-friction layer used. In the literature we find various systems: aluminium sheets with talc powder,[281,282] rubber,[239] teflon,[283] polyethylene sheets with molybdenum sulfide grease,[284] teflon sheets in combination with grease,[103,189,285] polished metal sheets,[286] composite systems like the MGA

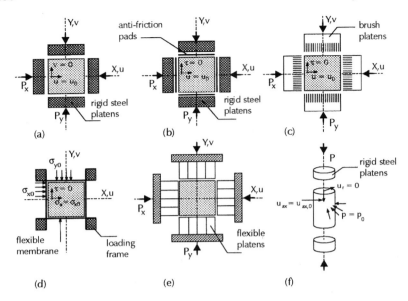

Figure 4.23 Various types of loading systems for compressive testing of concrete.[238] (Reprinted with permission of ACI, Detroit, MI.)

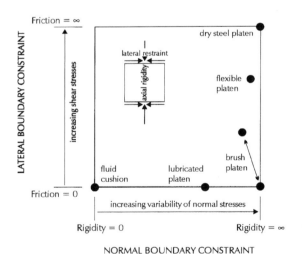

Figure 4.24 Classification of loading system according their frictional-restraint-reducing ability and their axial rigidity.[238]

pad,[100] and more recently stearic acid.[106,110,287] Teflon in combination with grease, or the application of stearic acid between specimen and steel loading platen both lead to extremely low coefficients of friction. For example, by

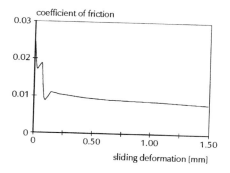

Figure 4.25 Coefficient of friction for the teflon platen used by Vonk.[89] (Reprinted by kind permission of Dr. Vonk.)

using two teflon sheets with some grease in between, Nojiri et al.[285] report coefficients of friction varying between 0.025 and 0.03. Vonk used a thin teflon sheet (0.05 mm thick) in combination with a thin layer of ordinary bearing grease (Molykote BR2 plus) between the teflon and the steel loading platen. The frictional restraint of this teflon intermediate layer varied with sliding distance, and is shown in Figure 4.25. The platen shows stick-slip behaviour, i.e., the initial coefficient of friction is about 0.03, which for larger deformations decreases to about 0.01. The results of uniaxial compression experiments carried out by Vonk indicate that this stick-slip behaviour may have affected the peak strength of concrete prisms of varying slenderness.[89] Similar results are obtained for stearic acid.

A great danger in applying lubricated platens is related to the thickness of the grease layer: if it becomes too thick, grease may be squeezed out in the lateral direction, thereby reversing the frictional restraint from inward to outward direction (see also Figure 3.18). The effect was noted before by Newman.[244] Thus, stick-slip behaviour and excessive grease quantities can nullify the effect of the intermediate layer. In particular, the grease is a large problem as the failure mode may be changed significantly as discussed in Section 3.2.4. Another disadvantage of intermediate layers is that their application in multiaxial experiments is restricted to multiaxial compressive states of stress. No tensile load can be applied.

The second type of friction-reducing loading platen is the so-called non-rigid platen. At present two variants have been developed. Hilsdorf[288] designed and used so-called brush-bearing platens. Later, brushes were used successfully by many other researchers too.[80,236,254] The other variant is the flexible platen developed by and extensively used at the Bundesanstalt für Materialprüfung (Federal Institute for Testing of materials, BAM) in Berlin.[251,289,290]

In brushes, the friction between specimen and loading platens is reduced due to bending of the separate brush rods. The maximum allowable loading is confined to the buckling-resistance of the individual brush rods.[276,288] An

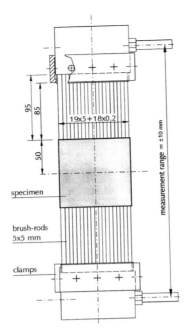

Figure 4.26 Example of brush-bearing platen.[80]

example of a brush platen is shown in Figure 4.26. A total of 361 steel needles
with a cross-section of 5 x 5 mm² were separated by a grid consisting of 0.2
mm thick phosphor bronze strips at the clamped side of the rods only. The
clamp construction is shown in Figure 4.26. The effective length of the rods
was 85 mm. As contact between the brush surface and the specimen determines
the effectiveness of load transfer and also of the uniformity of stress in the
specimen, the brushes had to be ground after several experiments.* Moreover,
because in these particular experiments the deformations were measured with
LVDTs mounted between two opposite base platen of the brushes (as shown
in Figure 4.26) contact effects had to be eliminated as much as possible.
Eventually, this led to the decision of using a thin capping between the brushes
and the specimen. Although this complicated the initiation of an experiment
somewhat, it proved a very reliable method. The frequent grinding of the brush
surfaces makes the application rather tedious. However, the frictional restraint
can be very low indeed, especially when the lateral deformations of the brush

---

* Contact effects develop if no full contact between loading platen and specimen exists. The
measured deformations are too large when the system of Figure 4.26 is used, because plastic
deformations in the contact zone must first lead to a better contact between the platen and the
specimen. The effect can be interpreted as a Hertzian contact problem between two spherical
surfaces with very large radius. Contact effects decrease at higher levels of applied load when
microcrack processes become more important.[80]

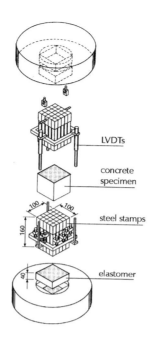

Figure 4.27 Flexible loading platen used by Schickert.[251, 289] (Reprinted by kind permission of Deutscher Ausschuss für Stahlbeton, Berlin.)

rods are limited. Therefore, by using brushes, the same peak strength is found when experiments are carried out on prisms with varying slenderness.[80] The brushes are well suited for applying tensile loads to a specimen. Obviously, the loading platens should be glued to the specimen in that case. In applying the glue (in general two-component, rapid-hardening epoxy resins are used) one should prevent the glue from entering the small space between neighbouring brush rods. Gluing the brush rods together would affect the performance of the brush considerably (in a negative sense).

The other type of non-rigid loading platen is the flexible platen developed at the BAM in Berlin. In Figure 4.27 the system is shown, including the fixation of LVDTs and extensometers for deformation measurements. Relatively large steel rods (cross-section 20×20 mm², length 160 mm, distance between the steel stamps is 0.1 mm) are placed on top of an elastomer pad (of 40 mm thickness) that is contained in a solid steel platen. The elastomer allows the steel rods to move more or less freely in lateral directions. Therefore, almost unrestrained testing is possible.

However, at the edges of the steel rods tensile stress concentrations will occur. This causes vertical tensile splitting cracks between the steel rods as shown in Figure 4.28a. The failed patterns of prismatic specimens with relatively low slenderness ($h/d < 1$), clearly showed a square grid of splitting cracks on the surface of the prism that has been in contact with the flexible platen (see

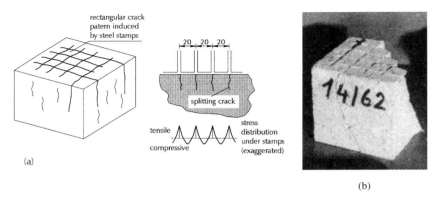

Figure 4.28 Splitting tensile cracking at the edges of the steel stamps in a flexible platen (a), and photograph of a fractured specimen, after Schickert[290] (b). (Figure (b) is reprinted by permission of Deutscher Ausschuss für Stahlbeton, Berlin.)

Figure 4.28b). This suggests that the observed failure mode is largely determined by the type of loading platen used. Thus, the square steel stamps basically act as small, rigid loading platens; triaxially stressed regions will develop in the specimen below the stamps as indicated in Figure 4.28a as well. It should be mentioned here that similar observations have been made for tests between brush-bearing platens. In that case, however, the area of the needles is considerably smaller than the size of the steel stamps of the flexible platens. Consequently, many possible sites of crack initiation under a brush are possible, and the material structure of the specimen will largely determine where a crack will develop.

In Figure 4.29 a drawing of a fractured cube is shown. The surface shown was in contact with the brush. Clearly, cracks develop under the loading platen, but the cracks are wider near the edges. This means that surface splitting may have occurred. Indeed, measurements shown in Figure 4.30 indicate that surface unloading occurs before the core of the specimen fails. Corrections were made for the contact effect that was mentioned before by replacing the initial part of the pre-peak stress-strain curve by a straight line up to the point where the maximum tangent is measured. After that the complete diagram is translated to the origin.[80]

In the above described experiment, the surface measurements were carried out using 60 mm long strain-gauges, the overall deformations were measured with LVDTs mounted between the brushes as shown before in Figure 4.26. The strain gauges show unloading, whereas the LVDTs registered a continuous increase of axial strain. This example would suggest that brushes are ideal for compressive experiments. However, there exists considerable doubt about brushes as well. The friction experiments carried out by Vonk[89] show that frictional restraint of brushes increases with increasing deformation (see Figure 3.20b). Consequently, as discussed in Section 3.2.4, brushes have a restraining

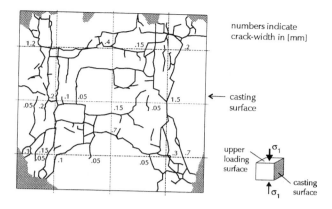

Figure 4.29 Crack pattern in the upper surface of a concrete cube, loaded between brushes.[80]

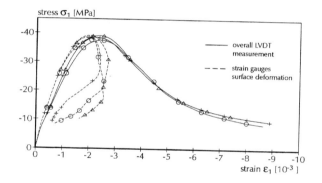

Figure 4.30 Difference between surface-strain measurements and overall strain measurements for a 100 mm cube loaded in uniaxial compression between brushes.[80]

effect in the softening regime. Moreover, as the lateral deformations increase, which occurs predominantly in the softening regime, the surface of the brush will become more and more spherical as shown in Figure 4.31. As a result, the specimen will be loaded non-uniformly, and this should be taken into account in the interpretation of the test results.

Now let us return to the flexible platens. Schickert[290] carried out an extensive test programme on prisms of varying slenderness loaded between flexible platens. A comparison was made with the response of prisms loaded between conventional rigid steel platens. The experiments were carried out under load-control in a stiff loading frame. As a consequence only the pre-peak behaviour could be determined. The most interesting result was the observed effect of slenderness on compressive strength. This effect was measured for four different concretes and mortars. In Figure 4.32 a comparison is shown of the strength of a specimen loaded between rigid platens and between flexible platens. The

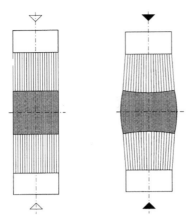

Figure 4.31 Deformation of brushes during a compression experiment. In the softening regime large lateral deformations cause the brush surface to become spherical, which may lead to non-uniform loading of the specimen.[80]

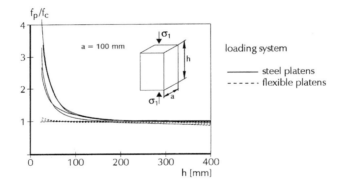

Figure 4.32 Effect of specimen slenderness on the compressive strength of concrete.[290] The area of the prisms was 100×100 mm^2. A comparison is made between prisms loaded between rigid platens and between flexible platens.

flexible platen results were normalised with respect to the strength of a 100 mm cube; the rigid platen results were normalised with respect to the strength of a prism with slenderness 2.5. The cross-sectional area of all prisms was 100 mm^2.

It is interesting to note that the prisms loaded between flexible platens showed an almost constant failure strength. Obviously, the friction reducing ability of the platens is good enough, and confining effects as found for rigid steel platens are completely removed. The confining action of rigid steel platens was explained in detail in Figure 3.18. When the rigid platen results are

TABLE 4.1
Relative Compressive Strength for Prisms of Varying Height Loaded Between Rigid Platens and Between Flexible Platens, for Four Different Concretes[290]

Loading system	Material[a]	Slenderness h/d								
		0.25	0.50	0.75	1.00	1.50	2.00	2.50	3.00	4.00
Rigid steel platen	B25/16	4.02	1.95	1.38	1.22	1.10	1.02	1.00	1.00	1.00
	B35/16	3.40	1.91	1.38	1.24	1.08	1.01	1.00	1.00	1.00
	B15/04	3.28	1.52	1.24	1.11	1.00	1.00	1.00	1.00	0.97
	B25/04	2.82	1.54	1.31	1.20	1.08	1.02	1.00	1.00	1.00
Flexible platen	B25/16	1.09	1.04	1.01	1.00	1.00	1.00	0.99	0.98	0.96
	B35/16	1.09	1.03	1.00	1.00	1.00	1.00	1.00	1.00	0.96
	B15/04	1.00	1.00	1.00	1.00	0.98	0.96	0.94	0.92	0.89
	B25/04	1.00	1.00	1.00	1.00	0.99	0.97	0.95	0.93	0.87

[a] Strength classification and maximum aggregate size are indicated in this column.

considered, we see that a four time increase of compressive strength is found for prisms with slenderness of 0.25. The curves for rigid platens level off between $h/d = 2.5$ and 3. For that slenderness, the end-effects do not seem to affect the compressive strength any more. However, one should be careful in extending the range of slenderness above $h/d = 4$, because in that case eccentricities, and eventually buckling may become increasingly more important. In Table 4.1 all relative strength results obtained by Schickert[290] are summarised. The main conclusion would be that for multiaxial experiments on cubes, flexible platens would give a reliable result, whereas in uniaxial compressive experiments one could decide to use rigid platens, but the specimen slenderness should then be between 2.5 and 3. However, a nagging feeling remains in view of the failure modes observed in the experiments. The mechanism proposed in Figure 4.28a suggests that when flexible platens are used, one could also conclude that the specimens all fail at the same splitting load, i.e., splitting near the edges of the steel stamps. This idea is reinforced when one considers the axial stress-strain diagrams for prisms of various slenderness and loaded between flexible platens, Figure 4.33. These graphs suggest that failure occurred at a more or less constant stress level, but the curvature of the stress-strain diagrams just before peak suggest that the real peak is not measured. It would be interesting to repeat some of these flexible platen experiments under displacement control.

At this stage one could raise the question: which load transmitting medium is most optimal for compressive testing, in particular when we are also interested in the post-peak response? This is not an easy question. At present, the RILEM Technical Committee 148 SSC ("Strain Softening of Concrete") is

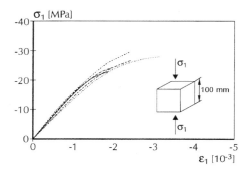

Figure 4.33 Stress-strain curves for concrete prisms of different height (B25/16) loaded between flexible platens.[290] (Reprinted by kind permission of Deutscher Ausschuss für Stahlbeton, Berlin.)

trying to clarify the problem. In a Round Robin test, the effect of both specimen slenderness and boundary restraint in uniaxial compression is investigated. Most of the effects described in this section and in Section 3.2.4 can be recognised in the results of the individual contributions. The comparison will lead to a recommendation for performing uniaxial compression tests for measuring strain softening, see Section 4.6. The committee's results indicate that no unique compressive softening diagram can be measured. Nevertheless, for practical purposes a proposal for a standard has been made. From a more scientific point of view, however, the best option still remains to perform experiments under different boundary conditions, in combination with either a macroscopic numerical model or a mesoscopic numerical model to explain the observed differences. This approach will be followed in Chapters 5 through 7 of this book.

b. Beam experiments

Frictional restraint plays an important role in beam experiments. In a standard beam test, such as a three-point-bend test, the specimen is placed on roller bearings and loaded to failure in a compressive machine. The situation is sketched in Figure 4.34. The rollers have to be designed very carefully to allow for truly frictionless lateral deformations in the beam. Unrestrained horizontal movement of the beam is of particular importance during crack propagation. If the effect is neglected, a higher load is needed to propagate a crack, and the fracture energy will be overestimated.

In the early days of fracture research of concrete, much attention was given to the so-called notch-sensitivity of the material. This implies that a variable external load is found when a notch (most common is to make a saw cut) of varying depth is present in the beam. Gjørv et al.[273] have shown that the notch-sensitivity of concrete beams depends on the frictional restraint in the supports. With increasing notch depth in a beam, the notch-sensitivity increases when the supports are restrained. In different words, a higher load must be applied to the

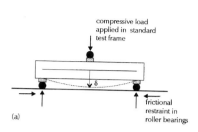
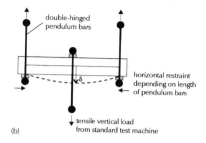

Figure 4.34 Three-point-bend test where the specimen is placed on rollers and loaded in a conventional compression machine (a), and the same beam loaded between pendulum bars (b).

beam when the supports cannot move freely. In Section 4.6 we will discuss the need for standard testing in fracture mechanics. Here it is mentioned that one of the geometries proposed for evaluating the fracture energy of concrete is the three-point-bend test. In a recent paper the effect of rolling supports on the fracture energy G_f was analyzed.[291] It was concluded that freely rolling supports were essential in G_f-tests: when the supports are fixed, errors up to 15% were measured (depending on the beam depth).

Frictional restraint in roller bearings is extremely hard to overcome, and perhaps the ideal roller bearing does not exist. An improved solution is to load the beams between pendulum bars in a tensile machine, rather than between rollers in a compressive machine. The application of pendulum bars was mentioned earlier in this chapter for four-point-shear tests. The same solution can be applied in three-point-bend tests as shown in Figure 4.34b. When the pendulums increase in length, the freedom to move in horizontal direction increases. However, the frictional resistance is not completely overcome in the hinges connected to the pendulum bars. Similar as with the uniaxial compression experiments described in the previous section, the best choice is to include the boundary restraint as an additional variable in the test programme. This was done in our four-point-shear experiment (Figure 4.7). Further understanding of the situation can then be obtained by analyzing the boundary condition effect using a numerical model, either at the macro-level or at the meso-level.

4.1.3.2 Boundary rotations

A second type of boundary condition that may cause significant errors, in particular in uniaxial tensile or uniaxial compressive experiments, is the allowable rotation of the loading system. When large localised fracture zones develop in a specimen, i.e., of size equal to the specimen size, the boundary rotations interact with the propagating fracture zone. Moreover, the (flexural) stiffness of the specimen itself will affect the behaviour as well, in particular when the loading platen rotations are prevented (to some extent).

Basically, three types of experiments can be distinguished as shown in Figure 4.35. In the first type (I), both loading platens cannot rotate and a

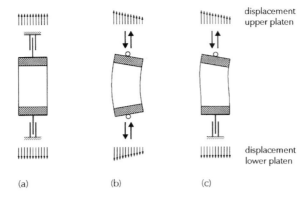

Figure 4.35 Tensile or compressive testing between fixed platens (a), between hinges (b), or between a fixed and a rotating platen (c).

constant displacement is applied to the specimen boundaries. In type II experiments, both loading platens are free to rotate, and the deformation at one side of the specimen can be larger than on the other side. The third possibility (type III) is a mixed set-up, in which one platen is fixed, and the other platen has some freedom to rotate. Situation three was the initial set-up of the four-point-shear experiment described in Section 4.1.2.2, but this was later changed to a symmetric machine between hinges (type II) as shown in Figure 4.7. The third variant is not very favourable because the state of stress is rather complex. The state of deformation or the state of stress in a specimen is better defined in type I and II experiments, but only before fracturing. As soon as large fracture zones develop, the situation changes dramatically, and highly non-uniform stress distributions develop, especially when the fracture zone is not immediately stress-free (as is the case for brittle, disordered materials like concrete). Using analytical or numerical tools, the difference between rotating and fixed loading platens can be analyzed. However, the accuracy of the outcome is highly dependent on the fracture law that is used. In Chapters 5 and 7 we will come back to such analyses. Here we would like to show the effects measured in laboratory scale experiments. In a first example we will show the effect of boundary rotations in tensile testing. Basically, the same phenomena can occur in compressive experiments, although slight differences exist. A second example will be given of an advanced tensile experiment carried out at the Politecnico di Torino.[292] This experiment will be explained here because it was analyzed with one of the numerical micromechanics models that will be described in Chapters 5 through 7. Moreover, it was quite helpful for a better understanding of tensile softening of concrete.

a. Boundary rotations in uniaxial tensile experiments

The first example of boundary rotation effects concerns a tensile test between rotating and fixed loading platens. Over a period of about 12 years,

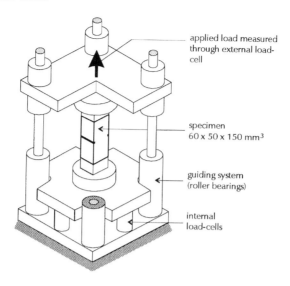

applied load measured through external load-cell

specimen
60 x 50 x 150 mm³

guiding system
(roller bearings)

internal
load-cells

Figure 4.36 Experimental set-up for deformation-controlled, uniaxial tensile tests in the Stevin laboratory.[84] (Reprinted with kind permission of Dr. Hordijk.)

numerous displacement-controlled tensile experiments have been carried out in the Stevin laboratory.[84,146,187,190,293-295] Since the first experiments, the tensile machine has been transformed several times. In the first tensile fatigue experiments,[188] cylindrical specimens were loaded between hinges, i.e., a type II experiment following Figure 4.35b. The main purpose of these hinges was to adjust the loading system and the specimen to avoid shear on the piston of the hydraulic actuator. Moreover, the hinges could be fixed as some of the fatigue tests implied a reversal of load from tension to compression. Thus, the hinges were not expected to act as hinges *during* the experiment. Results from true type II experiments where the platens were expected to rotate during the test can be found elsewhere.[271,295,296]

After the fatigue experiments, the set-up in the Stevin laboratory was modified to determine the parameters for the Fictitious Crack Model,[5] and a specially designed guiding frame was installed to ensure that the upper and lower loading platen of the machine would remain parallel throughout the experiment. The machine with fixed platens is shown in Figure 4.36. The rotational stiffness of the loading platens was estimated at 10^6 Nm/rad.[84] This value is for the combined rotational stiffness of the upper and lower loading platen. Small errors were found in the load measurement, four small load cells were installed below the lower loading platen as shown in Figure 4.36.[84] The reason was friction in the guiding system: the external load cell could not be used to measure the specimen load. The four internal load cells were placed below the cross-shaped lower loading platen. Note, however, that the load-cells affect the rotational stiffness of the lower loading platen.[57] In Figure 4.37, the

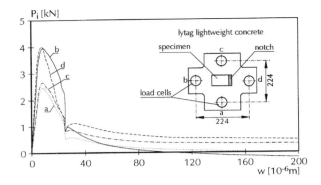

Figure 4.37 Load-distribution under the lower loading platen during a uniaxial tensile test. The load cells are numbered *a* through *d* and are drawn as circles. Note the non-uniform loading after the peak.[57]

four signals of the internal load cells are plotted separately for a tensile test on a lytag concrete specimen. In the inset, a top view of the lower loading platen is shown. The specimen is glued to the centre of the loading platen; it is a 200 mm high, 100 mm wide and 50 mm thick single-edge-notched specimen. We observe that the loads are largest in load-cells *b* and *d*. Because concrete is heterogeneous, crack initiation will always start from a certain location along the circumference of a specimen. In this example we used a single-edge-notched specimen, which helped to initiate the crack at a given location. However, in un-notched specimens or double-notched specimens crack initiation and propagation is non-uniform as well.[148] From the initiation point, the crack will propagate through the specimen's cross-section. In other words, the applied tensile load becomes gradually more eccentric. As a result, the distribution of loads over the four load cells will become highly non-uniform. This can be seen in Figure 4.37.

In particular the difference between load-cells *b* and *d* changes significantly just beyond peak. The load-cell measuring the highest load (*b*) will of course deform more than the load-cell at the other end (*d*). Thus, the lower loading platen rotates, although the rotational stiffness is still quite high, viz. $8.75.10^6$ Nm/rad. From a comparison with the earlier cited value, it must be concluded that the upper platen stiffness is considerably lower than the lower platen stiffness. It is almost impossible to design a machine with non-rotating platens. The decision of inserting the internal load-cells leads in principle to a type III machine. With hindsight this is of course all very easy to understand.

Now let us examine the difference in response between a specimen loaded in a type I machine and in a type II machine. Recently, an experiment was carried out in which cylindrical specimens of two different diameters (50 and 100 mm) and two different lengths (100 and 200 mm) were loaded between rotating and fixed-end platens.[294,295] The type I experiment was carried out in

(a)

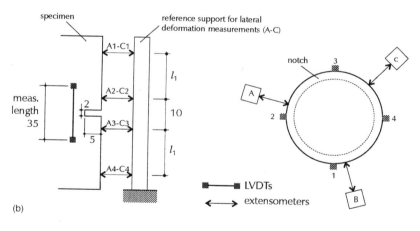

(b)

Figure 4.38 Experimental set-up for uniaxial tensile tests between freely rotating platens (a), and locations of LVDTs and extensometers for axial and lateral deformation measurement, respectively.[294]

the tensile machine of Figure 4.36. For the type II experiment between hinges, a new set-up was built (see Figure 4.38a).

The hinges in the machine were placed as closely as possible to the loading platen-specimen interface. The cylinders were glued to the steel loading platens using a commercial two component epoxy resin. In addition, the set-up was

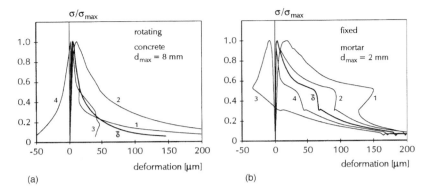

Figure 4.39 Experimental stress-crack-opening diagrams for concrete and mortar cylinders loaded between freely rotating (a) and fixed (b) end platens.[294]

designed to assure symmetric deformations in the axial direction of the specimen. In each test, four LVDTs were glued to the surface at $90°$ intervals as shown in Figure 4.38b. The average axial deformation measured with these four LVDTs was used as feed-back signal in the closed-loop system. The measuring length was 35 mm; the loading rate was equal to 0.08 µm/s. Moreover, 16 extensometers were used for measuring the lateral deformations of the cylinder. This was done at three positions along the circumference of the cylinder as indicated in Figure 4.38b. Lateral deformation profiles were measured for both the fixed and freely rotating boundary conditions. In Figure 4.39 results are shown of tensile tests on 100 mm diameter, 100 mm long cylinders loaded between fixed (a) and rotating (b) boundaries. In each graph the result of a single experiment is shown; the scatter was relatively small.[295] The examples shown here can be considered as representative for the difference between rotating and fixed platens. For each test the average axial deformation (thick solid line) is plotted against the average axial stress, and the individual measurements of the four LVDTs are shown as well. The numbers 1 through 4 in these graphs correspond to the numbering in Figure 4.38b. In the freely rotating case (Figure 4.39b), the fracture initiates near LVDT number 2, which gives the largest deformation (w_2) just beyond peak. At the other side of the specimen (LVDT no. 4) negative deformations develop; they increase in magnitude during further loading of the specimen. The loading on the specimen is highly non-uniform during the entire loading process: at one side of the cylinder the specimen shows an increasingly larger crack opening, whereas at the other side compressive deformations are measured.

The situation is somewhat more complex in the type I experiment. Again, due to the heterogeneity of the material, fracturing starts at the weakest spot along the circumference of the cylinder. As can be seen from Figure 4.39a, the largest deformation just beyond peak is measured with LVDT number 1. At the other side of the cylinder relatively small compressive deformations are mea-

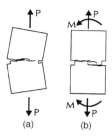 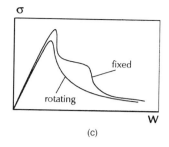

Figure 4.40 Effect of boundary conditions on softening: (a) freely rotating boundaries and (b) fixed boundaries. In (c), the stress-crack-opening diagrams are shown.[294]

sured. When w_1 has exceeded 150 μm, the trend reverses: w_1 decreases slightly, whereas w_3 starts to increase rapidly. The other two deformations are in between these extremes during the entire loading process. The average stress-crack opening diagram shows a bump. This particular shape of the σ-w diagram is caused by the non-uniform cracking of the heterogeneous material and the ability to stress-redistribution when fixed loading platens are used. In Figure 4.40, the difference between fixed and freely rotating boundaries is explained qualitatively. In Figure 4.40, (a) shows the cracking in a type II experiment between freely rotating platens, (b) shows fracturing of a specimen loaded between fixed platens, and (c) shows the two stress-crack opening diagrams. Under rotating boundaries, the experiments suggest crack opening from the weakest location. Under fixed boundaries this happens as well, but now because the specimen boundaries are forced to remain parallel during crack propagation, a closing bending moment develops as indicated in Figure 4.40b. Depending on the specimen size and shape, closure of the initial crack may occur as well. This means that there is also an interest in measuring cyclic response of concrete (Section 3.3.3). The bending moments prevent the crack from further extension, until at a given moment, the other side of the specimen starts to fracture. A situation has been created where two crack branches interact: the crack tips tend to avoid each other. This is similar to the particle scale in Section 3.3.1.

The bump has been observed by many researchers over the past years.[253,297-299] Since then, the above crack mechanisms and the ability to stress-redistribution in the experiment with fixed boundaries has been recognised as a solid physical explanation for the bump in the softening curve. The flexural stiffness of the specimen itself has an effect on the bump too.[298] In Figure 4.41 stress-crack opening diagrams for specimens of different slenderness are plotted. The material was a lightweight concrete containing Isol lightweight particles of size $4 < d < 8$ mm. The specimen lengths were 250, 125 and 50 mm at a constant cross-sectional area of 50×60 mm² (type A, B and C, respectively); in addition, a specimen with a smaller cross-section (40×50 mm²) and length l = 250 mm (type D) was tested as well. All specimens were double-notched, i.e.,

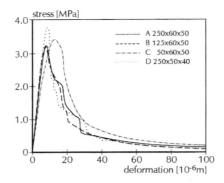

Figure 4.41 Effect of specimen size on the stress-crack opening behaviour in uniaxial tension.[84] Tendencies observed in these single tests were representative for the entire test series.

two 5 mm deep saw cuts were made, decreasing the net area to 50×50 mm² for the types A, B and C, and to 40×40 mm² for type D. The measuring length was 35 mm. In Figure 4.41, a comparison has been made between the stress-crack opening diagrams for these four different specimen sizes. All tests show a more or less significant bump in the softening branch. The bump seems to move upward when the specimen length decreases, as can be seen from a comparison of the type A, B and C geometries. For the type C specimen, the bump almost disappeared, but in this case the curvature of the pre-peak diagram increased, and moreover, the peak was more rounded. The maximum rotation in the fracture plane was determined for specimen types A, B and C.[84] In-plane and out-of-plane rotations were measured, and a maximum non-uniform crack opening could be determined in this way, see Figure 4.42. The maximum non-uniform crack openings were situated along a (more or less) straight line, which suggests a relation between the rotational stiffness (in this case flexural stiffness changes caused by different specimen length) and the maximum non-uniform crack opening.

Let us now return to the cylinder experiments between hinges and fixed platens. As mentioned, lateral deformation profiles were measured in these experiments, see Figure 4.38b. In Figure 4.43, these profiles are shown for two extreme experiments, namely for 50×200 mm Colton sandstone cylinders. The behaviour of Colton sandstone is comparable to that of normal concrete and mortar.[274] Only the lateral displacement along the location C in Figure 4.38b are shown. The other profiles are more or less similar; at least they indicate the same mechanism. The diagrams show the radial deformation profiles at different stages of average axial crack opening. In addition to the height of the specimen (on the vertical axis), the shape of the specimen is drawn to the right of each diagram. Here we also plot the locations where the radial deformations were measured. The radial deformations are relative to the radial deformations of point C4; the magnitude is given by the small scale bar at the top of each diagram. In Figure 4.43a we can identify the location of the bump

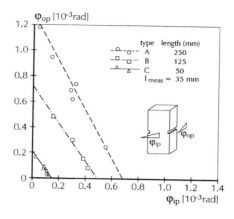

Figure 4.42 Maximum non-uniform crack opening for uniaxial tensile tests on three different specimen sizes. The net cross-sectional area was 50×50 mm² for all three specimen types.[84] (Reprinted by kind permission of Dr. Hordijk.)

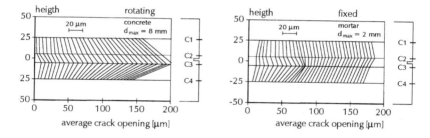

Figure 4.43 Lateral deformation profiles for Colton sandstone cylinder (50×200 mm) loaded between fixed (a) and rotating end-platens (b).[294]

in the test between fixed loading platens. Figure 4.43b shows the significant radial deformations for the test between hinges. The opening profile remains the same during the complete test. Basically these diagrams support the failure mechanisms presented in Figure 4.40. The specimen between freely rotating platens shows a continuous opening at one side of the specimen, whereas a sort of waggling effect is observed for the fixed specimen. Such behaviour was speculated earlier.[253] The experiments shown in Figures 4.41 and 4.42 indicate that the length of the specimen has an important effect too. It sets an upper limit to the allowable rotations in the fracture zone. However, an important observation follows from Figure 4.41: the total shape of the stress-crack opening diagram for the shortest specimen (type C) is significantly different from all other curves. Much of this impression is of course given by the low Young's modulus for this particular geometry. This may have been caused by end effects from the loading platens. One might be tempted to conclude that the

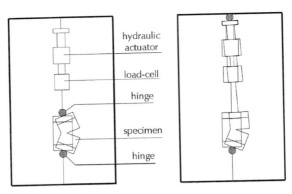

Figure 4.44 Effect of position of hinges in a uniaxial tensile test. Note the variation in lateral displacement of the specimen ends.

smallest specimen yields the best results, especially when the initial Young's modulus is corrected. The bump disappears almost completely for this geometry. However, the fracture process itself may have been affected by the shape of the specimen and boundary conditions under which it was loaded. In Chapters 5 and 7 we will clarify the crack growth processes in more detail. It will be shown that small specimens are needed to ensure a stable experiment. However, it will also be shown that the crack density in the specimen may change depending on the boundary rotations.

One last remark about tensile testing should be made here. In the fatigue experiments that were mentioned before,[188] the hinges were located at a relatively large distance from the specimen. The hinges were needed for adjusting the specimen in the test apparatus; the hinges were fixed during a test. In the tensile experiment of Figure 4.38, the hinges were placed as closely as possible to the specimen ends. Here the hinges were expected to rotate during the experiment. The two situations seem similar, but the demands for the respective experiments were completely different. However, assuming that in both cases rotations during the experiment were required, one can easily see that the response will be quite different depending on symmetry or asymmetry in the test set-up. Two extreme situations are given in Figure 4.44. In Figure 4.44a, the hinges are close to the loading platens, and the centres of the loading platens are forced to remain at the same position. In the situation of Figure 4.44b, the hydraulic actuator and load-cell are placed between the specimen and the upper hinge. Therefore, in this case, the centre of the upper loading platen can move in lateral direction. These differences should be included in numerical models when it is attempted to simulate the observed behaviour. Some examples will be given in Chapter 7. In this context it is important to realize that concrete (and rock) are heterogeneous materials. In most cases symmetry cannot be assumed in numerical simulations. The asymmetric failure modes shown in Figure 4.44 are the rule, rather than the exception, for materials such as concrete.

In uniaxial compression tests, the most common situation is a type II experiment. Most compressive machines have a spherical seat near one of the loading platens (e.g., Figure 4.18b); the other platen is usually fixed. This set-up is designed to allow for adjustment of the loading platen to the specimen, which is important, especially when the specimen is not exactly orthogonal and parallel. The experience with spherical seats is that no movement of the loading platen is possible during the test. Sometimes, the spherical seatings are designed such that they can be fixed during the experiment. In that case one assumes that a type I experiment is carried out. It may seem obvious that a type II compressive experiment is impossible. Buckling instabilities may occur depending on eccentricities in the load application or in the specimen itself (because concrete is a heterogeneous material). In uniaxial compression tests where the spherical seating was allowed to rotate during the test, a very eccentric mode of failure is found, in combination with an extremely steep descending branch.[89]

b. Maintaining uniformity of deformations in a three-jack system

From the previous section it will be clear that maintaining a uniform displacement is the main problem in fracture tests of heterogeneous materials. It also presents a contradiction. Why should a specimen fail at every point in the critical section at exactly the same moment? However, the Fictitious Crack Model *requires* that uniform fracture occurs. One might ask oneself if this *requirement* from the model is realistic: are we asking the material to behave as we would like it to behave? Recently it was proposed to maintain uniformity in the fracture zone by loading a specimen in a machine with three hydraulic actuators.[292,300,301] Three-jack systems are not new. Heilmann et al.[302] were the first to apply it in tensile testing. The experimental set-up of Carpinteri and Ferro[292] at the Politecnico di Torino is shown in Figure 4.45. Dog-bone-shaped specimens were loaded in vertical direction.

A central hydraulic actuator supplied the main load. During the experiment, the axial deformations were measured along the circumference of the critical cross-section of the specimen. As soon as an asymmetry of deformations (for example between front and back, or between left and right side of the specimen) was observed, the two other jacks would be activated. They were used to apply a counteracting bending moment such that the deformations would remain uniform over the measuring length during the complete experiment. In Figure 4.46, the result of an experiment is shown. The stress-crack opening diagrams measured with the individual LVDTs around the circumference of the specimen are located in a narrow band.

Essential is that fast electronics are available to control the experiment in this manner. Carpinteri and Ferro carried out numerous experiments on samples of different size. The results yielded information about the size-effect on tensile strength and fracture energy of concrete. In Section 7.2.2 the boundary conditions will be analyzed in detail using a numerical lattice model.

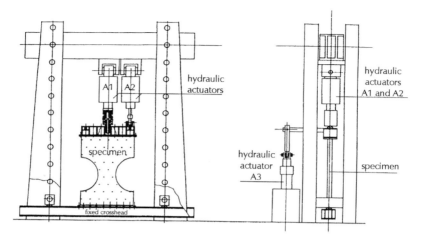

Figure 4.45 Tensile loading system with three hydraulic actuators developed at the Politecnico di Torino.[292] (Reprinted with kind permission from RILEM, Paris.)

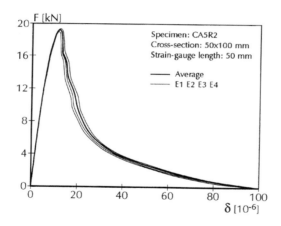

Figure 4.46 Stress-deformation diagrams for four individual LVDTs placed at the corners of a specimen loaded in the set-up of Figure 4.45.[292] (Reprinted with kind permission from RILEM, Paris.)

4.1.3.3 Platen interactions

Interactions between adjoining loading platens are of importance mainly in multiaxial experiments. In true triaxial experiments on cubes, or in biaxial tests on prismatic specimens, some of the load applied in a certain direction (let us assume σ_1) can be taken by the loading platens in the loading axes perpendicular to the σ_1 axis. The principle is shown in Figure 4.47. If rigid loading platens are used (i.e., without friction reducing measures as discussed in Section

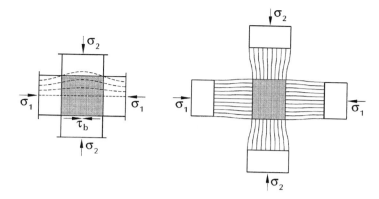

Figure 4.47 Platen interactions in a biaxial compression experiment between rigid steel platens (a), and between brushes (b).

4.1.3.1), the distribution of stress in a certain loading direction depends on the ratio of the Young's moduli of concrete specimen and the adjoining steel platens. Because the loading platens in lateral direction carry some of the load, an apparently higher load must be applied to fail the specimen. Therefore, the measures to reduce friction are of extreme importance in multiaxial experiments. When we consider the brushes of Figure 4.26, the acceptable lateral deformation is $18 \times 0.2 = 3.6$ mm. This value holds in theory because the loading applied to the specimen must be symmetric. Moreover, it is assumed that the spacing between the brush rods is exactly 0.2 mm, which is generally not true. The application of a certain loading system must always be considered very carefully, since otherwise highly undesirable loading situations will emerge.

Another example is the hydraulic cell of Figure 4.13. Here, a cylinder is loaded in axial direction through steel platens glued to the specimen ends, and by fluid pressure in lateral direction. In the experimental set-up it was decided to keep the steel platens outside the pressure vessel as shown in Figure 4.48a. Similarly, an undesirable situation would have developed when the platens were kept in the fluid reservoir. Differential deformations at the steel-concrete interface might have led to premature failure of the epoxy layer between the steel and the concrete, see Figure 4.48b. The differences in deformation are caused by the differences in Young's moduli of steel and concrete. Again, for materials such as sandstone the problems would have been even more substantial because of the low Young's modulus of such materials.

Finally, in the Colorado Cubical cell (Figure 4.21), leather pads having a circular opening were placed between the fluid cushions and the concrete (or rock) specimen. These leather pads were inserted to avoid contact between adjacent fluid cushions, which might have impaired the proper functioning of the loading device. Danger of contact between adjoining platen increases with increasing compressive deformations of the specimen. However, many corner

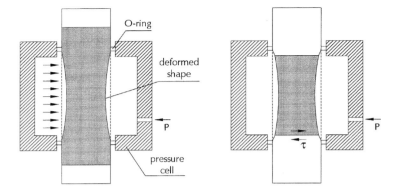

Figure 4.48 Hydraulic cell with axial steel platen outside the pressure vessel (a), and an alternative situation where the axial platens are situated inside the fluid reservoir (b).

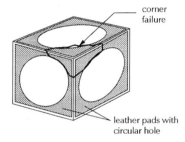

Figure 4.49 Possible undesirable effect of leather pads in the Colorado cubical cell. The pads were inserted to avoid contact between fluid cushions in adjoining loading directions.

failures were reported, which may have been caused by the differential stiffness of the fluid cushion and the leather pad as shown in Figure 4.49.

4.2 SPECIMEN SELECTION

Specimen selection and manufacture is of extreme importance in experimentation. Because of the rather rough character of concrete, this aspect is often neglected. However, when we start to measure in the micron range, a very good specimen manufacturing method is of utmost importance. Moreover, when accurate data are needed, for example for constitutive models for finite element analysis of reinforced concrete structures, an efficient and accurate specimen manufacturing method is of extreme importance. There are several decisions to make. First of all, the size and shape of the specimen should be accurate, and no problems should arise when the specimen is placed in the testing machine. No constraints should arise from inaccurate specimen dimensions as soon as loads are applied. Related is the decision about the surface of the specimen. Do we leave a casting skin, or do we prefer to use specimens where the skin has been removed by sawing a surface slice from the concrete? Then, of course, the specimen shape and size should allow study of the state of stress that we are interested in. In this book we have mainly discussed "simple" material experiments in which we tried to create a pure state of stress or deformation. The most common experiments are "simple" uniaxial compression, uniaxial tension and multiaxial states of stress. As a consequence of selecting a fracture mechanics based approach, there have also been efforts to design mode II, mode III and all type of mixed-mode experiments. In particular, many different mixed-mode I and II specimens have been proposed over the years. It is difficult to give a general solution to specimen selection and manufacturing methods. Of general interest is the decision on the size, which is related to the discussion about representative volume for particle materials such as concrete. The representative volume is of main interest if we intend to model the observed behaviour using continuum theories, as we discussed in the introduction to Chapter 3. For fracture, a continuum representation does not seem to make much sense, except when we limit our attention to the pre-peak behaviour. If this limitation is set, it is generally accepted that the smallest specimen dimension should at least be 3 to 5 times larger than the largest aggregate particle in the concrete (or rock). Thus, for a concrete containing 32 mm river gravel, the minimum specimen size should be between 100 and 150 mm. For dam concrete, where often large aggregates up to 150 mm are used, we would then end up with specimens with a minimum size of 450 to 750 mm. In the latter case, a lower compressive strength is usually found, but the size becomes relatively large, so that one should worry about the capacity of the testing apparatus. For high strength concrete, the above value of three to five times the largest aggregate particle should be debated. High strength concrete is generally more homogeneous than normal strength concrete. The matrix can be as strong as the aggregates. Therefore, the claim of minimum specimen size seems less severe as in the case of normal strength concrete. In fact it is believed that a substantial reduction of specimen size is possible in the case of high strength concrete. However, normal practice is to use the same standard size cubes and cylinders as for normal strength concrete. It would be interesting to investigate whether a size reduction of standard cubes and/or cylinders is possible for high strength concrete. In that case one should of course specify the range where smaller samples could be used. For many reasons this would be of interest, not in the last place in view of the (in many cases) insufficient capacity of the loading equipment when concretes with a compressive strength above 120 MPa are tested. For fracture tests we need not worry too much about the representative volume, especially in the approach that will be presented in Chapter 7, where we use meso-level models in which the heterogeneity of the material is directly incorporated. Nevertheless we will usually perform experiments on specimens where the above cited minimum sizes are maintained.

4.2.1 SPECIMEN SURFACE

The surface of the specimens should be considered as well. In many of the fracture experiments, where we are interested in extremely accurate deformation measurements, high demands exist for the specimen size and the condition of the surface of the specimen. This last point is of extreme importance when we are using optical crack-detection equipment. The standard manner to produce specimens is to cast them in a mould, and leave the specimen as it is. The most accurate surface smoothness can be obtained when steel moulds are used. Timber is generally less suitable, and can be used only a few times. Timber moulds will wear out after a number of castings, and the accuracy of size and shape of the specimen will decrease when play in the connections in the moulds increase.

A problem that will always remain when casted specimens are used is the casting surface. The smoothness of this surface is generally not sufficient. Bleeding water normally causes a rather rough surface. Moreover, the skin of the concrete in contact with the mould is normally very rich in cement, and fine particles are more abundant (wall effect).[41] Probably the best method for specimen preparation is to saw them from larger blocks. The surface layers, which are generally weaker than the core material, are removed from the specimens. As a rule of thumb, one should remove at least a layer with the same thickness as the largest aggregate particle, but removing thicker layers may be advantageous under some circumstances. When the surface smoothness after sawing is considered to be insufficient, a second treatment is needed. Grinding the surfaces with a diamond grinding wheel, or using carborundum powder is essential then. In that case a high surface smoothness can be obtained, quite easily around 1 µm. In the hydraulic experiments of Figure 4.13, an extremely high surface smoothness of the cylinders was needed because the seals of the pressure cell were in direct contact with the concrete. The high surface smoothness was obtained by grinding the specimens in a very accurate lathe which was normally fitted for metal working. After impregnating the surfaces with a low viscosity epoxy, the surfaces were polished a second time. Needless to say, the costs for each specimen were extraordinarily high, but the reward was a very small scatter in test results. More recently, the same procedure was followed for uniaxial compression experiments.[104] Again the specimens were sawn from larger blocks, and subsequently milled to the required smoothness.

In the case of compression tests, specimens should have a high surface smoothness to avoid contact difficulties with the loading platens. If the necessary equipment is not available, one can decide to apply a capping of rapid hardening cement or filled epoxy between the specimen and the loading platen. Of course, problems may arise when a capping is used in compression tests with lubricated platens. When brushes are used, one should avoid penetration of the capping material between the individual brush rods.

As far as the mechanical and physical loadings on the specimen are considered, the decision of selecting either cast or sawn specimens undoubtedly

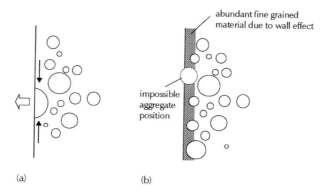

Figure 4.50 Splitting of aggregate segments located near the specimen's surface of sawn specimens (a), and comparison to the skin of concrete of cast concrete (b).

affects the result. For example, drying shrinkage will have a considerably different influence on the crack growth in the concrete casting skin. Compressive loading may cause a splitting type of loading on aggregate segments that remain visible at the surface after sawing or coring, see Figure 4.50. This last effect may be more or less important depending on stress-concentrations caused by the specific specimen geometry that was selected. Using the numerical tools that are presented in Chapters 5 through 7, such effects might readily be simulated in the near future. These affects are mentioned here solely for the sake of completeness; no fail-proof recipe for the specimen manufacture can be given.

4.2.2 ABOUT NOTCHES

As may have become clear, in general notched specimens are used in fracture experiments. In particular, in tensile softening experiments notched specimens are selected because they confine crack growth to a known location, and the placement of the control LVDTs is known from the specimen geometry. Most common is to either make a notch by means of a rotating diamond saw, or to directly cast it in (for example using thin metal or teflon plates). Concrete is rather notch insensitive,[273] in particular when large coarse aggregates are added to the mixture. In different words, the structure of the material itself causes tensile stress concentrations that can be even more severe than stress concentrations from the notch. This does not imply that no negative effects emerge from notches. As shown in Figure 4.51, stress redistributions around a notch are quite significant. In the example, a prism of size $100 \times 200 \times 50$ mm was pulled under uniform boundary displacement in the vertical direction.

Around the 15 mm deep, 5 mm wide notch, stress concentrations develop, whereas large stress relieved zones develop above and below the notch. In this

Figure 4.51 Stress distribution around a notch in a prism subjected to uniaxial tension.

example, the notch is assumed to be sawn into the concrete specimen. Concrete is considered as a continuum in this example.

Notches restrict the number of possibilities where a crack may nucleate. Thus, by chance a notch may be situated in a stronger or weaker part of the specimen, which may cause the observed strength to be either higher or lower. Until recently, the use of notches was the only reliable method to perform stable deformation controlled uniaxial tests. However, newly developed electronics allow for testing of un-notched specimens, see Section 4.1.2.1. In such a scheme, as mentioned before, many LVDTs are placed in a row along the specimen length. Dedicated electronics and software decide which LVDT is used for test-control. A fracture zone will nucleate from the weakest spot. Also a more favourable situation develops when circumferentially notched cylindrical specimens are used[294] because the crack can nucleate at the weakest spot along the circumference. This is especially the case when the cylinders are loaded between freely rotating loading platens (in all directions!). If circumferentially notched prisms are used (see Figure 4.52 where a comparison is made between cylinders and prisms), crack nucleation will occur from one of the corners. Stress concentrations will occur there. Note that the boundary conditions in the experiment play an important role in the crack-growth process as well, as was earlier discussed in Section 4.1.3.2. The situation will be elucidated further in Chapters 5 and 7.

It should be mentioned here that different size/scale effects on strength and fracture energy are found when either notched or un-notched specimens are used. The scale effect in tension was recently discussed in depth in a overview

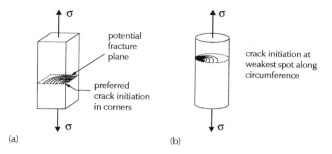

Figure 4.52 Crack nucleation and growth in circumferentially notched cylinders and prisms.

paper by Carpinteri et al.[303] We will now turn to this important aspect of fracture mechanics of concrete.

4.2.3 SIZE AND SCALE OF SPECIMENS AND MATERIALS

Much of the demands regarding the minimum size of the specimen to be tested stems from continuum mechanics. Quite clearly, concrete exhibits a large scale heterogeneity, which undoubtedly affects the test result. Normally, the minimum size of the specimen is linked to the maximum size of the largest material structural feature in the concrete, which is usually the maximum size of the coarse aggregates. As mentioned before, many researchers assume that a fivefold increase of specimen size over the largest aggregate size is sufficient to consider the specimen to be homogenous. The result can then be used to tune continuum models (constitutive equations). For fracture, unfortunately, the situation is not straightforward. During the fracture process, microcracks develop and grow to form macrocracks of size equal to the size of the specimen itself. Therefore, not only are the boundary conditions in the experiment of importance, but also the size of the structure in which the crack grows. Basically, during crack localization, the size of the specimen has an important effect on the elastic energy release in the areas outside the fracture band (see Chapter 5). It has been shown exhaustively that the strength of a specimen depends on its size, not only in tension but also in compression (Figure 4.32), although the situation is more complicated in compression. For uniaxial tension a result obtained by Carpinteri and Ferro[292] is reproduced in Figure 4.53.

With increasing size, the tensile strength decreases. The experimental results are compared to the so-called Multi-Fractal Scaling-Law (MFSL) developed by Carpinteri and co-workers. The size range of the experiments is described well with the model, but extrapolating the predictions from the model to larger sizes is still a haphazard state of affairs. In fact, the size range of most experiments does not exceed 1:32, and extrapolation to practical structural sizes is inevitable. Most unfortunately, however, the size effect laws that are most widely debated, i.e., the above mentioned Multi-Fractal-Scaling-Law (MFSL),[303] Bažant's Size-Effect Law (SEL),[304,305] and the weakest-link

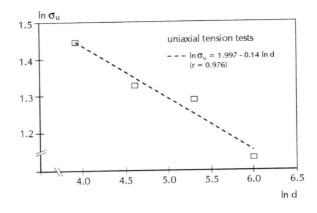

Figure 4.53 Size effect on tensile strength.[292] (Reprinted with kind permission from RILEM, Paris.)

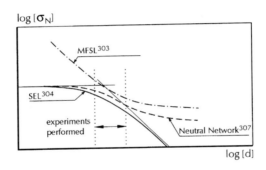

Figure 4.54 Comparison of Size-Effect Law (SEL),[304,304] Multifractal Scaling Law (MFSL, Carpinteri et al.),[303] and a Neural Network model.[307]

Weibull theory,[306] all lead to different extrapolations outside the range of observations. A comparison between the SEL, the MFSL and a Neural-Network-based best fit[307] of Bažant's test data obtained during the past few years at Northwestern University shows quite different tendencies for large sizes, see Figure 4.54.

First of all, the oldest size-effect law, the Weibull weakest-link theory, predicts a decreasing strength with increasing specimen volume. In a material sample, the chance for an imperfection is larger as the material volume increases, and a decreasing strength of the total volume is observed. The Weibull theory predicts a decreasing scatter with increasing specimen size. Second, the Bažant SEL describes (empirically) the transition from the theory of strength of materials (for small sizes) to the size effect predicted by Linear Elastic Fracture Mechanics. The latter limit is a straight line with slope -1/2, which stems from the fact that the stress intensity factor in LEFM depends on the square root of the crack length, see Section 5.3.2. Curiously enough, the LEFM limit predicts a zero strength for a certain structural size. The SEL is the gradual transition between the strength of materials curve and the LEFM curve as shown in Figure 4.54. The nominal strength of a structure is given by the empirical relation

$$\sigma_N = \frac{Bf_t}{\sqrt{1 + \lambda / \lambda_0}} \qquad (4.1)$$

where f_t is the tensile strength of the material, $\lambda = d/d_a$ is the relative size of the structure ($\lambda \geq n = 3$ for concrete), and B and λ_0 are constants for geometrically similar structures. The factor n relates to the crack band width $w_c = nd_a$ (where d_a is the maximum aggregate size, see Figure 3.35b). For concrete, n equals 3 according to Bažant, although different values can be found in different publications. The empirical constants B and λ_0 must be derived from experiments on specimens of similar geometry but different size (see Section 4.6.1).

The SEL has the same peculiar behaviour as LEFM, which is in conflict with Weibull's theory. Of course, the validity of Weibull's theory should be checked for notched specimens, where the stress-concentration around the notch directly affects the volume that must be taken into account, but the questionable LEFM limit is not solved by this observation. Many researchers doubt that the extrapolation to LEFM is valid for large sizes. In the MFSL, the experimental range is covered quite accurately (in fact it overlaps to a large extent with the SEL), but now the extrapolation to large sizes shows a different tendency than the SEL. The MFSL indicates an asymptotic strength limit for large sizes, which is in agreement with the Weibull prediction. The asymptote for large sizes seems physically more meaningful then the abrupt cut-off by the LEFM limit of SEL. Both the SEL and MFSL are (semi-) empirical formulations, and extrapolation outside the range of observations is of course a delicate affair. It is the author's opinion that much of the debate concerning the extrapolation for large sizes of each of these laws is completely irrelevant. First, a proper physical mechanism for the size effect should be found before reliable extrapolations can be made. Of course the main actors in the size-effect debate, Bažant and Carpinteri, would not agree with this point of view. However, it is stated here that a more physical based approach as advocated in this book will ultimately lead to the most reliable result. Quite interestingly, a recent application of neural network modelling by Arslan and Ince[307] to the size-effect experiments carried out by Bažant at Northwestern, indicate that his own test results would support the asymptote hypothesis of a lower strength limit at large sizes as predicted by the MFSL. This reinforces the idea that the SEL, which is based on earlier observations by Leicester[308] and Walsh,[309] is more based on ideas that have almost developed into a belief.

Because of the above mentioned problems in the size-effect laws available to date, it does not make much sense to deeply discuss these matters in this book. Rather we will continue on the path set out, namely to find a physical argument for strain softening. At present, we are trying to elaborate this approach and new large scale tensile experiments are being set up with the sole purpose of elucidating the physical mechanisms underlying size effects in concrete.[310]

4.2.4 SPECIMEN SHAPE

The specimen shape selected for a given experiment depends on a number of factors. Important are the manufacturing method, the state of stress and the failure mode to be achieved. In addition, other practical factors such as the wish to correspond to existing standard test geometries (in particular those geometries commonly used for determining the compressive strength and Young's modulus of concrete, such as cubes, cylinders or prisms) must be considered. Of course for research purposes, any imaginable shape would do as long as it supports the hypothesized state of stress.

For simple uniaxial tension, prismatic or cylindrical specimens are best suited. For displacement control in tension, notched specimens are needed, or a dedicated electronic system as discussed before. For applying the tensile load to the specimen it must either be glued to steel loading platens or held between specially designed grips. Commercially available, two-component epoxies are quite suitable for glueing the specimen. Sometimes, however, some experimentation is needed before a proper glueing procedure has been developed. If the epoxy is not strong enough, or simply to avoid problems in loading, the specimens are tapered at the ends, or they have a dog-bone shape. Mechanical grips are generally not useful in tensile experiments since through the lateral confinement subjected to the specimen, a biaxial tensile state of stress will develop in the specimen's ends. As clearly shown in Figure 3.83b the biaxial tensile strength is lower than the uniaxial tensile strength. Consequently, the specimen will fail at the grips. Some authors have proposed to use grips nevertheless, but then generally dog-bone shaped specimens are used that will always break in the middle section. As we have seen, notches cause a strong deviation from the pure uniaxial state of stress is a specimen. Of course, the deviations increase if in addition tapered or dog-bone shaped specimens are used. An example of the linear elastic stress distribution in tensile specimens with variable cross-section is shown in Figure 4.55. For dog-bone shaped specimens (Figure 4.55a), a strong variation of tensile stress is observed, whereas only stress concentrations are observed in the rounded corners in the variant of Figure 4.55b.

Note, however, that the stress concentrations in the corners of variant Figure 4.55b tend to force crack nucleation in those areas. For tensile tests the dog-bone shape of Figure 4.55a is more suited because only a marginal stress gradient develops. This shape was used by Ferro[301] in his tensile experiments, whereas the shape of Figure 4.55a is used in the current size effect tests in the Stevin laboratory.[310]

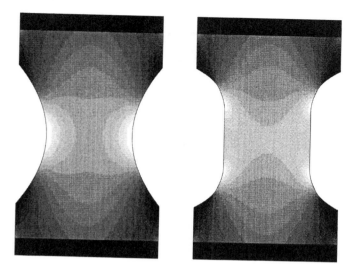

Figure 4.55 Comparison of linear elastic stress-distribution in two different types of dog-bone shaped specimens.

As was mentioned in the previous section, stress concentrations in corners and along edges are affecting the behaviour of prismatic specimens. One is not always free to select the most optimum shape for a given experiment. For example, when performing truly triaxial compression experiments, cubical specimens are the only possible choice. Shear bands in triaxial compression experiments tend to initiate in the corners of prismatic specimens used in the experiments[253,262,311] (see also Figure 3.99). Of course, questions can then be raised about the validity of the test result: to what extent is the mode of failure caused by the experimental set-up. Such a question might be answered by comparing results obtained from specimens of different shape with one another. For example, in Figure 4.56, the influence of specimen shape on the compressive strain softening diagram of concrete is shown. These results were obtained by one of the contributors[112] to the Round Robin experiment organized by RILEM committee 148SSC. The general observation from such tests is that the compressive strength tends to be slightly higher in case of cylinder tests, although the differences are very small. Again, stress concentrations in the corners of the prismatic specimens may cause cracks to nucleate at an earlier stage of loading, thereby reducing the peak load.

Some countries have decided to use cubical and prismatic specimens rather than cylindrical specimens for the determination of many strength parameters of concrete. As there are undoubtedly disadvantages in manufacturing cylindrical specimens, and the subsequent loading of the top surface (normally the casting surface of the cylinder is ground flat or sawn off), but other disadvantages blur the experiment using prismatic specimens. As we will see in Section

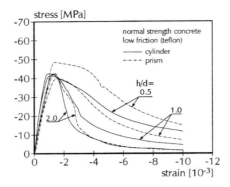

Figure 4.56 Comparison of compressive strain softening diagrams from prism and cylinder tests.[112]

4.6, the tendency is to define standard experiments for the determination of tensile and compressive softening, in spite of the fact that a unique curve can never be obtained. The standard tests are, however, quite valuable as they allow for a better comparison of experimental results obtained in different laboratories around the world. The only way of defining a softening diagram for computational purposes is a method of inverse modelling, as will be discussed in Chapter 6.

For the fracture modes II and III, and all mixed modes of loading, often dedicated specimen geometries are developed. In Section 3.4 several of the problems in shear testing, and the various geometries adopted were presented. In addition to the information given there, it should be mentioned that the various test geometries are often based on a linear-elastic, finite element analysis, comparable to those presented in Figures 4.51 and 4.55 for uniaxial tensile tests. Although such analyses are quite valuable, one should realise that the stress distribution in the specimen constantly changes as the fracturing process progresses. This makes the interpretation of fracture experiments a tedious job, and, in fact, it cannot be properly done without parallel analyses with a fracture model, even if the model is far from perfect. In Chapter 7 we will readdress these matters.

4.3 DATA-ACQUISITION

An important part of experiments is the collection of data. This should best be done in a rapid and reproducible manner, without interfering with the experiment. The result from any type of measuring device should be recorded and stored in some form. Traditionally, extensive manual labour was required for the data collection process, but much has changed since the introduction of digital computers. In the early days, mechanical devices were sometimes used to plot, for example, the load versus axial elongation of a specimen. Nowadays

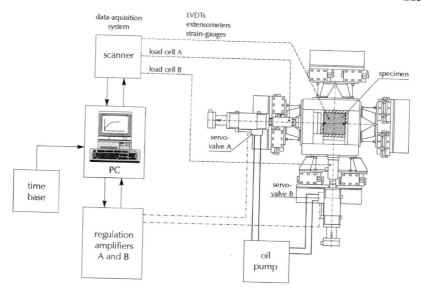

Figure 4.57 Test-control and data-acquisition system for the biaxial test frame developed at the Stevin Laboratory.[214]

electronics have taken over, and fast scanning of all connected transducers and other measuring equipment is possible. Temporarily halting the experiment for recording crack patterns and collecting other test data is not necessary any-more, as the speed of the data acquisition systems has been improved over the years. For example, it is possible now to store digital images in a computer for later processing. However, a drawback of the new methods is that the amount of data is increased by a few orders of magnitude as compared to the more traditional manners. Thus, handling the data and reducing them has become an important aspect. In this section, we do not aim at giving a complete overview of data collection methods, nor of brands of equipment that can be obtained on the market. Systems are improved continuously, and it is expected that their speed will increase further in the future. Instead, the principle of data collection is outlined.

Most of the complete systems to date are capable of both controlling the experiments and of collecting and storing data. An example of such a system is shown in Figure 4.57. Here, we show the set-up as it was developed for the biaxial test frame in use in the Stevin Laboratory. In the right part of the figure the biaxial test set-up is shown. The specimen is a double-edge-notched square plate and is visible in the centre of the loading frame. The actual loading situation was shown before, namely in Figure 3.71b. On the left hand side of the figure, the amplifiers for the control LVDTs and load-cells are shown. Through Analog/Digital convertors (A/D) these data are not only sent directly to the servo-controllers, but also to a Personal Computer which is the heart of

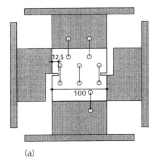 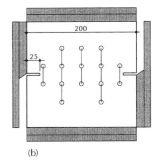

(a)   (b)

Figure 4.58 Example of extensometer pattern at the surface of a 200 × 200 mm double-edge-notched specimen for the biaxial loading frame of Figure 4.57.[214] (Reprinted by kind permission of Dr. Nooru-Mohamed.)

the total system. Also connected to the PC is a data-logger (scanner) that consisted in this particular example of four serial placed A/D convertors for measuring data from strain gauges and extensometers placed additionally on the specimen's surface. Figure 4.58 shows an example of extensometer positions at a 200 mm square specimen. This layout was identical at the rear side of the specimen. Finally, connected to the PC is a function generator, or linear time base (double system, one for each actuator). In the PC the load-path to be tested is constructed, and the requisite signals are used to control the experiment. As mentioned, additional strain gauges, extensometers and LVDTs can be placed on the specimen, and monitored during the experiment using a fast scanner. The scanner used in the experiment of Figure 4.57 was developed and built at the Instrumentation and Measuring Group of the Stevin Laboratory. In total 32 channels could be scanned, each having its own programmable amplifier. The scanner was set up such that four channels were scanned simultaneously, thereby reducing the total scan time (including the time for storing data on floppy disk) to 976 ms. This is still relatively slow, and recent developments allow for much faster storage of data. The inherent problem is that the amount of information from one single experiment is now so large that data reduction becomes an essential part of the task of the experimentalist. Because of the fast scan times, it is possible to load the specimen continuously. However, small errors are introduced because the loads and deformations are scanned at slightly different time steps. For reducing the error, it would be possible to measure the load several times between scanning the various deformation gauges. This would reduce the error to a minimum. In the biaxial experiment of Figure 4.57, a typical scan consisted of signals from 2 loads, 2 deformations that were also used as control signals, 6 to 20 extensometers, and 12 strain gauges. Moreover, the elapsed time from the beginning of the test was stored with each scan. Normal procedure nowadays is to store data in such a format that they can be conveniently handled by, for example, spread sheet programs and custom built graphics software.

4.4 CRACK DETECTION TECHNIQUES

A very important aspect of fracture testing is crack detection. Without having some insight in the physical mechanisms of crack growth, fracture mechanics becomes meaningless. If the underlying principles are unknown, how will it ever be possible to derive a sensible model? Therefore, in this section a number of crack detection techniques will be described. Quite a number of techniques are available to date, each with its own virtues and deficiencies. In general it can be stated that it is most profitable to combine different techniques in order to overcome the drawbacks of a single method. Viewing the problem from different angles is certainly quite advantageous.

Cracks are visible to the naked eye when the crack width is between 50 and 100 μm. Below that size, measures are needed for vizualization, such as magnification of the crack, or enhancement by some colouring method. Interior cracks are invisible during the experiments. However, they can still be measured, either by using acoustic techniques or by postmortem examination of the specimen's interior.

Crack detection techniques can be divided into direct and indirect methods, although the distinction is not always clear. Using most direct methods, it is possible to access information from the specimen's surface only, and mechanisms taking place in the specimen's interior remain undetected. More destructive methods are needed when information regarding interior crack processes is to be obtained, or alternatively, indirect methods should be used from which the requisite information can be deduced. In the following overview, most of the existing methods are mentioned, and relevant literature sources are given. For some of the methods, in particular for those that contribute to the main line of reasoning in this book, some more detail, and occasionally some results that are relevant to the discussion in other chapters, are included.

Indirect methods include compliance measurements and double cutting techniques, acoustic emission and ultrasonic pulse-velocity measurements, various kinds of interferometry techniques such as laser-speckle interferometry, Moiré interferometry and holographic interferometry, surface strain measurements using strain gauges or other types of extensometers, and infrared vibro-thermography. Next to these techniques, numerical modelling can be regarded as an indirect method as well. We will return to numerical modelling in subsequent chapters of the book, viz. Chapters 7 and 8.

Direct measurements include all optical and scanning microscopy techniques, as well as X-ray measurements. Impregnation, colouring (dyeing) techniques may enhance the objects (cracks) that are studied with these different viewing techniques. An overview of different techniques is given. In the overview, the distinction between direct and indirect techniques will not be used, but rather five main groups of crack detection techniques are distinguished, namely local surface deformation measurements, full-field, surface deformation measurement techniques, microscopy, acoustic techniques, and other methods.

Finally it should be realized that the resolution of the crack detection tool should correspond to the size/scale of the mechanism investigated. This is not always easy to recognize, especially when new unexplored areas are investigated. A simple example is that during softening large cracks must appear if the huge energy release is to be explained. Therefore, it does not seem realistic to use high resolution crack-detection techniques for visualising crack growth in the softening regime. Contrary to this, crack initiation is related to the nucleation and growth of very small cracks, and the highest resolution available must be strived for.

4.4.1 LOCAL SURFACE DEFORMATION MEASUREMENTS

Local surface measurements can be used to detect (in a rather coarse manner) locations where cracks develop. Because the strain gauges or extensometers always must have a certain length, preferably several times the maximum aggregate size if the continuum idea is maintained, the resolution of this technique is not very large. The method can be employed for measuring the width of the process zone (or fictitious crack, see Figure 3.32).[312] Another application might be, again with a rather low degree of resolution, to measure the extent of crack growth in fracture specimens. Non-uniform opening in tensile specimens can be demonstrated quite clearly in this way.[148] By following the rate of strain increase of strain gauges or extensometers mounted at different parts of the specimen, an idea of crack growth sequence can be obtained. The same can be done in uniaxial compression, but there strain gauges and/or extensometers should be attached over the entire specimen's surface. In Figure 4.59, for example, deformation readings at different locations around the circumference of a specimen are shown. The specimen was a single-edge-notched concrete plate of dimensions $200 \times 200 \times 50$ mm^3. The local deformation w_i is plotted against the overall deformation w measured with the control LVDTs (see inset of Figure 4.60; the tensile load is applied in the vertical direction whereas the loading platen remain parallel to one another throughout the experiment).

The measuring length of the local extensometers was 35 mm, that of the control LVDTs 65 mm. Because the cracking process is a highly localised phenomenon, the total deformations measured with each of these gauges is equal to the crack width, plus the elastic strain of the part of the specimen contained in the measuring length. Note that during crack growth the specimen unloads (softening!). Therefore, at larger crack openings, i.e., in the tail of the softening diagram, the contributions from the elastic deformations in the measuring length become negligibly small. The deformations w_i and w in Figure 4.59 and 4.60 were not corrected for the elastic deformations in the measuring length. In the inset of Figure 4.59, the cross-section of the specimen is shown, and the arrows indicate in which sequence an axial deformation larger than 10 μm was exceeded. For $w < 27$ μm, the fracturing of the specimen is highly non-uniform. At $w = 27$ μm a sharp drop in the tensile load is

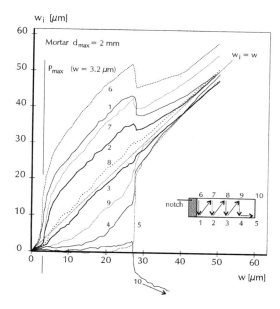

Figure 4.59 Local deformations w_i vs. global control deformation w in a uniaxial tensile test between non-rotating loading platens.[148]

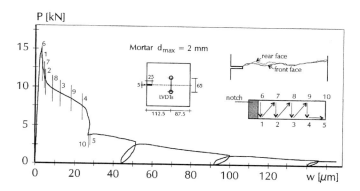

Figure 4.60 Load-average deformation diagram and crack pattern for the single-edge-notched tensile test of Figure 4.59.[148]

measured (Figure 4.60), and after that the differences in local deformations w_i gradually become smaller. Figure 4.60 shows how the sequence maps on top of the average stress-crack opening diagram. The results indicate that a zone of relatively large deformations gradually proceeds through the specimen's cross-section. During this process, local deformations are highly non-uniformly distributed, but uniformity is restored when the global deformation

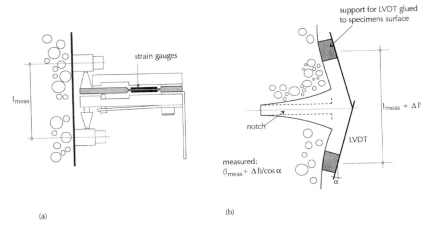

strain gauges

support for LVDT glued
to specimens surface

l_{meas}

$l_{meas} + \Delta l'$

notch

LVDT

measured:
$(l_{meas} + \Delta l)/\cos \alpha$

α

(a)

(b)

Figure 4.61 (a) Extensometer and (b) local rotations as source for errors.

$w > 27$ μm. The result indicates that in the descending branch at least two mechanisms are important: a growth mechanism in the part where a plateau is observed, and a more uniform degradation mechanism when the tail part of the curve is reached.

The extensometers used in the above experiments were developed by the Measurement and Instrumentation group of the Stevin Laboratory; they are shown schematically in Figure 4.61a. These extensometers are based on a half Wheatstone bridge, strain-gauge configuration. Bending of a thin metal strip in the extensometer is a measure for the displacement between the two measuring points that must be glued to the concrete surface. Because the extensometers are located at some distance from the surface, rotations of the measuring points might introduce an error in the displacement measurement as shown in Figure 4.61b. In tensile tests the rotations are generally negligibly small. In compressive experiments the effects might become more important, but this has to be considered from case to case. Rotations of measuring points do not blur the test result when strain gauges are glued directly on the specimen's surface.[187,271,312] It should be mentioned here that positioning the extensometers and/or strain gauges should be done with the greatest possible care.

In the Stevin Laboratory specially designed tools are used to fix the measuring points at their exact location, such that a perfect alignment is obtained. Again it is emphasized that accuracy in testing and measuring procedures is of the utmost importance for obtaining reliable and reproducible results. In concrete testing this aspect seems to be neglected too often. Disadvantage of the strain gauges is that they can be used only once and, more seriously, the gauges crack simultaneously with the specimen. Thus, the technique cannot be used for measuring the width of the crack. However, this might also be turned into an advantage because the fracturing of the gauge is a direct measure for crack

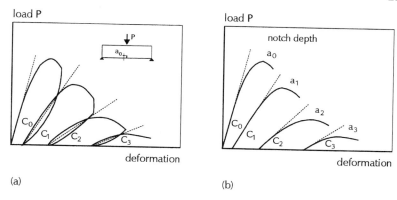

Figure 4.62 Compliance technique for determining an effective crack length in concrete beams: (a) through loading/unloading cycles and (b) by loading specimens having different initial notch lengths.

growth. Specially designed crack gauges (or crack foil) have been developed for this purpose. The gauges consist of a large number of parallel wires and the breaking of the subsequent wires is used as a measure for crack growth. The carrier for the wires is usually a thin foil of thickness not exceeding a few micrometres, such that the fracturing of the specimen is not affected. The device was used successfully in fracture experiments on asphalt concrete.[313] Note, however, that by using crack gauges, or any other surface strain measurement technique, information is obtained regarding surface crack processes only. The technique does not yield information on interior processes. Finally, it should be emphasized that deformation measurements using strain gauges, extensometers or LVDTs yields local information only. A complete view of the surface deformations can of course be obtained when the entire surface is covered with gauges, but this is not a very practical approach. In the next section full-field surface deformation measurement techniques will be presented.

In the early days of fracture mechanics applications to concrete, the effective crack length was estimated from deflection measurements in three-point bend tests using notched beams.[314,315] The idea is quite simple and it is shown schematically in Figure 4.62. The effective stiffness of a specimen decreases with increasing crack length. This can be shown by repeated load cycling, both in the pre-peak and post-peak regime (Figure 3.60). Alternatively, as shown in Figure 4.62b, one might decide to perform experiments on specimens with different initial notch size. When comparing with numerical computations (based on a linear elastic fracture mechanics approach), an effective crack length can be computed.[315] Others have proposed more analytic methods to compute the effective crack length.[314]

The effective crack length is not the crack that is directly visible to the naked eye, and generally it is assumed that a relatively large zone of distributed microcracks precedes the growth of the macroscopic crack. It is not unlikely that under repeated loading/unloading cycles, non-linear effects start to affect the effective stiffness of the specimen. These non-linear effects are not only caused by processes taking place in the microcrack zone, but frictional effects, crack branching from grain bridges and crack overlaps in the macroscopic crack will certainly have some effect as well. In experiments the difference between the effective crack length from compliance measurements and the observed macrocrack is defined as the process-zone length.[316] In their paper, Kobayashi and co-workers use a replica technique, which is a more sensitive technique to measure the extent of surface cracking (the replica technique is described in Section 4.4.4). The results indicated that the fracture process-zone increased in size as the crack grew. It should be mentioned that the macroscopic crack is determined from surface crack measurements only, which is not completely representative of the interior crack processes. For example through impregnation and cutting it can be shown that the interior crack front is curved[144,149,150] (Figure 3.39). We will return to these matters in Section 4.4.3. The effective crack can be interpreted as an approximate average crack length that should be taken into account when LEFM models are used for describing concrete fracture (Section 5.3.2). Thus, compliance methods are rather indirect methods, and the result depends to a large extent on model assumptions that must be made for computing the effective crack length. Therefore, process-zone lengths determined in this way are rather uncertain.

4.4.2 FULL-FIELD SURFACE DEFORMATION MEASUREMENT TECHNIQUES

Many different techniques can be used to visualize the surface deformation field of an entire specimen. Each technique has its own specific use and resolution, and also may affect the result to some extent. In this section we will discuss photo-elastic coating techniques, various interferometry methods and stereo-photogrammetry. By using optical microscopy, composite images of the entire specimen can of course also be obtained, but this is discussed in Section 4.4.3.

The use of surface strain measurements over a large part or the whole specimen has many advantages. Not only are crack growth processes measured in the area where the cracks are expected to grow, but in addition, other sources of strain concentrations are detected. Thus, one might get an impression of spurious crack growth in other areas as where cracks are supposed to grow. In compression experiments the stereo-photogrammetry technique[317] in particular can be quite helpful to elucidate the movement of blocks separated by shear bands. In some of the full field techniques it is necessary to attach reference points, measuring grids or other coatings to the specimen's surface. These measures might interfere with the fracture process, and should be carefully studied. If necessary (and possible!) the effects should be subtracted from the measuring data.

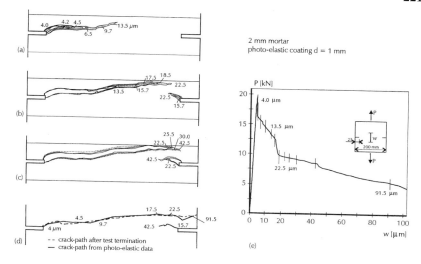

Figure 4.63 Photo-elastic coating technique for crack detection.[148] Isochromatics at various average crack openings (a–c), final crack path observed after the experiment (d) and load-crack opening diagram (e).

4.4.2.1 Photo-elastic coating techniques

Photo-elastic layers of reflecting foil or plates can be glued to the surface of a specimen, or alternatively, they can be cast as a fluid layer which solidifies in time. The layer has always a certain thickness, which is most uniform when plate material is glued to the specimen. The coatings that must be cast on the surface are definitely thicker in areas with pores, which will undoubtedly affect the result. The plate material is available at various thicknesses, which defines the accuracy of the method directly. Most applications of photo-elastic coatings are related to finding elastic solutions for elements with complex shapes and/or complicated boundary conditions. For example, in Chapter 3, we already saw the application of photo-elasticity to elucidate stress transfer mechanisms in systems of stacked particles (Figures 3.7c and 3.11c). For crack detection in brittle solids, the method has been considered as not being very effective, and has not been applied very much. After gluing a layer of photo-elastic coating on part of the specimen, crack growth can be detected when the specimen is illuminated under polarised light and viewed through a polarization filter.[148,318,319] Results shown in Figure 4.63 indicate crack growth from one of the notches in a double-edge-notched uniaxial tensile test between non-rotating loading platens.

At peak ($w = 4.0$ μm) first crack growth is detected. The stress-crack opening diagram indicates a steep drop just after the peak, leading to a sort of plateau, see Figure 4.63e. At $w = 13.5$ μm the end of the plateau is reached, and as can be seen from Figure 4.63a, the crack has reached the centre of the specimen. It should be mentioned that in this particular example, photo-elastic coatings were glued both at the front and rear of the specimen, and crack growth was almost identical at both sides. At 22.5 μm, we see a second drop in the load-crack-opening diagram. After this second drop, the full ligament between the two notches is cracked, although the crack consists of two separate overlapping branches. In fact, the mode of failure is similar to what was reported in Section 3.3.1.2 at the level of the particle structure of the concrete. The crack-path indicated by the photo-elastic coating technique corresponds quite well with the actual crack path that was observed after completing the experiment (Figure 4.63d). Note that the tail of the softening diagram in Figure 4.63e is higher than normally observed in mortar (compare to Figure 3.49). This difference can be explained from the bridging of the crack by the photo-elastic coating. The stress-crack-opening diagram should be corrected for this effect, but this was not attempted because the photo-elastic experiments were used as a first trial to explain — qualitatively — softening mechanisms in concrete. Based on these observations, which date back to 1988, it was concluded that two distinct mechanisms are involved in concrete fracture, namely first a process called perimeter cracking where surface cracks grow and penetrate into the specimens interior, and second, a process referred to as flexural ligament failure. The second stage coincides with the tail of the softening diagram and indicates a very slow stable fracture process where the two overlapping cracks join together as explained in Section 3.3.1.2. It should be pointed out that the crack tends to align with the edge of the photo-elastic coating, see Figure 4.63c. However, other results show that the same mechanism occurs when larger areas of the specimen are covered with a 0.25 mm thick coating.[147,318]

4.4.2.2 Interferometry techniques

Other, more accurate, full-field strain measurement methods include Moiré interferometry and holographic interferometry and speckle interferometry. The interferometry techniques can be used to measure the surface deformations of specimens under load. Traditionally, visible light was used to illuminate the specimens, but more accurate results are obtained nowadays by using monochromatic laser beams. Moiré interferometry is a method where two grids, one obtained from the unloaded specimen, the second from the loaded specimen, interfere. From the interference patterns crack growth can be derived.[320] Basically, results from Moiré interferometry experiments confirm the photo-elastic coating results that were shown in Figure 4.63. Surface crack growth processes were completed in the steep part of the softening diagram. The disadvantage of the photo-elastic method, where a coating has to be applied to the specimen's surface, does not apply to the interferometry techniques. It should be noted that the cracking is indicated by subsequent strain-contours in the Moiré method, quite similar to the images obtained in photo-elastic coating experiments.

More accurate than the Moiré interferometry method is perhaps laser interferometry[321] and laser holographic interferometry.[172,322,323] A resolution of

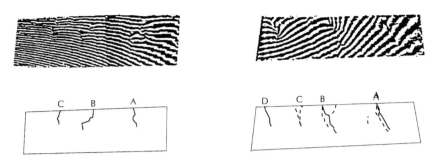

Figure 4.64 Fringes observed in laser holography study of fibre-reinforced composites.[172] (Reprinted by kind permission of Chapman & Hall, Andover, U.K.)

0.63 μm is claimed, i.e., equal to the wavelength of the He-Ne laser used in the experiments.[323] The principle of the technique is similar to Moiré interferometry, and fringe patterns are obtained by superimposing the image of the distorted specimen on a reference pattern. An example of cracking measured using laser holography in a fibre reinforced composite is shown in Figure 4.64. Cracks can be detected as discontinuities in the fringe patterns. A disadvantage of the laser interferometry experiments is that a very stable experimental set-up, free of vibrations, is needed. Otherwise the resolution of the technique decreases by several orders of magnitudes, or the results might become useless altogether. The interferometry experiments are quite contradictory regarding the possible extent of the fracture process zone in concrete, i.e., using the original definition of "a cloud of microcracks advancing in front of a stress-free macrocrack". This was clearly demonstrated by Mindess,[143] who carried out a thorough survey of crack detection techniques with the goal of finding a decisive answer to the size of the "microcrack cloud". His survey was so contradictory, so many different lengths were found in different experiments, that it seems more realistic to find different explanations for softening in tension, as for example attempted in Section 3.3.1.2. The results of Section 3.3.1.2 point to bridging mechanisms in the main cracks, a mechanism that has been confirmed in numerical analyses carried out at the meso-level, see Section 7.2.1.

4.4.2.3 Stereo-photogrammetry

The above techniques have been used mainly for visualizing cracking in tension. In compression, many distributed cracks nucleate and grow throughout the specimen's volume, and seem to develop into a localized deformation zone in the post-peak regime. The photo-elastic and interferometry methods are either too sensitive or not sensitive enough to register all the individual cracks. In fact, the isochromatics or fringes of multi-crack patterns become rather messy, and it is difficult to base clear-cut conclusions on such observa-

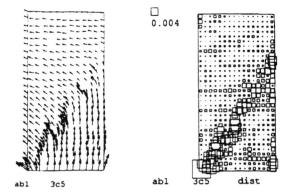

Figure 4.65 Stereo-photogrammetry applied to concrete, after Torrenti et al.[325] Development of shear band in prism subjected to uniaxial compression between non-rotating end-platens, and aluminium sheets as friction reducing measure: (a) block displacements, and (b) distortion. (Reprinted by kind permission of Chapman & Hall, Andover, U.K.)

tions. Another technique, which has been successfully applied, in particular for studying strain localization in compression, is the stereo-photogrammetry method. This technique, developed by Desrues,[317] was used by Torrenti et al.[324,325] to study the development of shear fracture in uniaxial compression, including the effect of boundary rotations (see also Section 4.6). A noted disadvantage of stereo-photogrammetry is that data processing is a cumbersome and time-consuming activity. The stereo-photogrammetry technique requires specimens to be loaded in plane stress. Consequently, relatively thin specimens are used (for example, Torrenti used prisms of height × width × thickness = 120 × 60 × 20 mm). During loading, a sequence of photographic images is taken from the specimen's surface. Stereo-comparison of these photographs yields the incremental displacement field between two images. Numerical analysis is needed to translate the results to a strain field. An example of results from photogrammetry observations is shown in Figure 4.65. The concrete specimen was loaded between fixed (non-rotating) loading platens, using a system of 5 x 5 mm and 2 mm thick aluminium squares placed between the concrete and the steel loading platen as friction-reducing measure. Clearly visible is how a shear band develops, and how the relatively intact parts of the specimen interact. It should be mentioned that at present the stereo-photogrammetry technique can only be used for large deformation measurements like shear band development in concretes and soils. For measurement of isolated microcracking, impregnation and/or colouring using dye in combination with direct (optical) observations seems to be more effective, see Section 4.4.3.

4.4.3 MICROSCOPY, IMPREGNATION AND IMAGE ANALYSIS

Isochromatics and fringes give a rather approximate view of crack extension only. In spite of the high resolution obtained, in particular in the laser interferometry based methods, direct observation still seems the best method. However, clear disadvantages exist here too. First of all, with conventional optical microscopes, the field of view is of course very small and, in general, the distance between the optics and specimen surface is rather small too, which makes direct observations of cracking under load rather tedious. However, in first applications, small magnifying crack microscopes have been used to detect the advance of the crack tip, for example under 10X magnification. When the specimen is illuminated using light in the visible wavelength-range, the resolution is always smaller than the wavelength of the light used. Using a long-distance microscope, it is possible to view specimens from a larger distance. This option has been developed in the Stevin Laboratory and will be described below.[52] The normal optical microscopes are mainly suitable for measuring in-plane deformation and crack growth. Using a con-focal microscope, however, it is possible to explore the out-of-plane dimensions as well.[326,327]

The optical microscopes have the advantage that wet specimens can be viewed. This becomes a problem when conventional Scanning Electron Microscopes (SEM) are used. Examples of SEM applications are from Tait and Garrett[328] and more recently Krstulovic-Opera.[329] Using SEM, specimens must be dried first and vacuum-coated with a gold or palladium coating before the specimen can be viewed in the microscope. Of course the magnification is much higher now, but it is generally believed that removal of the free water in the cement affects the structure of the material quite substantially. For example, drying shrinkage may cause additional cracking. Such cracks, which are merely the result of the imaging technique, must obviously be prevented. Recently, Environmental Scanning Electron Microscopes (ESEM) have been applied to view concrete and cement micro-structures. The great advantage of ESEM is that a much lower vacuum is needed. Therefore, a substantial part of the surface moisture remains in the specimen and allows for studying of fracture processes under wet conditions. As the technique is not very widespread yet, limited data is available to date. An example is the application of ESEM in studying healing mechanisms in rock salt.[330] An alternative technique to circumvent the moisture problems in SEM viewing is to make replicas of crack patterns, and view the replicas under the microscope rather than using the specimens directly.[316,331,332] A print (replica) of the cracked surface of a specimen is made. The replica is made of an acetylcellulose film. The film is dissolved in methyl acetate, and removed from the surface after the solvent has evaporated.[331] Small particles can be torn from the specimen's surface when removing the replica film. This may disturb the observation of microcracking. Therefore, several replica's must be made, and a good sample is selected before coating and viewing in the SEM. By using the replica technique, crack bridging (which was discussed in detail in Section 3.3.1 [Figure 3.43]) was confirmed.[332] The advantage of the replica technique is of course that the fracture patterns are not affected by the viewing technique as in SEM. Moreover, it is fairly easy to obtain replicas from the same specimen at different levels of loading, and gain some insight in crack growth processes. The resolution is high, i.e., better than 0.5 μm.[331] However, again only surface crack processes can be studied.

It will be obvious that under large magnification, the area that can be viewed reduces significantly in size. As mentioned, when normal optical microscopes are used, high resolution of larger parts of the specimen's surface can be obtained by scanning a larger area of the specimen, simply by moving the microscope in front of the loaded specimen. An example of a remote-controlled, long-distance optical microscope is shown in Figure 4.66. The set-up was used for viewing crack growth in splitting tensile experiments. The experiments were designed especially for studying the interface between aggregate and matrix, and were discussed earlier in Section 2.2.3. The microscope operates at a distance of 150 to 300 mm from the specimen's surface. The optics, which resemble the construction of deep-space telescopes, are connected to a cradle, which is fixed to a stable floor stand. The cradle with optics can be translated in three orthogonal directions using stepper-motors, each having a step-size smaller than 1 μm. The resolution of the microscope is 1.1 μm. Connected to the microscope is a CCD camera with electronic shutter, which allows for scanning the surface under constant illumination. This last aspect is quite important if mosaics of the scanned surface are made. As shown in the scheme of Figure 4.66b, the image from the CCD camera is fed into a variable scan frame grabber which is located in a fast PC. The same PC is used for controlling the stepper motors, which are connected through an indexer. The total set-up is controlled with a dedicated software package developed in the Stevin Laboratory. The specimen is illuminated by means of two fibre-optic light sources, which are connected to the same cradle as the microscope. Thus, the light source travels with the microscope to guarantee adequate illumination throughout scanning. The CCD images have a size of 756×581 pixels with a grey value from 0 (black) to 255 (white). The images are stored on the hard disk of the PC, and after completing the experiment, mosaics of the scanned areas can be made. An example of a partial scan (3 images at a resolution of 3.65 μm²/pixel) near the interface of a granite aggregate embedded in cement paste is shown in Figure 4.67.

A crack containing a small overlap is observed to run into the interface in Figure 4.67. The overlap is a small version of the earlier examples of Figures 3.42 and 4.63. The software developed for scanning the specimen's surface has been designed such that only those locations where previously cracks were detected are scanned over and over again, plus an additional area around the supposed crack tip. In Figures 3.45, 3.46 and 3.80 examples of crack patterns obtained with the long distance optical microscope were shown. Recently, the technique was adopted for measuring the roughness of surface cracks.[333] Important in this application is the processing of the CCD image to obtain the skeleton of the crack structure, which is needed for applying the box-counting method[333] for the determination of the fractal dimension of the crack structure, see Figure 4.68. For obtaining the thresholded damage pattern of Figure 4.68b, a commercial image processing program was adapted to fit the purpose.

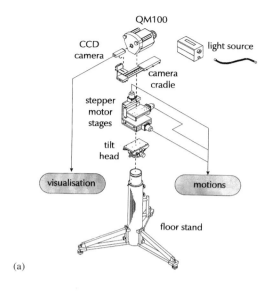

(a)

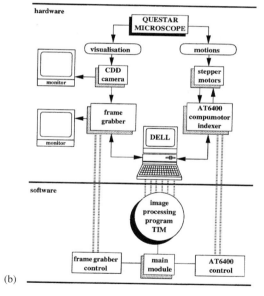

(b)

Figure 4.66 Overview of the long-distance optical microscope (a) and image processing system (b). The boxes visualization and motions refer to the connections with the frame grabber and the indexer, respectively.[52] The microscope is also shown in Figure 4.8.

It should be mentioned that cracks sometimes must be enhanced by impregnating or colouring in order to separate them from the material structure. In the recent experiments with the scanning optical microscope (Figure 4.67) light-coloured aggregates were embedded in a matrix of white cement. The contrast

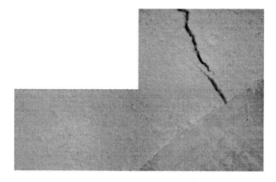

Figure 4.67 Example of a mosaic image, showing an example of crack face bridging near the interface between cement matrix and a granite cylindrical aggregate.[52]

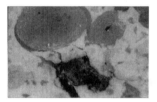 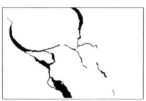 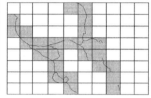

Figure 4.68 Scanned microscope image (a), thresholded damage patterns (b), and application of the box-counting method to thinned, skeletonized cracks (c).[333]

between cracks and material structure is optimized in this manner, which very much helps to enhance the crack structure using an image processing program. Optimizing the illumination of the specimen in order to enhance shading in the crack is very important. No method can be given, but trial and error and experimentation with different light sources will eventually give the best result. Dye penetrant might be used as well,[34,149,150,334] or fluorescent epoxy.[89,92,144,146,311,335] The fracture patterns shown in Figures 3.37 and 3.42 (as well as those of Plates 1 through 9) were obtained using vacuum impregnation of fluorescent epoxy resin. Using such a technique, the internal cracking can be visualised after the specimen has been cut into slices. Impregnation is done under vacuum and, again, moisture must be removed in order to allow saturation of the crack structure with epoxy or dye. The best option is of course to impregnate the specimen under load,[149] but in most cases impregnation is carried out after the specimen has been loaded to a prescribed point in the stress-deformation diagram. After that, the specimen is inserted in a basin of dye or fluorescenting epoxy, and is placed in a vacuum chamber. In the impregnation tests described in Section 3.3.1, vacuum was maintained for about 45 minutes, which corresponds to the pot-life of the low viscosity (110 mpas at 23°C) epoxy that was used. After complete hardening of the epoxy, the

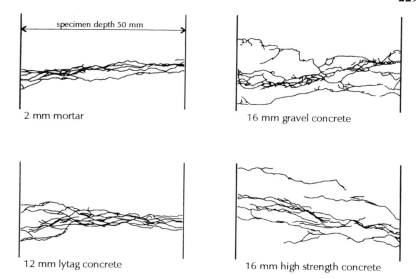

specimen depth 50 mm

2 mm mortar

16 mm gravel concrete

12 mm lytag concrete

16 mm high strength concrete

Figure 4.69 Projection of internal cracks on a single plane. Results from impregnation experiments[144-146] are shown. For four different materials the global curvatures in the crack are shown in this way, resembling the width of the fracture zone detected using AE source location techniques.

specimens were cut in slices, and internal cracks could be viewed and photographed under ultraviolet light.

Another appropriate material used for impregnating concrete and rock samples under load is molten Wood's metal.[336-338] Wood's metal (Cerrosafe®) is a non-wetting liquid for minerals with a relatively low melting point (between 70 and 88°C). It is solid at room temperature. The material can be impregnated under pressure at the temperatures cited above, and after cooling down, a good view of the material's pore-structure and crack structure is obtained. At a pressure of 10 MPa the liquid metal penetrates into cracks and pores as small as 0.08 μm. The sample can be viewed under high or low magnification using SEM and/or other optical means. It should be mentioned that the impregnation techniques reveal only the cracks and porosity that are interconnected. Sometimes, small channels are made into the specimen to facilitate impregnation of colouring,[334] but in most applications, the dye penetrant must be introduced through the outer surface. This implies that only the cracks and porosity that are connected to the outer surface can be detected.

Results of impregnation experiments were discussed in Section 3.3.1 in detail. An additional result is presented in Figure 4.69. The internal crack profiles from the various slices into which a specimen was cut were projected on the back plane of the specimen. The width of the crack band that is found in this manner is a measure for the global curvatures in the crack plane. Results are shown for four different concretes, i.e., the same concretes as mentioned in Section 3.3.1, and clearly the band width increases with increasing maximum aggregate size. We have to bear in mind that the cracks grow through the lightweight lytag particles, and crack roughness is determined here solely by the sand particles with $d_{max} = 4$ mm. Therefore, the crack band in lytag concrete is only slightly wider than in the 2 mm mortar. For the normal and high strength 16 mm concretes, the differences are marginal, indicating that the material structure is decisive rather than the strength of the material. These results are in good agreement with Acoustic Emission (AE) source-location measurements presented in the next section.

For visualizing fast crack growth in, for example, impact tests, images should be gathered at an extremely rapid rate. Typically the sampling rate should be lower than 1 µs. Developments in these directions are ongoing as CCD camera's and frame grabbers are developed with increased sampling rate.[339] Typically, the standard CCD image of 512×512 pixels is split into many images containing less pixels, for example all the way down to 32×32 pixels. These low resolution images are sampled one by one at a fast rate. The sampling rate could increase up to 10,000 images/s. This is at the limit of what is needed for measuring the speed with which a crack grows in a specimen under impact tensile loading.

4.4.4 ACOUSTIC TECHNIQUES

No information of the fracture process taking place in the specimen's interior is revealed when surface crack detection techniques are used (except for the step-wise impregnation procedure). Many researchers support the idea that surface processes are not representative for the complete fracture process taking place. In fact, impregnation experiments like those presented in Section 3.3.1 reveal that curved crack fronts develop and, moreover, the same experiments indicate that isolated microcracking takes place in the specimen's interior. Acoustic techniques seem a suitable tool for gathering information about interior crack processes in a specimen *under load*. Two different approaches can be distinguished, namely the acoustic emission technique where the elastic waves generated by the rapid release of energy from crack processes are measured with piezo-electric transducers, or alternatively the ultrasonic pulse technique in which the deflection and reflection of acoustic signals by discontinuities in the specimen are studied in order to obtain insight in the internal crack structure of the material. As far as acoustic emission (AE) is concerned, measure the number and intensity of acoustic emissions.[340-343] Typical parameters are the peak amplitude, signal duration and signal energy as shown in Figure 4.70. Most common is to define a threshold in order to limit the amount of data that is generated in an experiment. Also, quite importantly, the duration of an event must be set to decide between individual events.

In general, the main measurement in an experiment concerns the number of events, or hit rate, and correlation of this number to fracture energy or other toughness characteristics. Quite surprisingly, the number of events is, after an initial increase around peak stress, at a constant high level when it is deter-

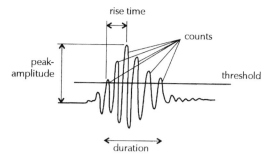

Figure 4.70 Typical AE event and definitions.

mined in uniaxial tensile tests.[341] This would confirm the observations from Section 3.3.1, where crack growth is presented as a continuous process in which macro-crack growth is followed by tensile failure of crack face bridges that are initially left in the wake of the macro-crack. Some authors claim that frictional effects are the major cause for the long tail of the softening diagram in tension.[84,165] In that case, clearly different AE pulses (with different amplitude and/or duration) would be expected to occur, but none of this has been found to date. In the bridging mechanism (Figure 3.43) such a shift in failure mode is not needed as this model is characterised by an on-going mode I fracture process (Figures 3.46 and 3.48).

Many different transducers are sometimes placed on a specimen, in order to determine the location of the AE sources.[345-347] Source location relies on a rather exact knowledge of the speed of sound waves through the material and, for heterogeneous materials such as concrete, this is obviously not a constant. Therefore, the determination of AE sources is approximately precise up to several millimetres. The main problem in AE studies is, however, that the physical mechanism underlying the acoustic signals is not at all clear (see also the previous paragraph). Next to simple crack growth in mode I (opening), frictional contact and sliding of crack faces may occur, or even hitherto unimagined mechanisms. Ohtsu[346] developed a technique to discriminate between mode I and mode II AE events. This so-called Moment Tensor Analysis was applied in a number of "mixed-mode I and II", or rather biaxial tension-shear experiments (in the test set up of Figure 3.71b).[347] The experiments confirmed the growth of curved cracks from the notches as shown in Section 3.4. The comparison of the locations of the subsequent AE events and the final crack pattern in the specimen was quite favourable as shown in Figure 4.71. As will be shown in Section 7.3.2, the curved crack paths can be simulated by means of a numerical lattice model, using a local mode I criterion only. The equipment available at the time these tests were carried out did not allow for full 3D source location. Recent 3D source-location experiments carried out at Northwestern University[342] (Figure 4.72) confirmed the three-dimensional curved crack front as obtained from impregnation experiments.[144,149,150]

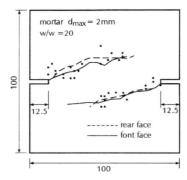

Figure 4.71 Comparison between AE source location measurement and real crack path in a biaxial tensile/shear test ($w/w_s = 2.0$) carried out in the biaxial test frame of the Stevin Laboratory.[347]

Figure 4.72 AE source locations in coarse mortar specimen with $d_{max} = 5$ mm (a), and in fine mortar specimen with $d_{max} = 1$ mm (b).[342] (Reprinted with kind permission of Aedificatio Verlag GmbH, Freiburg.)

The general conclusion from source location experiments in mode I fracture tests of concrete is that the fracture zone width increases with increasing particle size of the concrete (natural aggregates). This was shown in Figure 4.69 for impregnation experiments. In Figure 4.72 the effect is shown from AE experiments by Landis and Shah;[342] the same observation was made earlier by Nomura et al.[167] For both a coarse mortar ($d_{max} = 5$ mm) and a fine mortar ($d_{max} = 1$ mm) results of AE source-location experiments are shown in Figure 4.72. Clearly, the width of the fracture zone increases as can be seen from the frontal view of the beam.

Finally, the ultrasonic pulse technique is worth mentioning. In rock mechanics this technique is frequently used for oil reservoir detection, and for gathering information about the composition of the earth's crust (acoustic tomography). In laboratory fracture experiments, the technique can possibly be used to locate cracks. For example, in experiments the departure and arrival times at a great number of different transducers is determined, and through reverse analysis the location and width of a crack should be determined. The technique is quite demanding as far as computational effort is concerned.[348]

Berthaud[349] studied the change of wave velocities in uniaxial compression tests on concrete, and tried to correlate the results to damage. Crack growth in concrete in compression is highly anisotropic, which is confirmed by different velocities in different directions. Ultrasonic pulse velocity tests are also used in more practical circumstances to determine the damage in (reinforced) concrete structures. It should be noted, however, that the method is highly sensitive to the moisture content in the material structure.[350]

4.4.5 OTHER TECHNIQUES

Before closing this section on crack detection techniques, a few other methods that are used and developed by a number of researchers are mentioned. These methods are difficult to group under the four above categories, which does not mean that they are not equally important. The first is the so-called double-cutting or multi-cutting technique, a truly destructive method as the specimen is demolished step by step, although this is done in quite an ingenious way. The other two techniques can be categorized as non-destructive: they yield information of the interior fracture process without cutting the specimen. They are the X-ray technique and infrared thermography.

The multi-cutting or double-cutting technique was originally developed for measuring the process zone length in ceramics.[351] Recently, the method was applied to concrete.[352] The idea is that the compliance of a specimen with an earlier developed fracture zone will change when some of the bridging stress is removed. In an experiment this can be done by making a saw cut in the direction of the main crack. Hu[353] used wedge-splitting tests in which the crack zone was guided along a straight path from the notch by making two 2 mm deep side grooves at the specimen's surface. The principle of the technique is shown in Figure 4.73. A load-deformation diagram is shown, indicating the increased compliance C_u beyond the peak. At this stage a fracture zone or process zone with length a_{fpz} has developed in the specimen, which is consecutively cut to remove the bridging stress. At each cutting the compliance C_p is determined and plotted against the cutting length. Comparison of the compliance curve C obtained from uncracked specimens with different notch size a gives a measure of the process zone length x as the difference between the total process zone length a_{fpz} minus the length of the stress-free crack a_{sfc}. From these experiments it was concluded that the fracture process zone is not a material constant, but depends on strain gradients in the geometry investigated,[352,353] which was concluded earlier from uniaxial tensile experiments.[182] This supports the idea that fracture processes are very much dependent on size and boundary conditions in the experiments, and points in the direction of meso-level modelling as will be demonstrated in Chapter 7.

A variant of the above technique[353] was used earlier by Foote et al.[354] After a process zone was created in a specimen, the part of the specimen containing the process zone was cut perpendicularly to the direction of the crack into several small beams. The flexural strength of each of these beams was used as a measure for fracture process zone propagation. It will be obvious that cutting techniques rely on the assumption that no additional damage is done by the consecutive cutting, a point which is unfortunately hard to prove.

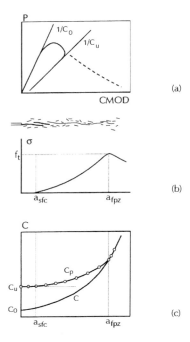

Figure 4.73 Multi-cutting technique for the determination of fracture process zone length.[353]

X-ray techniques were pioneered at Cornell University in the 1960s, but only the largest cracks could be made visible. More recently, the technique was used in Japan for studying internal crack growth around reinforcing bars by Otsuka.[355] For visualizing cracks in concrete a contrast medium is needed. Experiments with different types of barium sulphate and organic iodine showed that an aqueous solution of the latter (440 mg/ml) could best be used for crack detection. Injection holes are essential to bring the contrast medium into the specimen's interior. The results of Otsuka clearly showed the growth of fine hair cracks from the lugs at the reinforcing bars. Not only pull-out specimens were considered, but also lapped splices were investigated. Meso-level analyses of steel-concrete bond are presented in Chapter 8.

The last technique mentioned here is infrared thermography which has been developed and applied by Luong[356] for studying damage processes in concrete and rock. The technique is non-destructive. Infrared thermography uses a photovoltaic detector in combination with advanced electronics to detect the energy radiated by the specimen. All matter radiates energy in the form of charged particles that are accelerated constantly. The temperature of the materials increases when the acceleration of the particles is larger. The detection system translates the radiated energy into a real-time thermal picture, in which temperature variations of 0.2°C are visible. Microcracking is a highly efficient heat production mechanism (from ceramics studies it is for example known that the temperature at crack tips can increase up to 500°C). Luong suggested that the energy dissipation due to microcracking could best be studied under load-reversals. Therefore, he changed the name of the method to vibrothermography. In the experiments, the load is gradually increased, but halted at various levels. At these levels a vibrating load of 100 Hz is applied, and the heat distribution is examined. The spatial resolution of cracking and failure is not very large, and the method seems best suited for studies in the catastrophic regime when the size of cracks or shear bands has become of the same order as the specimen's dimensions.

4.4.6 CONCLUDING REMARKS

Crack detection techniques are considered an important aspect of fracture mechanics. As a matter of fact, fracture mechanics becomes meaningless if fracture mechanisms are neglected. Theories break down if they are not based on sound physical mechanisms. It is shown in this section that many different crack detection techniques are available to date. Many other views on fracture mechanisms in concrete exist in literature, and here we have tried to summarize the work that supports the view that fracture is a progressive process that is different in each distinct structure of varying size and boundary conditions. Thus, fracture energy according to Hillerborg and the associated definition of a fracture process zone are rather meaningless because the details of the process change with every new structure that is considered. This conclusion is supported by findings of Mindess.[143] From his survey of literature it can be concluded that a fixed size of a process-zone cannot be retrieved from all the information available to date. Rather, the process zone is dependent on geometry and boundary conditions. Of course, classes of problems can be defined where the details of the fracture process are quite similar. For example, crack growth in bending or in uniaxial tension seem quite comparable, and it is no surprise that a softening diagram from a tensile test can be used to analyze the flexural problem. However, it is asserted here that under more complicated loadings, such as the shear problem discussed in Section 3.4, which is further worked out in section 7.3, can be described quite accurately when the hypothesis of a macroscopic fracture energy and constant fracture process zone is abandoned. Instead, an approach based on a simple strength criterion and meso-level analysis of more complicated loading cases shows that a mode I criterion suffices.

4.5 THEORY OF EXPERIMENTAL DESIGNS

The theory of experimental designs is an important tool in experimentation. Not only is the theory suited for systematic survey of experimental parameters (it reduces the chance of drawing biased conclusions), but also the models are a useful tool when it comes to estimating parameters for numerical models. After a short outline of the theoretical background, an example is presented, which is related to the manufacture of specimens (i.e., a relation with Section 4.2 is established).

A full description of the theory of experimental design will not be given here, more comprehensive descriptions can be found in Mendenhall[357] or in Cochran and Cox.[358] The theory of experimental design found its basis in the agricultural sciences, where it was developed by Fisher in 1919 as tool for the design of experiments having a large number of variables. Usually, an experiment is designed such that only those parameters that are expected to have some influence on the experimental result are varied. Care should be taken that the validity of the conclusions is not affected by (unknown) variables that were not included in the experiment. Mostly used are so-called linear models, which means that the experimental results are described in terms of a model which is linear in the parameters β_i, and that contains k variables following,

$$y = \beta_0 + \beta_1 x_1 + \beta_2 x_2 + \ldots\ldots + \beta_k x_k + e \tag{4.2}$$

where y is a vector of size n containing the experimental results. The variables x_i are dependent on the levels of the influence factors and may either be quantitative, or by means of dummy variables (for example coding by 1 and 0), may also be qualitative. The vector e contains all random errors, which are assumed to be mutually independent and which have a normal distribution with expectation $E(e) = 0$ and variance $var(e) = \sigma_0^2$. In shortened form, Equation 4.2 may be written as

$$y = X\beta + e \tag{4.3}$$

where X is the design matrix that contains all information regarding the treatment combinations (experimental variable combinations). The purpose of the experiments is to estimate the variables β_i by means of the method of least squares. The estimate b of β is determined through

$$b = (X'X)^{-1} Xy \tag{4.4}$$

The vector b is of size $(k-1)$, and the expectation is $E(b) = \beta$. In case of an orthogonal experimental design, $(X'X)$ reduces to a diagonal matrix. The random error can be estimated from the sum of squares of deviations of the observations according to

$$\sigma_0^2 = \frac{KS_r}{n-k-1} \tag{4.5}$$

where the difference between the number of observations (n) and the number of parameters ($k+1$) is the number of degrees of freedom ($n-(k+1)$). The number of degrees of freedom should in the ideal case be as large as possible. The sum of squares KS_r can be written as,

$$KS_r = (y - \bar{y})/(y - \bar{y})$$

where the bar indicates the estimate of the experimental result y.

The variances and covariances of the estimates b_i of β_i are fully determined by the matrix $(X'X)^{-1} \cdot \sigma_0^2$. The estimate for the variance and covariance of b_i (and b_j) is given by,

$$\bar{\sigma}^2(b_i) = \left[(X'X)^{-1}\right]_{ii} \bar{\sigma}_0^2 \tag{4.6a}$$

and

$$\overline{cov}(b_i, b_j) = \left[(X'X)^{-1}\right]_{ij} \bar{\sigma}_0^2 \tag{4.6b}$$

The bar above the symbols for variance and covariance makes the distinction between the estimate and the real value. Further it can be proved that

$$\frac{(b_i - \beta_i)}{\bar{\sigma}(b_i)} \tag{4.7}$$

has a students-t distribution, with $n-(k+1)$ degrees of freedom. The $(1-\alpha)$ confidence intervals for the β_i can now be determined through

$$\beta_i \pm t_{\alpha/2} \cdot \bar{\sigma}(b_i) \tag{4.8}$$

In addition, hypotheses concerning the parameters b_j or linear combinations of the form

$$\bar{l} = a_0 b_0 + a_1 b_1 + \ldots + a_k b_k = a'b \tag{4.9}$$

may be investigated.

For example, the hypothesis H_0 that $(k-g)$ variables are equal to zero,

$$H_0 : \beta_{g+1} = \beta_{g+2} = \ldots = \beta_k = 0, \qquad (g < k) \tag{4.10}$$

may be checked using an F-distribution with $(k-g)$ and $n-(k+1)$ degrees of freedom according to

$$F = \frac{(KS_{r0} - KS_r)/(k-g)}{KS_r/(n-k-1)} \tag{4.11}$$

where KS_{r0} is the sum of squares of deviations of the observations for the reduced model with $(k-g)$ values of β_i equal to zero, and KS_r is the sum of squares of the original model. When the value of F surpasses a critical value, the hypothesis H_0 must be rejected. In that case it must be concluded that the parameters β_i $(g+1 \leq i \leq k)$ have a significant influence on the experimental result.

The confidence limits for linear combinations $l = a'b$ may be determined as well. They are fully described by,

$$var(\bar{l}) = a'(X'X)^{-1}a\bar{\sigma}_0^2 \tag{4.12}$$

The procedure may now be summarized as follows. For a linear model containing all main effects and higher-order-factor interactions in the b_i are determined using Equation 4.4, and the random error is estimated with Equation 4.5. In general only two-factor interactions are considered (indicating the combined effect of two main factors). Higher order interactions (i.e., higher than two) are generally not significant, or remain undetected because the scope of the test series is too limited (mostly because of practical reasons). Using the F-test, it can be investigated whether the model can be simplified. Confidence limits for the b_i or linear combinations for the b_i are determined. For the above statistical analysis, which involves a number of multiplications and inversion of large matrices, standard computer programs are available.

Let us now consider an example. In a series of displacement-controlled uniaxial compression tests, the effect of the manufacturing method on the compressive strength of the specimens is investigated. The specimens (100 mm cubes) are sawn from eight prisms that are casted in four batches (designated 1 through 4). From each prism 6 cubes are sawn, and consequently three different positions of the cubes can be distinguished as shown in the inset of Figure 4.74. Furthermore, a casting direction always has to be chosen. The prisms were cast into the vertical direction, and it was expected that the direction of loading with respect to the direction of casting would have some effect on the experimental outcome as well. Thus, three variables had to be taken into consideration: the difference between batches (or stated differently, is the method of casting reproducible?), the difference in position of the cubes in the original prism, and the direction of loading (indicating possible effects of initial anisotropy). In total, 48 specimens were manufactured and loaded in uniaxial compression. A statistical analysis according to the above theory of experimental design was carried out in which the peak compressive stress was taken as response variable. The linear model is given by

$$\sigma_{1p} = Ey = \Sigma\beta_i x_i \qquad (i = 0,....,17) \tag{4.13}$$

The variables x_i represent the level of the influence factors: the loading direction with respect to the direction of casting ($x_1 = -1$ for perpendicular loading, $x_1 = 1$ for parallel loading, see inset in Figure 4.74 and Table 4.2), the position of the cube (x_2, x_3), the batch (x_4, x_5, x_6), and a number of two-factor interactions (x_7, x_8, ... , x_{17}). Because all factors are qualitative, values for the x_i must be selected following a coding system (as indicated above for the value of x_1) in order to perform the necessary computations, see Table 4.2. The x_i should be selected such that rows in the design matrix X are independent. The

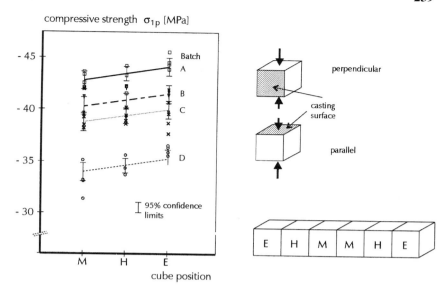

Figure 4.74 Effect of cube position on the uniaxial compressive strength.[80]

x_i values shown in Table 4.2 satisfy this requirement. Text books on statistical designs generally contain numerous coding schemes for the $x_{i.}$[357,358]

The mean strength of all tests is given by the estimate b_0 of β_0. The number of degrees of freedom n available for estimating the error in this particular example is 29, namely 47 (i.e., the number of successful tests, one test failed) minus 1 (average strength b_0) minus 17 (for the x_i). In Table 4.2 next to the overview of the main factors and two-factor interactions, the values of the x_i, the estimated values of b_i, the standard deviations, and the P-values are shown. The P-value indicates whether or not a factor is significant. The significant values according to a student's-t distribution (95% interval) are underlined. They are: the linear term for differences in the position of the cube (x_2), the three terms (x_4, x_5, x_6) for differences between batches, and a two-factor interaction ($x_{13} = x_2 x_5$). This two-factor interaction describes the combined effect of loading direction and batch. The reduced model, containing only the significant factors can then be formulated as

$$\overline{\sigma}_{1p} = \overline{y} = -39.56 - 0.61x_2 - 1.25x_4 + 0.55x_5 - 0.67x_6 - 0.55x_2 x_5$$

$$(4.14)$$

The estimate for the random error is

$$\overline{\sigma}_0^2 = 1.51 \; MPa \quad (v = 29)$$

The average uniaxial compressive strength determined from the 100 mm cube tests is $\sigma_{1p,100} = b_0 = -39.56$ MPa. The effect of cube position on uniaxial

TABLE 4.2
Statistical Analysis of the Effect of Various Factors Related to the Manufacturing of Concrete Test Specimens on the Uniaxial Compressive Strength[80]

Factor and levels	Value of x_i	Estimated value b_i [MPa]	St. dev. (var/σ_0^2)	P-value ($\alpha = .05$)
Mean	$x_0 = 1$	$b_0 = -39.56$	0.0215	0.000
Loading dir. D (perpend/parallel, see Figure 4.74)	$x_1 = -1,1$	$b_1 = 0.11$	0.0215	0.541
Cube position P (E,H,M, Figure 4.74)	$x_2 = 1,0,-1$	$b_2 = -0.61$	0.0312	0.009
Position P2	$x_3 = 3x_2^2-2$	$b_3 = -0.08$	0.0111	0.532
Batch B	$x_4 = -3,-1,1,3$	$b_4 = -1.25$	0.0044	0.000
Batch B2	$x_5 = (x_4^2 -5)/4$	$b_5 = 0.55$	0.0215	0.005
Batch B3	$x_6 = (5x_4^3-41x_4)/12$	$b_6 = -0.67$	0.0042	0.000
Two-factor interactions				
DP	$x_7 = x_1x_2$	$b_7 = -0.05$	0.0313	0.800
DP2	$x_8 = x_1x_3$	$b_8 = 0.07$	0.0111	0.588
DB	$x_9 = x_1x_4$	$b_9 = 0.10$	0.0044	0.240
DB2	$x_{10} = x_1x_5$	$b_{10} = 0.12$	0.0215	0.498
DB3	$x_{11} = x_1x_6$	$b_{11} = -0.07$	0.0042	0.413
PB	$x_{12} = x_2x_4$	$b_{12} = 0.12$	0.0062	0.230
PB2	$x_{13} = x_2x_5$	$b_{13} = -0.55$	0.0312	0.016
PB3	$x_{14} = x_2x_6$	$b_{14} = -0.03$	0.0062	0.773
P2B	$x_{15} = x_3x_4$	$b_{15} = -0.05$	0.0023	0.444
P2B2	$x_{16} = x_3x_5$	$b_{16} = -0.08$	0.0111	0.555
P2B3	$x_{17} = x_3x_6$	$b_{17} = -0.01$	0.0021	0.812

compressive strength is shown in Figure 4.74. The results indicate that stronger specimens are found when the cubes are obtained from so-called "end-positions" of the prisms. An explanation for this behaviour might be that the end platens from the moulds in which the prisms were cast had a stiffening effect. Thus, when the prisms are cast, the ends will be better compacted. Recently, it was found that specimens cast in stiffer moulds yield a higher strength.[359]

The results also indicate that differences between batches exist. This is of course a well-known fact. It shows that when valid conclusions are to be drawn, a systematic experimental design will help. If the factor "batches" was not taken into account, the significance of the factor "position" might not have been found at all, because the "error" would have become too large. Practitioners and even some researchers are tempted to neglect such detailed analyses under the "slogan" that concrete is a material with large scatter anyway.

Moreover, often it is argued that the conditions under which the material is manufactured in the laboratory do not correspond to the situation at building sites. The point that is made here is that through careful experimentation, fundamental physical mechanisms *can* be evaluated much easier, which might lead to improved analysis models and/or materials with improved properties.

The compressive strength is of course but a single response parameter measured in a displacement-controlled test. The complete stress-deformation curve is measured, and by taking the stress at different prescribed levels of axial deformation, one might apply the theory of linear models to obtain the best fit of the complete stress-deformation curve as well. When the area under the curve is integrated, one obtains a measure for the fracture energy in compression. Using the same linear model of Equation 4.13, the fracture energy computed up to an axial strain of 0.6% was used as response variable. In addition to the factors mentioned in Equation 4.14, now also the direction of loading with respect to the direction of casting was found to be significant. The two-factor interaction x_{13} was not significant when the fracture energy was taken as response variable. Indeed, from the stress-strain diagrams of Figure 3.9 it can be observed that initial anisotropy has an important effect. An explanation in terms of microcrack growth was given in Section 3.2.2.

It should be mentioned that the number of experiments to be carried out can be greatly reduced by using so-called fractional factorial designs. This means that not every main factor and higher order interaction can be estimated, but given a careful planning of the experiment it is possible to find at least the most important main factor effects and, if necessary, a number of two-factor inter-actions as well. It must be judged of course against the greatly reduced experimental effort needed whether such a fractional design is suitable for the problem that is addressed. Some effects in the fractional designs may coincide, and cannot be estimated, but in most cases it is clear which factors might be left out. Carefully conducted preliminary experiments might be helpful in deciding which factors can be left out. Fractional designs are treated in detail in the book by Cochran and Cox,[358] where references to further sources may be found as well.

The theory of experimental design is suitable for estimating parameters of numerical models as well (Chapter 6). In that case, numerical simulations of a given structure are carried out. A number of model parameters is varied according to a careful design, similar as shown here for the experiments. The statistical analysis is similar to the one described in this section.

4.6 THE NEED FOR STANDARD TESTING

Most of the material presented in this chapter was related to fundamental aspects of testing. Of course, most of these matters are important in standard testing as well, but then a single procedure is prescribed which must be followed each time a test is performed. These regulations concern the manufacturing procedure of the specimens, the shape and size of specimens, the testing machine and boundary conditions, the loading rate, the way of presenting the results, etc. It is not our intention to give detailed information on current standards here because they may vary from country to country. Rather, we will address a number of recent efforts to develop standardized tests for strain-softening in uniaxial compression and in uniaxial tension. There can be several reasons for organisations, practitioners and researchers to decide on a standard test:

1. for the determination of certain parameters needed in an accepted model that is used in building practice
2. as a means for comparing experimental results obtained in different laboratories
3. as a benchmark problem for numerical models
4. as a means for quality control

The first reason is most important. When a model has become an accepted tool for structural analysis in engineering practice, a test must be designed such that the relevant model parameters can be measured. As may have become clear from Chapter 3 and parts of Chapter 4, and as will be further elaborated in Chapters 6 and 7, the direct measurement of parameters for fracture models is not possible, and always an inverse analysis of experimental results must be carried out. Nevertheless, for applications a standard test method should be agreed upon. Very slowly now, progress is being made for a standardized test for measuring mode I fracture properties of concrete. In the rock mechanics community also efforts are underway to define a test standard, which is, incidentally, quite different from the approach followed in concrete mechanics. As we will see in Section 4.6.1, the definition of a standard is hampered by the fact that not a single model exists which is accepted by all researchers at this moment. More seriously however seems the fact that the development in fracture mechanics of concrete are grossly ignored by practical engineers, which may stem from the above fact that no generally accepted model exists, but perhaps also from the fact that the current generation of practitioners has not received any education in fracture mechanics.

As long as the situation remains as sketched above, there is still no reason not to decide on a standard test method. Difficulties in comparing experimental results obtained in different laboratories are a complicating factor in a fast development of new models. Everyone tries his own experiment, carried out under widely varied boundary conditions — which are in most cases not even clearly described — and no basis for comparison exists. Of course, it is an excellent development that new experiments are designed all the time, but next to those innovating efforts, there is room for standard testing as a comparative means between all the different test-series. The three remaining reasons mentioned above are of sufficient importance to introduce standard tests for fracture energy and softening in uniaxial tension and compression. A standard test presented in such a context might be more easily accepted at present, as it is not necessary to favour one of the existing models above all the other models. As long as no clear insight exists in all the boundary and size effects on fracture parameters of concrete, it seems a rather pointless effort to propose a theory for practical use and to develop a test for measuring the relevant parameters. As will be discussed in Chapters 5 and 6, not only material parameters are to be defined but, in many models, additional parameters are needed that have no physical basis whatsoever. Such model parameters originate in most cases from the chosen discretization technique.

Let us now summarise which efforts are undertaken to define a standardized test for concrete fracture. First in Section 4.6.1, we address efforts that aim at measuring tensile fracture properties of concrete (mode I), whereas in Section 4.6.2 the approach for measuring compressive softening is outlined.

4.6.1 FRACTURE ENERGY AND TENSILE SOFTENING

At present, three different draft recommendations for fracture testing have been published by RILEM.[67,360,361] Next to these developments, a fourth draft for fracture testing of autoclaved aerated concrete (AAC) has been proposed,[362] in which a geometry is suggested that differs substantially from the first three draft recommendations. Moreover, some researchers would favour a method based on the methods for fracture testing suggested by the International Society of Rock Mechanics (ISRM), see Ouchterlony.[363] The RILEM draft recommendations and the recommendation for AAR are based on Hillerborg's Fictitious Crack Model,[5] the Jenq and Shah[364] two-parameter model (which is an effective crack model based on classical, linear fracture mechanics, see Section 5.3.2), and Bažant's size-effect law.[304] The developments in rock mechanics are based on linear elastic fracture mechanics (Section 5.3.2) for which a critical stress intensity factor must be determined. Thus, all these various drafts were set up with a particular theoretical fracture model in mind. The controversy between the different proposals will be evident. Recently, Elices and Planas[368] reported that through measurement of only the tensile strength and the initial slope of the tensile softening diagram (Figure 4.75), it is possible to determine the parameters of all two-parameter models.[365] Let us now first examine which test geometries and procedures have been suggested for the determination of the softening parameters of Figure 4.75.

Let us first consider the Fictitious Crack Model, for which the complete tensile softening diagram is an essential input parameter. Hillerborg et al.[5] and Petersson[142] indicate that — as understood at that time — the uniaxial tensile test between fixed end-platens is the best option for measuring the complete tensile softening diagram and the associated tensile fracture energy G_f. However, because the performance of the test is rather complicated, it was proposed to use a three-point-bend beam instead, see RILEM.[67] In Figure 4.76 the actual beam geometry is shown, and in Table 4.3 the size of the beams needed for

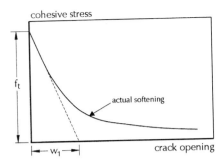

Figure 4.75 Most important parameters in a tensile softening diagram to be determined in a standardized test.[368]

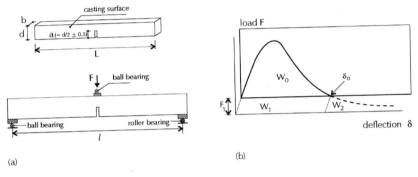

Figure 4.76 Three-point-bend beam recommended for measuring the fracture energy of concrete, after RILEM.[67] (Reprinted with kind permission from RILEM, Paris.)

concretes made with different types of aggregates is indicated. The beam should be supported by a roller and a ball, and the load must be applied through a ball bearing as well. Effects from torsion and frictional slip are supposedly circumvented in this way. The tensile strength, which is an essential parameter in all fracture models cannot be directly measured from the three-point-bend test, but has to be determined through an additional test, for example by the tensile splitting tests (Brazilian test). The beam must be loaded in stable displacement control, and after correcting the area under the softening curve for effects from self-weight, the fracture energy G_f of concrete can be determined (see Figure 4.76b) following,

$$G_f = \frac{W}{b(d-a)} \tag{4.15}$$

where $W = W_0 + 2F_1 \delta_0$ (see Figure 4.76b). According to Petersson,[142] $W_2 \approx W_1$. Quite unfortunately, the fracture energy determined in this manner is size

TABLE 4.3
Suggested Beam Size for the Three-Point-Bend Test for Measuring Fracture Energy of Concrete, after RILEM[67]

d_{max} [mm]	Depth d [mm]	Width b [mm]	Length L [mm]	Span ℓ [mm]
1–16	100 ± 5	100 ± 5	840 ± 10	800 ± 5
16.1–32	200 ± 5	100 ± 5	1190 ± 10	1130 ± 5
32.1–48	300 ± 5	150 ± 5	1450 ± 10	1385 ± 5
48.1–64	400 ± 5	200 ± 5	1640 ± 10	1600 ± 5

dependent, i.e., an increase of fracture energy is measured with increasing beam size.[366] Elices and co-workers have identified different sources of error which may have caused the size effect.[291,366,367] The sources of error mentioned are hysteresis in the loading machine (note that the hydraulic cylinder must unload in the softening regime), bulk dissipation in parts of the beam under the loading point and above the two supports, as well as frictional restraint from the rollers used as support and loading point (see also Section 4.1.3.1). Moreover, errors are caused by cutting of the tail of the softening diagram, for example because the self-weight of the beam causes an abrupt end of the experiment (the self-weight gives an additional load on the beam and must be considered as a partial load-control, which is of course unstable), see Figure 4.76b. For very large beams the test tends to become unstable at an earlier stage as compared to smaller test specimens. Self-weight correction can be done by using additional spring supports that counter-balance the self-weight of the specimen during the test, or alternatively by using a longer beam which is supported such that the effect of self-weight is eliminated at the point of crack nucleation and growth. This procedure was used by Elices et al.[366] Also, in uniaxial tensile tests the tail is cut for practical reasons because the test duration would be very long indeed when the test would be carried on until full separation. The total crack opening before the specimen load has been reduced to zero again can be as large as 1000 μm in a uniaxial tensile test (note: the variation in maximum crack opening can be very large as well, up to several hundreds percent). The error of cutting the tail at 400 μm rather than at 600 μm is estimated at 10%, whereas additional errors up to 10% may be caused by erroneous measurement of the axial load at very large crack openings.[84] The fracture energy is not the most important parameter in the tensile softening diagram,[368] but of much more importance are the tensile strength f_t and the initial slope $d\sigma/dw$ of the softening diagram (as will be further elaborated in Section 6.2). Estimating the peak stress f_t and the slope from a three-point-bend test can only be done through inverse analysis as will be discussed in Section 6.2. Such procedures lead to best estimates for the required parameters based on the experimental data available, and generally it is rather debatable whether the findings can be extrapolated to regimes outside the range of experimental observations.

One last remark should be made. Hillerborg and Petersson suggested that the uniaxial tensile test would be the most superior test to measure the softening parameters. The interpretation of the uniaxial tensile test is however quite complicated. As was already pointed out in Section 4.1.3.2, rotations of the loading platens may cause a significantly different fracture response of a specimen. The tensile strength is affected, as well as the fracture energy, but also the shape of the softening diagram, as was shown in Figures 4.39 and 4.40. This means that the required softening parameters cannot be derived from the supposedly superior uniaxial tensile test. However, effects from bulk dissipation at the supports (i.e., glue-platens in a uniaxial tensile test), and problems with self-weight are much smaller, if not non-existent, in the tensile test. Therefore, the opinion of some researchers, i.e., because of the rotational problems of the uniaxial tensile test, the three-point-bend test should be preferred, is not shared by the author.

A standard test as described above should not only give the correct material parameters, but test performance should be as simple as possible as well. The specimen should be easy to handle (preferably a low self-weight), and there should be no risk of breaking the specimen during handling. In the case of the three-point-bend test described above, the net stress at the notch of the standard beams according to Table 4.3 is normally not higher than 20% of the failure stress.[369]

The standardized test for measuring the fracture energy of autoclaved aerated concrete, which was proposed recently,[362] deviates from the above procedure. A wedge-splitting test is proposed, rather than a three-point-bend beam. The wedge splitting test was proposed earlier by Tschegg and Linsbauer.[370,371] A modified version was developed and tested by Brühwiler.[171] In Figure 4.77 the specimen and loading procedure are clarified. A standard cube is modified as shown in Figure 4.77a. Two steel platens are attached to the top (Figure 4.77b) and, through a wedging device, the splitting force is applied. The applied horizontal splitting force is related to the vertical compressive load through

$$P_s = \frac{P_v}{2\tan\alpha} \cdot \frac{(1 - \mu.\tan\alpha)}{(1 + \mu.ctg\alpha)} \approx \frac{P_v}{2.\tan\alpha} \cdot \frac{1}{(1 + \mu.ctg\alpha)} \qquad (4.16)$$

where α is the angle between the side of the wedge and the vertical load line and μ is the coefficient of friction between the wedge and the rollers that are used to guide the wedge. With a frictional restraint of 0.1 to 0.5% and a wedge angle $\alpha = 15°$, the contribution of frictional forces on the splitting force P_s is approximately 0.4 to 1.9%.[315] Because of these low values, it is suggested to neglect the effect of friction in a first approximation of the fracture energy. The test must be carried out in stable displacement control, and the crack opening displacement (COD) is measured at the top of the specimen as indicated in

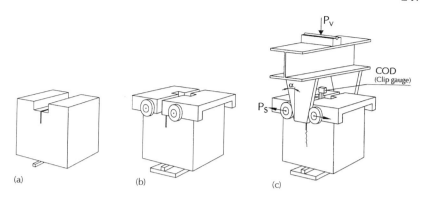

Figure 4.77 Wedge splitting test.[171,315] Reprinted by kind permission of Chapman & Hall, Andover, U.K.)

Figure 4.77c. From the measured load-COD curve, the tensile strain-softening diagram must be determined through an inverse analysis, for example by using the program Softfit.[372] These methods will be further elucidated in Chapter 6. It should be mentioned that the wedge splitting test can also be performed on cylinders that have been modified in a way similar as shown in Figure 4.77 for the standard cube. The great advantage of the wedge splitting test is of course that the specimen geometry is related to common standard test specimens used for measuring the compressive and splitting tensile test (Brazilian test, not to be confused with the above splitting tensile test).

Finally, here are some words about the draft recommendations for the determination of the parameters needed in the Jenq and Shah two-parameter model and Bažant's Size Effect Law. In both draft recommendations, a three-point-bend test is proposed. The size and shape are only slightly different from the first RILEM draft recommendation for measuring fracture energy based on the Fictitious Crack Model. The proposal for the measurement of the two-parameters of the Jenq and Shah model comes closest to the original RILEM 50 FMC proposal. The dimensions of the three-point-bend test are shown in Table 4.4 and Figure 4.78. The beam should again be tested in stable displacement control with the crack mouth opening displacement (CMOD) as control variable. At about 95% of the peak load, in the descending branch, the specimen must be unloaded manually, and reloaded again as soon as the load has reduced to zero. Subsequently, a number of such unloading-reloading cycles must be carried out. A typical result is shown in Figure 4.78b. The effective crack length is computed from the compliance decrease. In the draft recommendation, the various formulas are then given to compute the critical stress intensity factor K_{Ic}^s and the critical crack tip opening displacement $CTOD_c$. Thus, basically, the parameters for the effective crack model must be derived using a compliance technique (see also Section 4.4.1).

TABLE 4.4
Suggested Beam Size for the Cyclic Three-Point-Bend Test for Measuring the Two Parameters of the Jenq and Shah Model, after RILEM[360]

d_{max} [mm]	Depth d [mm]	Width b [mm]	Length L [mm]	Span ℓ [mm]
1-25	150 ± 5	80 ± 5	700 ± 10	600 ± 5
25.1-50	250 ± 5	150 ± 5	1100 ± 10	1000 ± 5

The third draft recommendation for the determination of the parameters used in Bažant's Size-effect Law[304] is based on three-point bending tests as well. The fracture energy is computed from the maximum loads measured in three geometrically similar beams. The shape of the beams is shown in Figure 4.79. Limits are set to the dimensions of the beams as follows. The length over depth ratio l/d should be at least 2.5. The notch depth to beam depth ratio a_0/d should be between 0.15 and 0.5. The notch width should be as small as possible, and should never exceed $0.5d_{max}$ (where d_{max} is the maximum aggregate size of the concrete). Moreover, the minimum width b and depth d should not be smaller than $3d_{max}$. Beams of at least three different sizes should be tested, i.e., with depths $d_1, d_2, ... d_n$ and span $l_1, l_2, ... l_n$. The smallest depth d_1 should not be larger than $5d_{max}$, whereas the largest depth d_n should not be smaller than $10d_{max}$. The ratio of d_n/d_1 should be larger than 4. All specimens should have the same thickness b. Thus, specimens that are geometrically similar in two dimensions must be used, which means that the ratio's l/d, a_0/d

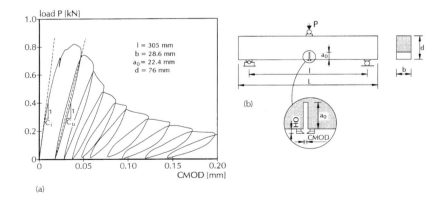

(a)

(b)

Figure 4.78 Typical response curve needed for the measurement of the two parameters in the Jenq and Shah model (a) and three-point-bend specimen (b), after RILEM.[360] (Reprinted with kind permission from RILEM, Paris.)

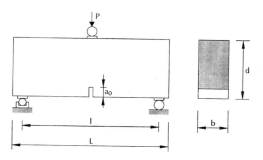

Figure 4.79 Three-point-bend specimen for the determination of the parameters needed in Bažant's Size-Effect Law, after RILEM.[361] (Reprinted with kind permission from RILEM, Paris.)

and L/d are the same for all specimens. Through a best fit of the size effect law through the corrected maximum loads determined in the beam tests, the fracture energy can finally be computed from equations described in the draft recommendation. Recently, an alternative method was proposed, in which beams of a single size are needed only.[373]

As can be seen from the above, the experimental procedures to be followed differ for each draft recommendation. The most complicated test seems to be the three-point bend test for the determination of the two parameters for the Jenq and Shah model, mainly because quite a number of load-cycles must be carried out. Also rather complicated is Bažant's test on three different specimen sizes, although much of this objection has been removed with his latest proposal.[373] The beam tests all have the disadvantage that they become unstable for very large sizes. A correction for the self-weight of the beam is essential, but none of the proposals contains such correction factors. The wedge splitting test seems to be most practical because it is based on existing standard test specimens (cubes or cylinders). The main objection here might be that the frictional component in estimating the horizontal splitting load is not quite known, but then frictional restraint in the supports affects the results from the three-point-bend tests as well. If the wedge splitting test is standardized it would be recommended to design a standard splitting device in which the splitting load can be measured directly, for example using a double working miniature actuator containing two load cells.[52] Typical for all standard tests seems that the stress-crack opening relation defined following Figure 4.75 can only be deduced from an inverse analysis. None of the procedures allows for a direct measurement of the required parameters. The fact that the most important reason for introducing a standardized test has not been resolved to date (namely reason [1] mentioned in the beginning of Section 4.6), means that the best choice at this moment seems the most simple test which can yield a reliable result. The wedge splitting test seems to be a good candidate, as it is based on familiar specimen geometries, and it could easily be used as a means for comparison of different laboratory investigations.

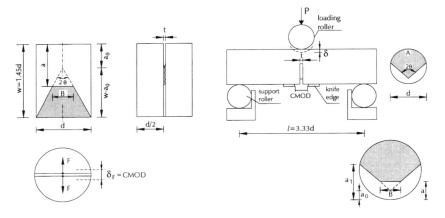

Figure 4.80 Short rod specimen (a) and chevron-notched beam (b) for measuring fracture toughness of rock.[363] (Reprinted by kind permission of the publisher, A.A. Balkema, Rotterdam, The Netherlands.)

As mentioned in the introduction to this section, in rock mechanics a standardized test for fracture toughness measurement must be defined. The model on which all efforts are based is conventional, linear elastic fracture mechanics. In Figure 4.80 the two proposed specimen geometries are shown. Both the short rod specimen (Figure 4.80a) and the chevron bend specimen (Figure 4.80b) are based on standard cores. A chevron shaped notch is used in order to stabilize the crack growth in the specimens. From the experiments a critical stress intensity factor is derived. Again numerical calibration is essential in order to retrieve the essential parameters. For the chevron bend specimen the (level 1) fracture toughness is calculated from

$$K_{CB} = \frac{A_{min} \cdot F_{max}}{D^{1.5}} \qquad (4.17)$$

where F_{max} is the maximum load the specimen can sustain, and $A_{min} = [1.835 + 7.15 a_0/D + 9.85 (a_0/D)^2].S/D$. A second level fracture toughness can be computed under correction for non-linear behaviour. Equation 4.17 was used by Alexander[61] to determine the interface fracture toughness, see Section 2.2.3.

4.6.2 COMPRESSIVE STRENGTH AND SOFTENING

The determination of the compressive strength of concrete has been standardized a long time ago, and accepted methods include the familiar cube and cylinder tests. Much of the concrete structural codes are based on compressive strength only. The tests are usually carried out in load-control and at a prescribed loading rate. The specimens are loaded through stiff steel platens, and normally one of the platens in the machine can rotate before loading is applied in order to overcome geometrical imperfections of the specimen. Sometimes it is advised to use a thin filler between the specimen and loading platen (capping). This seems particularly important if tests are conducted in a machine

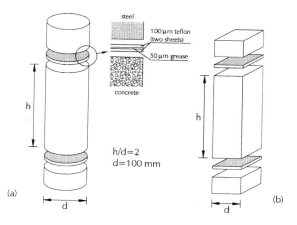

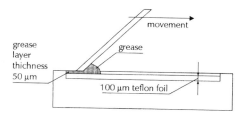

Figure 4.81 Standard test for measuring softening of concrete in uniaxial compression.[375]

Figure 4.82 Device for applying a constant amount of grease on a 100 μm thick teflon foil.

where initial adjustment of the loading platen is not possible. In RILEM[374] a number of demands concerning the test equipment are summarized. The steel loading platens are the worst choice in view of the boundary restraint introduced during the test, as was discussed in Sections 3.2.4 and 4.1.3.1. For the determination of the maximum compressive stress (with low scatter) the application of friction reducing measures seems to be a large improvement. Because there is interest in the determination of the full softening behaviour of concrete under compression — for example for the analysis of rotational capacity of reinforced concrete structures — the RILEM committee 148 SSC (Strain Softening of Concrete) aims at a definition of a standardized test for compressive softening. Based on the report of the committee,[375] it is proposed to determine the compressive softening curve in the future from uniaxial compressive tests between teflon platens, as shown in Figure 4.81.

The suggested specimen slenderness is $h/d = 2$, with $d = 100$ mm, as indicated. Both cylinders or prisms can be used. Very good results can be obtained with a system of two 100 μm thick teflon sheets with a 50 μm thick layer of simple bearing grease in between. The application of the grease is quite critical, and it is important to apply a constant quantity. This can be done using the device of Figure 4.82. A scraper is moved over a steel block containing a

150 μm deep void where first a teflon sheet with the same size as the specimen's cross-section is placed. A 50 μm thick layer of grease can be applied quite easily in this way. Again, no test method can be prescribed from which a unique compressive softening diagram can be measured. Rather, the policy is followed where a procedure has been suggested that gives reliable results with the smallest possible scatter. Of course, the procedure is still open to debate as no draft recommendation has been published to date. However, based on experience, it can be stated that the above compression test seems to give the most reliable result with present day experimental tools.

Chapter 5

MODELLING AS A TOOL FOR TEST INTERPRETATION

Knowledge of fracture mechanisms *is essential for fracture* mechanics, *but not vice versa !!*

5.1 INTRODUCTION

As may have become clear from the preceding chapters, a material may be modelled at various levels. According to the specific application that we have in mind the material can be modelled as a continuum, at the particle level, or even at the level of the individual cement grains or Calcium Silicate Hydrates (CSH). In the extreme case we might altogether descend to the molecular or atomic level (or even below that, as many physicists have done). For modelling fracture of concrete, we will confine ourselves to the meso-level and the macro-level. Only for setting the stage, i.e., when fracture laws for lattice models are presented, we will discuss shortly the theoretical strength of materials based on considerations of a regular atomic lattice. From such a scheme the theoretical cleavage strength of materials can be computed, but for various reasons these sometimes very high values can never be reached in practice.

The idea of modelling is that we would like to have a (numerical) tool that can describe in sufficient detail the observations of which some have been presented in Chapters 3 and 4. With such a model we would like to have the ability to make predictions outside the domain that was used for tuning. Of course, the term "sufficient detail" is rather vague. In principle, we would like to forecast the fracture response of structures that were not explored before by means of experiments. Important for making predictions is that the model parameters can be determined independently of specimen geometry and boundary conditions; otherwise they are not real material properties. This is of course the main problem. Many empirical formulations exist that are applicable only to the range of experiments to which they were fitted, for example the size-effect laws mentioned in Section 4.2.3. What follows in this chapter is mainly an exposure of models that are used for modelling localised cracks in concrete. This implies that cracks are modelled preferentially as a real discontinuity.

The modelling task can be separated in a number of sub-tasks. First, we have to select the level of representation. Historically, engineering mechanics has dealt with a continuum representation of the world. In other words, the state variables are nominal stress and strain. As shown in Chapter 3, the notion of strain is rather debatable for fracture, which means that other ways have to be found to simulate localised cracking in a continuum environment. Moreover, the old idea of uniform stress is open to dispute; as we have seen in Chapters 3 and 4 fracture zones usually develop asymmetrically in laboratory specimens, even though the initial set-up was symmetric and the specimen was supposedly loaded uniformly. The heterogeneity of the concrete is to blame for continuum approaches that break down for this material. In Section 5.2 we will review the various strategies that have been followed in the past, including smeared and discrete crack approaches. Lately, so-called higher order continuum theories have been developed to allow for including mesh-independent, localised deformation zones in a structure.[376] These models have become rather complicated, and only a limited number of researchers are capable of fully understanding all the details and implications of such models. However, if models become too complicated, even in a research environment, we should search for simpler tools that are easy to handle and that are still capable of capturing the essence of localised fracture. In 1989, at a summer school in Cargése (Corsica), theoretical physicists specialised in statistical mechanics tried to convince engineers of the potentials of a lattice approach to fracture.[377] The lattice technique was not accepted by many, but it seemed quite appropriate for our goal, i.e., to have a tool that can be used for interpreting and designing laboratory experiments. Lattice models will be described in Section 5.2 as well, as an alternative for continuum approaches. Thus, as a first step it must be decided how to discretize the problem: as a continuum or as a lattice.

The second step is to decide whether the heterogeneity of the concrete should be modelled directly or indirectly. In a continuum model, the material structure effects are not included directly, but rather the effect is hidden in the constitutive law. However, if a relation should be made between material structure and fracture response, it would be more appropriate to map the structure of the material under consideration directly into the model. Mapping the material structure on the structure (e.g., laboratory specimen) directly, has most likely the advantage that a more simple criterion for fracture can be used. In Section 5.2, we will discuss several approaches of mapping a material structure on a continuum finite element mesh or on top of a lattice.

In summary, in Section 5.2, the representation of concrete in a model will be described. Most of this is pure geometry. Thus far nothing has been said about a possible rule (law) for making fractures, but it will be evident that this law depends on the level of discretization. In Sections 5.3 and 5.4, several fracture laws will be presented. We will distinguish between fracture laws for continuum based models, and fracture laws for lattice type models. As far as the first category is concerned, we will review classical strength of materials theory, classical Linear Elastic Fracture Mechanics (LEFM), the Dugdale-Barenblatt plastic crack tip model, and the non-linear fracture mechanics (Fictitious Crack Model of Hillerborg and co-workers). Finally, we will discuss possible fracture laws for lattice type models. In the last models, however, the fracture law becomes very simple indeed, as we only need to specify a rule for breaking a linear truss element (i.e., spring or bar) or beam element. Note however that at the level of the individual lattice beams, continuum representations are used again. Always a lower bound seems to exist where continuum ideas must be included in the formulations.

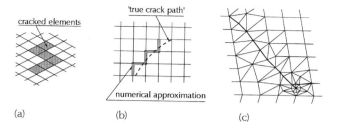

Figure 5.1 Crack modelling concepts in continuum finite element models: (a) smeared cracking, (b) discrete cracking along prescribed mesh lines, and (c) remeshing.

5.2 REPRESENTING THE MATERIAL

5.2.1 CONTINUUM OR LATTICE

In the first applications of numerical models for the analysis of reinforced concrete structures, the finite element method was adopted.[378] In this model, cracking was introduced as a smeared phenomenon, see Figure 5.1a. A crack was not considered as a discrete displacement jump, but rather the properties of the finite elements where a crack was supposed to develop were changed according to some continuum law. The idea was that through careful modelling and laboratory measurements, such as those presented in Chapter 3, a universal valid (macroscopic) constitutive law could be developed for the material. In combination with the powerful finite element technique, this would give a tool to analyze any concrete structure, and possibly also a model for designing new reinforced concrete structures, simply through computation. This was also the goal in the Dutch Concrete Mechanics project that has been run since the mid 1970s.[379-382] The experiments that have been done in the past decade, and that are presented in Chapters 3 and 4, do not give much faith that a unique and universal constitutive law can be derived for concrete (and other brittle disordered materials like rock and ceramics). Especially the post-peak behaviour is difficult to capture of course, although some researchers have faith that some day they will find a good continuum representation.[383] There is not so much to say about continuum representation. Making finite element meshes and setting up the equations for solving the boundary problem fall beyond the scope of this book. For the continuum representation, a good definition of the constitutive laws is of extreme importance. Some related matters will be discussed in Section 5.3. So, let us save our comments for that section.

The crack methodology that is adopted in the continuum finite element approach is directly related to the fracture law that is selected. Next to the above-mentioned smeared crack method, other crack methodologies have been used by various researchers, namely the discrete crack method without remeshing and the discrete crack method with remeshing. In these discrete approaches, we do not rely so much on the constitutive formulation of the complete stress-strain behaviour of concrete, but rather through the selection of appropriate interface elements and crack-tip elements, fracture can be simulated.[384-386] The disadvantage of the discrete crack method with a fixed finite element mesh is that the crack path is restricted to boundaries between the finite elements. In an exaggerated example, the crack would have to make right angles as shown in Figure 5.1b. This would have to be interpreted as an approximation of the real crack path, but will undoubtedly add new numerical problems. As far as the fracture law is concerned, a criterion for crack initiation, as well as for propagation, must be specified. These will be discussed in Sections 5.3.2 through 5.3.4.

A more appropriate discrete crack method is the approach developed by Ingraffea and Saouma[386] and Carpinteri and Valente.[387-388] In this approach, the crack shape dictates the mesh lay out. Thus, in each load step it is determined in which direction the crack will propagate, and subsequently the finite element mesh is adjusted to the new crack path, see Figure 5.1c. Basically, also a crack initiation and propagation criterion are needed. In Ingraffea and Saouma's model, Linear Elastic Fracture Mechanics is used, and the crack tip singularity must be computed and compared to a critical stress intensity factor as will be described in Section 5.3.2. Always, at the beginning of an analysis, a starter notch must be defined. For example, in a simulation of hydraulic fracture propagation, starter cracks with a quarter circular shape and radius $r = 12.5$ mm were used.[389] It is still not completely clear how large this notch should be and what direction it should have. The direction of crack growth depends on minimizing or maximizing another parameter. Valente's programme is based on a process-zone model, i.e., softening is included in the formulation.

All three continuum approaches mentioned above rely on a continuum description of the material outside the crack zone. This behaviour outside the localisation zone is sometimes referred to as "bulk behaviour".[390] In general it is assumed, at least for tensile cracking, that localisation occurs in the post-peak regime. The tensile strength of the concrete is the crack initiation criterion.[5] The bulk behaviour should then be modelled using a pre-peak stress-strain law, as is done in the Fictitious Crack Model that will be presented in Section 5.3.4. For compression, it is not clear at this moment whether the same approach will work. Some researchers claim that localisation in compression starts at the peak,[324] but confusion with the sub-critical cracking before peak, as was discussed in Section 3.2.3, should be avoided.

The alternative representation would be the discretization of the continuum in a lattice of truss or frame elements, i.e., bars or beams. This approach is not new, and dates back to 1941, when Hrennikoff[391] proposed the discretization of a continuum in a truss as a possible solution for solving (or at least approximating) problems in elasticity. The bars should be placed in a certain pattern, and elastic properties, suitable to the type of problem, should be given to the bars, see Figure 5.2.

Hrennikoff showed that for plane-stress problems a planar framework composed of bars having resistance against change of length would suffice. In order

Figure 5.2 Triangular truss used by Hrennikoff.[391] The dashed line shows the deformed mesh when the middle node is moved upwards.

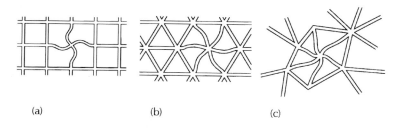

(a)   (b)   (c)

Figure 5.3 (a) Regular square lattice, (b) regular triangular, and (c) random triangular lattice of beam elements.

to come to the exact elasticity solution, the bars should be made infinitesimally small, but this of course would make the problem unsolvable and we would regain the original differential equation. Thus, it should be remembered that the method is an approximation of the exact solution only, but then this remark holds for every numerical solution. It is just a matter of handling things in the right manner and using judgement when a solution is obtained. The disadvantage of Hrennikoff's approach is that his trusses all had a Poisson's ratio of 1/3, which makes the method very appropriate for modelling steel structures. For concrete, with its variable Poisson's ratio (see Section 3.2.3), it seems less appropriate.

As mentioned in the introduction, theoretical physicists (i.e., from the field of statistical mechanics) were the first to model fracture in a lattice.[392-395] Instead of using bars, beam elements are used. In their original approach, Herrmann and colleagues used beams in a regular square configuration as shown in Figure 5.3a. The lattice model of Figure 5.3b was adapted for modelling fracture in concrete and sandstone laboratory-scale specimens.[162,274,396]

The first thing is to replace the regular square lattice by a regular triangular lattice of beam elements as shown in Figure 5.3b. In this way correct values of Poisson's ratios for concrete (and sandstone) are computed. The Poisson's ratio of a lattice depends on the shape of the lattice, the cross-section of the beams, and the elastic constants of the beams. The regular square lattice has a Poisson's ratio $\nu = 0$, which is not very suitable for the class of materials discussed here.

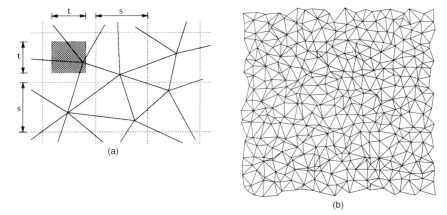

Figure 5.4 Construction of a 2D random lattice (a) and example of generated lattice (b).

For the regular triangular and random triangular lattice more freedom exists to obtain a given combination of overall Young's modulus and Poisson's ratio. In Chapter 6 we will readdress these matters. Some work also has to be done using random triangular lattices.[53,397,398] The random lattice was based on an approach originally proposed by Mourkazel and Herrmann.[399] The random lattice is shown in Figure 5.3c. The construction of the lattice is explained in Figure 5.4. Starting point is a regular square grid of size s. In each box of the grid, a point is selected at random. The randomness can be changed by selecting in each box a sub-box with a smaller size t than the grid box. Next, the three points that are closest to one another are connected with beam elements. The selection of the closest neighbours is done through trials using circles of various diameters.

The 2D random lattice was recently adapted to a general 3D version.[400] Note that the random lattice is not isotropic, i.e., the elastic properties are not the same in all directions, see for example the analysis of a hole in a plate in Section 5.3.2. Schlangen and Garboczi[401] have attempted to homogenise the random lattice of Figure 5.3c by prescribing different cross-sectional areas to different beams. As a result, some beams get a negative cross-section, which is physically not acceptable. The main reason for making the homogenised random lattice is that it has the same properties in all directions. At present, one of my former students, Schlangen, continues to work along these lines. Jirásek and Bažant[402] also published a random particle model which strongly resembles the lattice type models. In regular particle models a strong directional bias was found and they concluded that a randomly generated particle model would be more favourable.

Fracture in a lattice can be simulated quite easily by removing a beam as soon as it has exceeded a certain strength criterion. Various fracture laws can be used and several possibilities are presented in Section 5.4. An extreme case of a lattice is a model representing bonds between atoms. An example of an

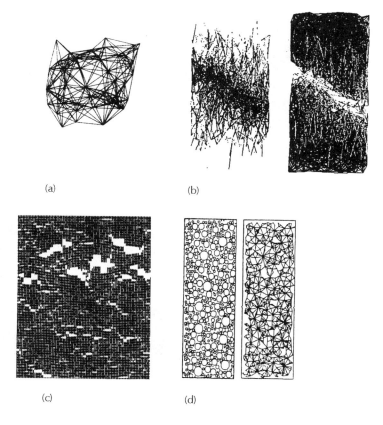

Figure 5.5 Examples of lattice models for simulating fracture in heterogeneous media: (a) Burt and Dougill,[403] (b) Berg and Svensson,[406] (c) Termonia and Meakin,[409] and (d) Bažant et al.[405] (Figures (a) and (d) are reprinted by permission of ASCE, New York. Figure (c) is reprinted from *Nature*, with kind permission from Macmillan Magazines, Ltd.)

atomic simulation is given in Section 5.4 as well. The breaking law should be adjusted of course to resemble the bonding strength between real atoms. In this way the ideal theoretical strength of an atomic lattice can be calculated.

The use of bar or beam models for concrete fracture is certainly not new. Burt and Dougill,[403] Zubelewicz and Bažant,[404] Bažant et al.[405] and Berg and Svensson[406] used, with varying success, lattice type models for simulating fracture in heterogeneous materials. Even applications for modelling fibre composites have been found.[408] In Figure 5.5a-d some examples are shown.

5.2.2 MAPPING THE MATERIAL STRUCTURE

The lattice models described above do not have very much in common with concrete or other disordered materials. Somehow, the material structure has to be included in the analysis. The most direct way is to generate the material

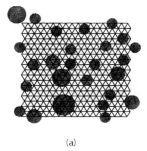

(a)   (b)

Figure 5.6 Regular triangular lattice projected on top of a generated particle structure of concrete (a), and lattice defined by the particle stack (b).

structure with a dedicated computer program, and next to superimpose the lattice model on top of the generated material structure as shown in Figure 5.6a. This is basically the approach followed in the lattice model that was developed in the Stevin Laboratory.[396,410] The computer-generated material structure can also be replaced by an image of a section through a real concrete (or any other material).[411] Subsequently, the same lattice overlay procedure as described above can be followed. Another way of interpreting a lattice is by assuming that the lattice beams or bars are the connections between the centres of neighbouring particles in a grain model as sketched in Figure 5.6b.[405,412-415] If a particle stack is made from particles of identical size, a regular triangular lattice is obtained (in two dimensions). However, if all the particles are of different size, the lattice elements have varying length and a random lattice is obtained. Beranek and Hobbelman[413] worked with a regular lattice and assumed that failure would take place in the interface between the particles, i.e., in the middle of the lattice element connecting the two centres of these particles. Similar assumptions were made by others,[402,205] but instead of using a regular particle stack, a random particle size distribution was chosen. Recently, it was shown that, at least for tensile fracture, it is not very important which approach is chosen.[398]

The first alternative, where a particle structure is projected on top of a lattice, seems most realistic. In this approach the size of the individual lattice beams must be smaller than the size of the smallest aggregate particle included in the model. Thus, contrary to the particle models where the beam length of the lattice equals the particle size in the material, the lattice beams always have a smaller length than the particles, and the solution (probably) depends less on the size of the mesh. As mentioned, first a particle structure of concrete must be generated. As we have seen in Chapter 2 (Figure 2.31), concrete must be considered as a multi-scale material. Essentially, aggregates are embedded in a more or less homogeneous cement matrix. A bond zone with deviating properties exists between the particle and the matrix material as shown in Section 2.2.2. The matrix itself consists of sand embedded in cement paste, and is thus a composite as well, with a distinct softening behaviour as shown in Figure 3.49. Preferentially, the fracture law should be as simple as possible, and an elastic, perfectly brittle law is proposed (Section 5.4). This implies that particles of very small size should be included in the model too. However, this increases the required computational effort and in general a lower limit for the particle size is set. The implications of making this lower limit will be shown in Section 6.3.3. A very dense particle structure of concrete is obtained using a Fuller distribution,

$$p = 100\sqrt{D/d_{max}} \tag{5.1}$$

where p denotes the percentage by weight of particles passing a sieve with aperture diameter D; d_{max} is the size of the largest particle. Walraven[218] derived a cumulative distribution function describing the probability that an arbitrary point in the concrete body, lying in an intersection plane, is located in an intersection circle of diameter $D < D_0$,

$$P_c(D < D_0) = P_k * \left(1.065 D_0^{0.5} d_{max}^{-0.5} - 0.053 D_0^4 d_{max}^{-4} + \right.$$
$$\left. -0.012 D_0^6 d_{max}^{-6} - 0.0045 D_0^8 d_{max}^{-8} + 0.0025 D_0^{10} d_{max}^{-10}\right) \tag{5.2}$$

Using this equation the circle diameters in a cross-section through a concrete sample can be generated. In Figure 5.7 an example of a generated particle structure is shown.

The aggregate volume P_k is 75% of the total concrete volume. The largest particle size is 16 mm, and the particles smaller than 2 mm have not been drawn. In Table 5.1, the number of particles of any given size are listed. Obviously, more small particles are present than large particles. For generating the particle structure of Figure 5.7 first the largest particles were placed randomly in the predefined area. If two particles would overlap, a new trial was made. Next, the smaller particles were placed, and this procedure was contin-

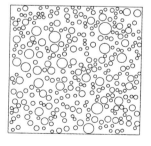

Figure 5.7 Generated particle structure of concrete according to a Fuller distribution and the cumulative distribution function Equation 5.2.[410]

TABLE 5.1

Particle Distribution in the Grain Structure of Figure 5.7[410]

D [mm]	1	2	3	4	5	6	7	8	9	10	11	12	13	14	15	16
Number of particles	1316	226	82	40	23	14	10	7	5	4	3	2	2	1	1	0

ued until, finally, all particles were used. Note, that an additional criterion is introduced, defining the minimum distance between two neighbouring particles A and B. The minimum distance is defined as $1.1 * (D_A + D_B)/2$ according to Hsu.[49] The method can also be used directly in a three-dimensional space. Schlangen[156] shows an example of a 3D particle distribution. A similar method using a finite element programme was followed earlier by Roelfstra and Wittmann.[416]

The next step is to overlay the generated particle structure with a lattice (or a finite element mesh). In the lattice model, simply a lattice of size $\ell < D_{min}/3$ is projected on top of the generated material structure and different properties are defined for beam elements appearing in different phases of the material. For example, in Figure 5.8 a generated regular triangular lattice is projected on top of a material structure. In Figure 5.8b it can be seen how different properties are assigned to the various beam elements. In this example, aggregate properties are assigned to the beams falling inside the circular aggregate particle, matrix properties to those beams falling in the matrix material, and bond properties are assigned to the beams falling over the boundary of the circular aggregate. In Chapter 6 the parameter identification procedure for the lattice model will be outlined.

Roelfstra, Sadouhi, and Wittmann[416] divided (in their 2D model) the generated particle structure in triangular 3-noded elements, see Figure 5.9a. Wang[417] used a similar method. De Schutter and Taerwe[418] generated a mesh based on

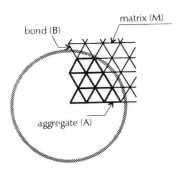

Figure 5.8 Triangular lattice projected on top of a generated particle structure of concrete (a), and assigning different properties to the lattice beams in aggregate, matrix and bond zones (b).[410]

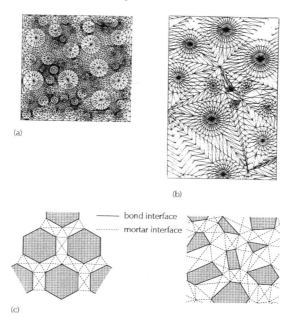

Figure 5.9 Finite element meshes in particle composites: (a) Roelfstra, Sadouhi, and Wittmann,[416] (b) De Schutter and Taerwe,[418] and (c) Vonk.[89] (Figures (a) and (b) are reprinted by kind permission of RILEM, Paris. Figure (c) is reprinted by permission of Dr. Vonk.)

Delaunay triangulation, and subsequently define a gravel particle in each of the triangular areas, taking into account the given grading curve and gravel content (see Figure 5.9b). Another approach was followed by Vonk,[89] who distinguished hexagonal aggregate particles embedded in a matrix, and subsequently deformed the hexagon randomly as shown in Figure 5.9c. Note that the method proposed by Vonk leads to rather angular aggregates, which may have some effect on the properties of the interface between aggregate and matrix. In most models, circular or smooth rounded aggregates are generated. In that case the material structure resembles concrete containing rounded river gravel.

There is an argument that the above approaches do not really reflect the material structure and the associated mechanical mechanisms of concrete. As we have seen in Section 3.2.2, when concrete is loaded in compression, particle interactions cannot be neglected. In particular in biaxial compression, and other extensile load-paths, out-of-plane splitting occurs, supporting the particle stack mechanism of Figure 3.7. In the 2D models presented thus far the aggregates are essentially cylinders: they have the same cross-section over the thickness of the model that is analyzed. Thus, the 3D splitting effect that was discussed in Chapter 3 is excluded a priori. A simple way to overcome this deficiency, at least in the lattice model, it to generate a lattice in 3D consisting of two parallel layers of either a regular or random triangular lattice. An alternative is

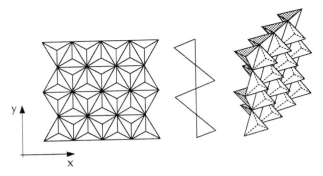

y↑

x

Figure 5.10 Parallel regular triangular lattice suitable for simulating 3D effects.[414]

to build a lattice from tetrahedrons, in which out-of-plane behaviour can be simulated as well. The lattice beams correspond to the edges of the tetrahedrons. In Figure 5.10 a small 4 × 4 lattice constructed from tetrahedrons is shown. In this model out-of-plane failure can be simulated when the lattice is loaded in biaxial compression as will be shown in Section 7.4. Note, however, that in this approach the aggregates are still cylinders, and that the 3D effect is caused by the lattice configuration. Of course, when a regular square lattice would be used still no out-of-plane action would occur.

5.3 FRACTURE LAWS FOR CONTINUUM-BASED MODELS

In this section fracture laws for continuum based models are described. The concrete is modelled as a continuum, and all the relevant material properties must be included in the mathematical formulation. Basically, a general 3D finite element programme would require a 3D constitutive law, with state variables nominal stress and strain, and (if applicable) by specifying the higher order terms and the internal length scale in the higher order continua.[376] We will not delve in all the details of a fully 3D constitutive law. There are excellent text books that describe such models in sufficient detail.[376,419,420] Here, we will discuss only failure criteria, namely the classical strength of materials theory (including three dimensional failure contours), classical Linear Elastic Fracture Mechanics (LEFM), the Dugdale-Barenblatt plastic crack tip zone and the non-linear fracture models designed specifically for concrete such as the Fictitious Crack Model. In Section 5.4 possible fracture laws for lattice type models are presented, but these are linked to the classical strength theories of Section 5.3.1 and to the linear elastic fracture mechanics presented in Section 5.3.2. As an example, some issues regarding micromechanical modelling of concrete fracture are included in Section 5.3.2.4. It is difficult to make a rigorous division between all topics, and it was felt that the micromechanical LEFM example could best be included in this chapter.

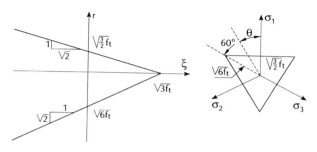

Figure 5.11 Tensile failure criterion of Rankine.

5.3.1 CLASSICAL STRENGTH OF MATERIALS LAW: BRITTLE FAILURE

The simplest way of introducing fracture is through the definition of a limit strength. As shown in Section 3.5, the strength of concrete is dependent on the state of stress (i.e., biaxial or triaxial). In addition there are signs of path dependency, in particular in the post-peak regime.[80] Underlying the strength of the material are micro-mechanical processes, which are, however, neglected when we limit ourselves to the macro-level. In this section we will discuss a number of strength-based macroscopic failure criteria. Most of these criteria are three-dimensional, and some knowledge of principal stresses and invariants is necessary. In the Appendix at the end of the book a short overview of these matters is given.

The simplest possible failure criterion is the maximum tensile strength criterion (Rankine, 1876). The material will fail in a brittle manner as soon as a maximum tensile strength is exceeded,

$$\sigma_1 = f_t \qquad \sigma_2 = f_t \qquad \sigma_3 = f_t \qquad (5.3)$$

where f_t is the tensile strength of the material, which should be determined from a simple uniaxial tension test. In the deviatoric plane (see Appendix for explanation) this criterion forms a triangular failure contour, and in the rendulic (or meridian) plane, two straight lines are found as shown in Figure 5.11. The criterion can be expressed in terms of the invariants.[420] Strict application of this criterion implies that for a uniaxial or multiaxial compression test on concrete, failure will never occur as no tensile stresses can develop. Of course, the situation changes when the exact boundary conditions are taken into account in the analysis. For example, in compression experiments shear acting at the specimen ends may cause tensile stress concentrations, which might eventually lead to fracture. Tensile stress concentrations also appear when the heterogeneity of the material is included, i.e., when the material is modelled at the particle level, as discussed in Sections 5.2.2 and 5.4 and Chapters 6 and 7.

For metals, often maximum shear-stress criteria are used, like those proposed by Tresca in 1864 and Von Mises in 1912. The second and third invariant

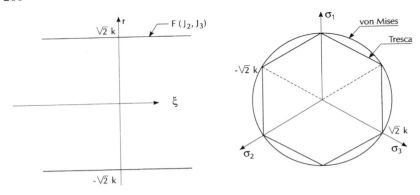

Figure 5.12 Comparison of the Tresca and Von Mises limit surfaces.

of the deviatoric stress tensor (see Appendix) is used as a fracture criterion, which for Von Mises leads to

$$f(J_2) = J_2 - k^2 = 0 \qquad (5.4)$$

where k is the maximum strength under pure shear. The difference between the Trecsa and the Von Mises criterion is that the Tresca criterion has a hexagonal shape in the deviatoric plane, whereas the Von Mises criterion is an ellipse. The difference is caused by the fact that in the Tresca criterion the maximum shear stress is used as a failure criterion, whereas Von Mises takes the maximum octahedral shear stress as criterion. The two failure surfaces are compared in Figure 5.12. We will not go into further detail because for concrete a definite effect of the first invariant I_1 is observed (Section 3.5). The Tresca and Von Mises criteria are quite suitable for metal plasticity.

Because the concrete failure locus depends on I_1, as was shown in Section 3.5 (more specifically, Figures 3.85 through 3.87), the J_2 criterium only is not very suitable for concrete. There are two well known simple failure criteria which include both I_1 and J_2 dependency, namely the Mohr-Coulomb and Drucker-Prager criterion. The Mohr-Coulomb criterion can be interpreted as the limit for the Mohr-circles. Basically, it is assumed that the material will fail when the maximum shear stress is exceeded as shown in Figure 5.13.

The maximum shear stress is dependent on the normal load acting on the shear plane and the angle of internal friction. Thus, the material will fail when

$$|\tau| = c - \sigma \tan \varphi \qquad (5.5)$$

where c is the cohesion and φ is the internal friction angle. Graphically the criterion is shown in Figure 5.14. In the meridian plane the criterion consists of two straight meridians, in the deviatoric plane the locus has a hexagonal shape, and in the biaxial stress plane the contour of Figure 5.14c is obtained.

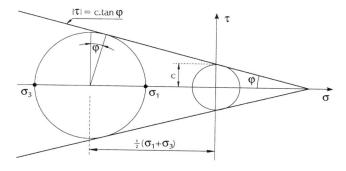

Figure 5.13 Mohr-Coulomb criterion as bounding surface for Mohr's circles. The cohesion c and angle of internal friction φ are indicated.

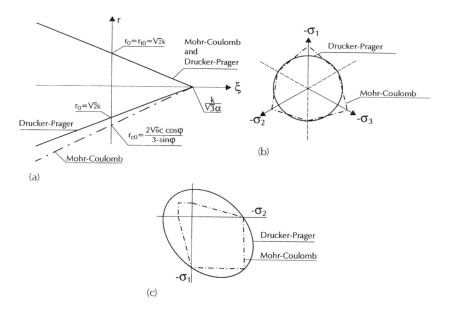

Figure 5.14 Comparison of the Drucker-Prager and Mohr-Coulomb limit surfaces in (a) the rendulic plane, (b) the deviatoric section and (c) the σ_1-σ_2 plane.

The biaxial failure locus resembles quite well the experimental biaxial failure envelope for concrete (see Figure 3.83a). When the Mohr-Coulomb criterion is expressed in terms of principal stress, we obtain

$$\sigma_1 \frac{1 + sin\,\varphi}{2c\,cos\,\varphi} - \sigma_3 \frac{1 - sin\,\varphi}{2c\,cos\,\varphi} = 1 \qquad (5.6)$$

with $\sigma_1 < \sigma_2 < \sigma_3$, and compression is negative.

The Mohr-Coulomb criterion can also be written in terms of the invariants I_1 and J_2, or, alternatively, in cylinder coordinates. Because the original criterion is based on principal stresses, like the Tresca criterion, the formulation in stress invariants becomes rather complicated and is not included here. For further information, see Chen.[420]

The Drucker-Prager criterion is, in this respect, somewhat simpler. It can be regarded as a smooth approximation of the Mohr-Coulomb criterion, and it is conveniently expressed in the invariants I_1 and J_2 following,

$$f(I_1, J_2) = \alpha I_1 + \sqrt{J_2} - k = 0 \tag{5.7}$$

With $\xi = I_1/3$ and $r = \sqrt{2J_2}$, the following expression in cylinder coordinates is obtained,

$$f(\xi, r) = \sqrt{6}\alpha\xi + r - \sqrt{2}k = 0 \tag{5.8}$$

where α and k are constants. When $\alpha = 0$ the I_1-dependency is omitted and Equation 5.7 changes into the Von Mises criterion. There is also a similarity between the Tresca and Mohr-Coulomb criteria. Tresca can simply be retrieved by removing the I_1 dependency in the Mohr-Coulomb criterion. The similarity between the Mohr-Coulomb and Drucker-Prager criteria can be seen in Figure 5.13. By selecting proper values of α and k, the Drucker-Prager circle can, in the deviatoric plane, either fall inside the hexagonal Mohr-Coulomb criterion, or lie outside the hexagon. In Figure 5.14b only the case where the Drucker-Prager circle encloses the Mohr-Coulomb criterion is shown. For this specific case

$$\alpha = \frac{2\sin\varphi}{\sqrt{3}(3 - \sin\varphi)} \qquad k = \frac{6c\cos\varphi}{\sqrt{3}(3 - \sin\varphi)} \tag{5.9}$$

The two criteria coincide then exactly along the compressive meridian ($\theta = \pi/3$).

The failure criteria discussed so far all have in common the fact that they are based on a single hypothesized parameter such as the shear strength of the material. The Tresca and Von Mises criteria are not in agreement with the experimental observations for concrete that were presented in Section 3.5. The Mohr-Coulomb and Drucker-Prager criteria come closer, but both limit surfaces neglect the effect of the intermediate principal stress which is quite important for concrete (see for example Figure 3.88). The fit can be improved by adding more parameters to the model, as has been done by Lade,[421] Ottosen,[422] Willam and Warnke,[250] and Podgorsky.[423]

The Lade criterion was originally developed for soils, but with a simple adjustment it can be applied to cohesive materials as well, and a good fit to compressive data for concrete can be obtained. The major advantage is that the model contains only three parameters. However, the best fit is obtained with

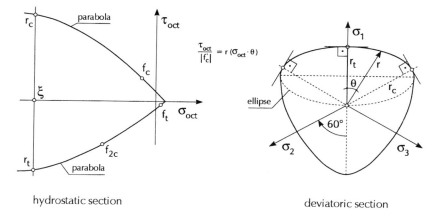

Figure 5.15 Construction of the Willam and Warnke[250] failure surface, after CEB.[424]

the Willam and Warnke five-parameter model. In this model, parabolic equations are used for describing the shape of the tensile and compressive meridians in the rendulic plane, whereas the deviatoric shape of the limit surface is approximated by using ellipses as shown in Figure 5.15. The Willam and Warnke criterion includes all the main characteristics of the failure contour for concrete. The criterion depends on three invariants, I_1, J_2 and J_3, and the surface is convex (which makes it suitable for numerical models, in which in general vertexes should be avoided as much as possible). Moreover, the shape in the deviatoric plane has a threefold symmetry and changes from rounded triangular for low hydrostatic stress to circular at high hydrostatic pressures. The meridians are parabolic and open-ended in the high compression space, but closed in the triaxial tensile regime. The tensile ($\theta = 0$) and compressive ($\theta = \pi/3$) meridians are given by

$$\theta = 0 \qquad r_t = a_0 + a_1 \left(\frac{\sigma_{oct}}{|f_c|} \right) + a_2 \left(\frac{\sigma_{oct}}{|f_c|} \right)^2 \tag{5.10a}$$

$$\theta = \pi/3 \qquad r_c = b_0 + b_1 \left(\frac{\sigma_{oct}}{|f_c|} \right) + b_2 \left(\frac{\sigma_{oct}}{|f_c|} \right)^2 \tag{5.10b}$$

The locus in the deviatoric plane is then constructed following

$$r(\theta) = \frac{2r_c \left(r_c^2 - r_t^2 \right) \cos \theta + r_c \left(2r_t - r_c \right) \sqrt{4 \left(r_c^2 - r_t^2 \right) \cos^2 \theta + 5r_t^2 - 4r_t r_c}}{4 \left(r_c^2 - r_t^2 \right) \cos^2 \theta + \left(r_c - 2r_t \right)^2} \tag{5.11}$$

with

$$\cos \theta = \frac{2\sigma_1 - \sigma_2 - \sigma_3}{\sqrt{2\left[(\sigma_1 - \sigma_2)^2 + (\sigma_2 - \sigma_3)^2 + (\sigma_3 - \sigma_1)^2\right]}} \qquad (5.12)$$

The parameters in the model, a_0, a_1, a_2, b_0, b_1 and b_2 can be determined from five basic experiments, namely a uniaxial compression test (f_c), a uniaxial tensile test (f_t), an equi-biaxial compression test (f_{2c}) and two tests in the high triaxial compressive regime. One of the triaxial data points should be located on the tensile meridian, the other on the compressive meridian. They are needed for fitting the parabolic meridians. The highest point in the triaxial compression regime limits the applicability of the model. The criterion cannot be extrapolated beyond these limits. The six parameters are as follows:

$$a_0 = \frac{2}{3}\alpha_u a_1 - \frac{4}{9}\alpha_u^2 a_2 + \left(\frac{2}{15}\right)^{0.5}\alpha_u$$

$$a_1 = \frac{1}{3}(2\alpha_u - \alpha_z)a_2 + \left(\frac{6}{5}\right)^{0.5}\frac{\alpha_z - \alpha_u}{2\alpha_u + \alpha_z}$$

$$a_2 = \frac{\left(\frac{6}{5}\right)^{0.5}\xi(\alpha_z - \alpha_u) - \left(\frac{6}{5}\right)^{0.5}\alpha_z\alpha_u + \rho_1(2\alpha_u + \alpha_z)}{(2\alpha_u + \alpha_z)\left(\xi^2 - \frac{2}{3}\alpha_u\xi + \frac{1}{3}\alpha_z\xi - \frac{2}{9}\alpha_z\alpha_u\right)}$$

$$b_0 = -b_1\xi_0 - b_2\xi_0^2 \qquad b_1 = \left(\xi + \frac{1}{3}\right)b_2 + \frac{\left(\frac{6}{5}\right)^{0.5} - 3\rho_2}{3\xi - 1}$$

$$b_2 = \frac{\rho_2\left(\xi_0 + \frac{1}{3}\right) - \left(\frac{2}{15}\right)^{0.5}(\xi_0 + \xi)}{(\xi + \xi_0)\left(\xi - \frac{1}{3}\right)\left(\xi_0 + \frac{1}{3}\right)} \qquad (5.13)$$

where

$$\xi_0 = \frac{-a_1 - \sqrt{a_1^2 - 4a_0 a_2}}{2a_2} \qquad (5.14)$$

As mentioned, for the determination of these six parameters, five material "constants" are needed: f_c, f_t, f_{2c}, and a combination of τ_{oct} ($\theta = 0$), τ_{oct} ($\theta = \pi/3$)

for one selected value of σ_{oct}. This value of σ_{oct} sets a limit to the application of the model and may not be exceeded. The model of Willam and Warnke was fitted to data of Schickert and Winkler.[251] The following five material constants were found

$$\alpha_z = \frac{f_t}{|f_c|} = 0.1 \qquad \alpha_u = \frac{|f_{2c}|}{|f_c|} = 1.16$$

$$\xi = \frac{-\sigma_{oct}}{|f_c|} = 1.5 \qquad (5.15)$$

$$\rho_1 = \left(\frac{3}{5}\right)\frac{|\tau_{oct}|}{|f_c|} \quad (for\ \theta = 0) \qquad \rho_2 = \left(\frac{3}{5}\right)\frac{|\tau_{oct}|}{|f_c|} \quad (for\ \theta = \pi/3)$$

In Figure 5.16 a comparison is shown between the data of Schickert and Winkler and the Willam and Warnke model. Using the same model parameters (Equation 5.15), a comparison was made with data from a different set of truly triaxial tests.[80] This latter comparison was shown in Figure 3.88 and was quite favourable. It must be concluded that the Willam and Warnke model can describe quite adequately the three-dimensional failure locus of concrete. Again by selecting specific values for the a_i and b_i parameters, some of the simpler models may be retrieved. For example when $a_0 = b_0$, and $a_1 = b_1 = a_2 = b_2 = 0$, the Von Mises criterion (Equation 5.4) is obtained. When $a_0 = b_0$, $a_1 = b_1$ and $a_2 = b_2 = 0$ is selected, the criterion becomes equivalent to the Drucker-Prager model, Equation 5.7. Both the Willam and Warnke model, and the four

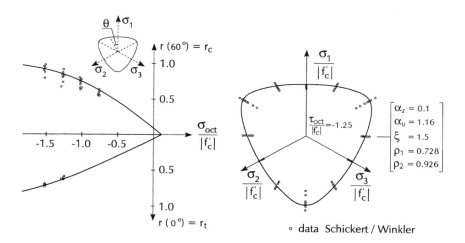

Figure 5.16 Comparison of the Willam and Warnke[250] failure surface with experimental data of Schickert and Winkler,[251] after CEB.[424]

parameter Ottosen[422] model (which is not discussed here) were recommended by the CEB.[424]

The disadvantage of the failure criteria that are based on a close fit of experimentally determined limit surfaces is that the underlying physics are completely neglected. As we have seen in Section 3.5, fracture mechanisms change under different states of multiaxial stress. This is not reflected in the above descriptions. For example, fracture response along the tensile meridian seems more brittle for an extended range of confining pressure than for stress states along the compressive meridian. Failure along the tensile meridian ($\theta = 0$) occurs in a single dominant direction, whereas along the compressive meridian failure occurs in the two least confined directions and is more ductile. This should have an effect on the shape of the limit surface. Using micromechanics models, such behaviour should be predictable and the shape of the limit surface could be better understood using such tools. In Section 7.4 an example is shown of a simulation of the biaxial failure contour of concrete using a lattice type model. Specifically, the ability of such models to simulate out-of-plane fracture under biaxial compression is examined. Biaxial compression is a typical example of extensile failure, see Figure A6 in the appendix.

5.3.2 LINEAR ELASTIC FRACTURE MECHANICS

Fracture mechanics is a branch of mechanics that evolved when it was recognised that imperfections in the material structure have an effect on the strength of the material or structure. Griffith[1] was the first to develop a theory for brittle fracture that was based on crack growth. The role of local imperfections on the stress-field may be quite obvious already from an analysis of the stress distribution around a circular hole (with radius $r = a$) in a plate of infinite size loaded in tension (S) at infinity. The solution is quite well known and can be found in Timoshenko and Goodier.[425] In Figure 5.17, the plate with a hole is shown, as well as the coordinate system. In the same figure the minimum and maximum stresses at the edge of the hole are indicated as well. The elastic solution is as follows:

$$\sigma_r = \frac{S}{2}\left(1 - \frac{a^2}{r^2}\right) + \frac{S}{2}\left(1 + 3\frac{a^4}{r^4} - 4\frac{a^2}{r^2}\right)\cos\theta \tag{5.16a}$$

$$\sigma_\theta = \frac{S}{2}\left(1 + \frac{a^2}{r^2}\right) - \frac{S}{2}\left(1 + 3\frac{a^4}{r^4}\right)\cos 2\theta \tag{5.16b}$$

$$\tau_{r\theta} = -\frac{S}{2}\left(1 - 3\frac{a^4}{r^4} + 2\frac{a^2}{r^2}\right)\sin 2\theta \tag{5.16c}$$

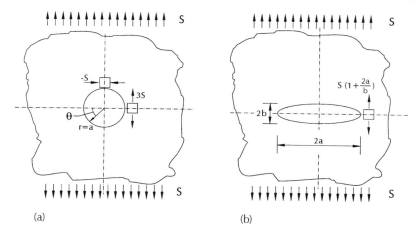

Figure 5.17 Stresses around a circular hole in an infinite plate subjected to uniaxial tension (a). In (b) the situation for an elliptical hole is shown.

At the edge of the hole, $r = a$, and Equation 5.16 reduces to

$$\sigma_r = \tau_{r\theta} = 0$$

$$\sigma_\theta = S - 2S \cos 2\theta \qquad (5.17)$$

The maximum circumferential stress ($\sigma_{\theta,max} = 3S$) is found for $\theta = \pi/2$ and $\theta = 3\pi/2$; the minimum circumferential stress is compressive and appears above and below the hole, $\sigma_{\theta,min} = -S$ (for $\theta = 0$ and π).

The state of stress around the hole can be approximated by means of the lattice model described in Section 5.2.1. By using a random lattice based on a grid size $s = 1$ mm (see Figure 5.4), a plate of size 75 * 75 mm² and unit thickness, containing a hole with a diameter of 12.5 mm was analyzed. The plate was loaded in uniaxial tension as shown in Figure 5.18a. The nodes of the random lattice at the edge of the hole were translated such that they fell exactly on the circumference of a circle of diameter $2r = 12.5$ mm. In Figure 5.18b, the elastic solution of Equation 5.17 is compared with the stresses around the edge of the hole computed with the random lattice. The agreement is quite satisfactory. Of course there are deviations because the material was discretized, but the approximation should improve when the grid size s is reduced. As mentioned in Section 5.2.1, the stiffness of a random lattice may deviate depending on the direction of loading. For the random lattice of Figure 5.18a, the vertical stiffness was 33 GPa, and the horizontal stiffness was 32.5 GPa. Again, the differences diminish as the grid size decreases.

For an elliptical hole (with semi-axes a and b) in an infinite plate loaded in tension, the maximum circumferential tensile stress is equal to $\sigma_{\theta,max} = S *$

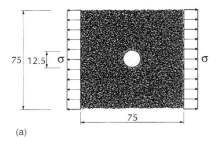

(a)

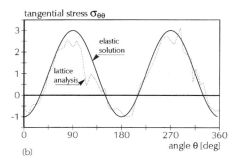

(b)

Figure 5.18 Plate with circular hole loaded in uniaxial tension. The plate has been modelled using a random lattice with $s = 1$ mm (a). In (b) the tangential stress σ_θ around the hole from the lattice analysis is compared to the elastic solution Equation 5.17.

$(1 + 2a/b)$, see Figure 5.17b. When b approaches zero, a slit-like imperfection is obtained, and the maximum tensile circumferential stress goes to infinity. Note that any inclusion with a differential stiffness as the surrounding matrix may act as stress concentrator in the way described above. Other stress-concentrators in real materials are pores and pre-existing cracks. In the case of slit-like (ellipsoidal) cracks, the normal elasticity solution leads to a stress singularity at the tips. For this case fracture mechanics must be applied.

In Linear Elastic Fracture Mechanics (LEFM) two approaches may be followed, namely the Griffith theory where a critical energy release rate is used as a propagation criterion for slit-like cracks, or the Irwin approach, which is based on a critical stress intensity factor as crack propagation criterion. The stress intensity factor approach is more appealing as it is based on conventional stress analysis. There are several text books where both theories are worked out in great detail, e.g., Broek,[426] Ewalds and Wanhill,[427] and Anderson.[428] Most of these books are written for metal fracture; no general text book exists for concrete. It is also not our intention to write a detailed textbook on fracture mechanics, but rather to show some of the concepts and to apply these for elucidating fracture mechanisms in concrete. Because Irwin's theory is very much related to the elastic stress analysis shown above, it will be treated first,

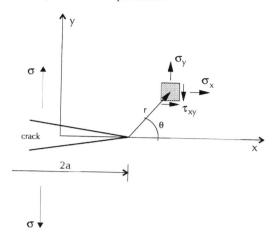

Figure 5.19 Definition of stresses in the vicinity of a crack tip.

followed by a discussion on the Griffith theory. We will give an example of crack growth in a tensile bar and, as another example, try to elucidate the shape of the compressive stress-strain curve on the basis of micro-mechanical arguments.

5.3.2.1 Irwin's model

Irwin proposed a model based on near stress distribution around a planar slit of length $2a$ as shown in Figure 5.19. When the plate is loaded in uniform tension (plane-stress), the stresses σ_x, σ_y and τ_{xy} can be expressed as follows,

$$\sigma_x = \frac{\sigma\sqrt{\pi a}}{\sqrt{2\pi r}} \cos\frac{\theta}{2}\left(1 - \sin\frac{\theta}{2}\sin\frac{3\theta}{2}\right)$$

$$\sigma_y = \frac{\sigma\sqrt{\pi a}}{\sqrt{2\pi r}} \cos\frac{\theta}{2}\left(1 + \sin\frac{\theta}{2}\sin\frac{3\theta}{2}\right) \qquad (5.18)$$

$$\tau_{xy} = \frac{\sigma\sqrt{\pi a}}{\sqrt{2\pi r}} \sin\frac{\theta}{2}\cos\frac{\theta}{2}\cos\frac{3\theta}{2}$$

All three stresses have a $r^{-1/2}$ singularity, i.e., they tend to infinity near the crack tip. The functions can all be seen as a product of a geometrical position $1/\sqrt{2\pi r}\,.f(r)$ and a factor $\sigma\sqrt{\pi a}$. This latter factor is the product of the far field stress and the semi-crack length a, and determines the magnitude of the stress concentration at the crack tip. This factor is called the stress intensity factor

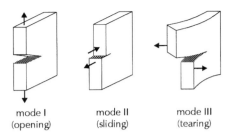

mode I   mode II   mode III
(opening)   (sliding)   (tearing)

Figure 5.20 Three principal crack modes: (a) mode I or tensile opening mode, (b) Mode II or in-plane shear mode and (c) mode III or out-of-plane shear.

$$K_{Ic} = \sigma\sqrt{\pi a} \tag{5.19}$$

in mode I (tension). The above is valid in plane-stress only. For plane-strain we must also consider $\sigma_z = v(\sigma_x + \sigma_y)$, where v is the Poisson's ratio.

Thus, in general form Equation 5.18 can be written as

$$\sigma_{ij} = \frac{K_I}{\sqrt{2\pi r}} f_{ij}(\theta) \tag{5.20}$$

This set of equations defines the near-tip stress field under mode I loading. Essentially it is the first term of a series expansion and is valid only in the vicinity of the crack tip. Next to the mode I (tensile) loading, the crack tip can undergo two other loadings, namely mode II (in plane shear) and mode III (out-of-plane shear). These three crack modes are shown schematically in Figure 5.20. In principle the opening mode I is the most important mode found in practice. Next to these basic fracture modes, mixed combinations of the three opening and sliding modes may occur too. For example, mixed mode I and II implies simultaneous crack opening and in-plane sliding. The respective stress intensity factors of the three basic modes are designated K_I, K_{II} and K_{III}. For concrete, mode I seems most important too, in particular when the material is modelled at the meso-level. At the macro-level, LEFM is not unrestrictedly applicable.

The crack is now assumed to propagate when the stress intensity factor K_I exceeds a critical value K_{Ic}, which for a given material, must be derived from experiments. In the sixties, as well as more recently,[315] there were some attempts to measure K_{Ic} for mortar and concrete.[429] In the latter paper, values of K_{Ic} ranging from 1.31 MPa $\sqrt{m}$ to 2.21 MPa $\sqrt{m}$ are mentioned for different concrete qualities. The main problem of concrete is, as may be obvious from Chapter 3, that a single crack will never propagate but it will always be accompanied by micro-cracking, crack branching and crack bridging (Figure 3.46 is a very clear illustration). For very large specimens (or strucures) it is normally assumed that the crack branching and bridging effects have a minor effect only and a good approximation of K_{Ic} can be obtained when the ratio

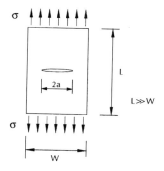

Figure 5.21 Centre-cracked plate.

between characteristic specimen size and characteristic material dimension (for example the size of the largest aggregate) is sufficiently large. But here the arguments brought forward in the introduction to Chapter 3 must be kept in mind. Alternatively, one might include all the branching and bridging effects in the analysis, and use LEFM as presented here at the micro-level.[164] If we do not want to go in so much detail and assume the crack to be a single entity, we must use non-linear fracture theories like the Fictitious Crack Model[5] or the crack band model.[140] We will return to these matters in Section 5.3.4.

The above solution is valid only when a single crack is contained in a plate of infinite size. When critical stress intensity factors must be measured in the laboratory, a first problem is experienced. The specimen is always of finite size, and Equation 5.20 must be adjusted. A general form for a modified formulation is

$$K_I = \sigma\sqrt{\pi a}\ Y\!\left(\frac{a}{W}\right) \tag{5.21}$$

where the geometrical factor $Y(a/W)$ must be determined by stress analysis. W is the finite specimen width in which the crack of size $2a$ is contained. Only a few closed-form analytical solutions exist for specimens of finite size. Most solutions are numerical approximations. For example, for the centre-cracked specimen of Figure 5.21,

$$Y\!\left(\frac{a}{W}\right) = 1 + 0.256\left(\frac{a}{W}\right) - 1.152\left(\frac{a}{W}\right)^2 + 12.200\left(\frac{a}{W}\right)^3 \tag{5.22}$$

This approximation by a power series has an accuracy of 0.5% for $a/W \le 0.35$.

Of interest further on in this chapter are the geometrical functions $Y(a/W)$ for a Single-Edge-Notched (SEN) plate loaded in uniaxial tension between fixed grips or between freely rotating boundaries, see Figure 5.22. For the tensile SEN plate between rotating boundaries, the geometrical factor is given in the stress intensity handbook of Tada et al.,[430]

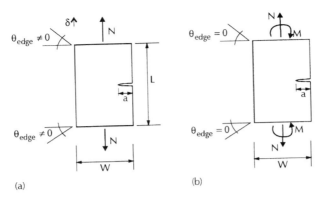

Figure 5.22 SEN plate between freely rotating boundaries ($\theta_{edge} \neq 0$, (a)) and between fixed boundaries ($\theta_{edge} = 0$, (b)).

$$Y\left(\frac{a}{W}\right) = \sqrt{\pi}\left(\frac{2W}{\pi a}\tan\frac{\pi a}{2W}\right)^{1/2}\left[\frac{0.752 + 2.02\left(\dfrac{a}{W}\right) + 0.37\left(1 - \sin\dfrac{\pi a}{2W}\right)^3}{\cos\left(\dfrac{\pi a}{2W}\right)}\right] \quad (5.23)$$

This equation is valid for all relative crack lengths (a/W) with an accuracy better than 0.5%. The other stress intensity factor that we will need is for the SEN plate between fixed boundaries as shown in Figure 5.22b, i.e., $\theta_{edge} = 0$. In this case an eccentricity is introduced by the growing crack, and a bending moment will develop in the specimen as indicated in Figure 5.22b.

The determination of the stress intensity factor can only be done through numerical integration, and was done by Marchand et al.[431] The solution is quite complicated and reads as follows:

$$Y(\xi) = \frac{F_1(\xi)\left(1 - \dfrac{F_2(\xi)C_{12}(\xi)}{F_1(\xi)\left[12L/W + C_{22}(\xi)\right]}\right)}{\left(1 + \dfrac{W}{L}\left[C_{11}(\xi) - \dfrac{C_{12}(\xi)^2}{\left(12L/W + C_{22}(\xi)\right)}\right]\right)} \quad (5.24)$$

where $\xi = a/W$, and F_1 and F_2 are Tada's functions for normal load (Equation 5.22) and pure bending (Equation 5.25), respectively. The C_{ij} are dimensionless crack compliances that contain all the information regarding the M, N, δ, θ relations. The C_{ij} are computed using numerical integration. For these details, the reader is referred to the original paper.[431] Here the results are displayed graphically in Figure 5.23.

A basic assumption in the analysis is that the length L of the plate should be larger than the decay length for local stress disturbances. This implies that

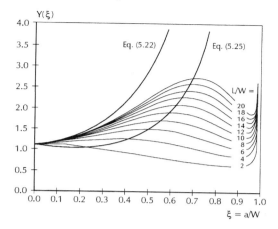

Figure 5.23 Geometrical factor (Equation 5.24) solution after Marchand et al.[431] (Reprinted by kind permission of Kluwer Academic Publishers.)

for small cracks ($a/W \ll 1$) the analysis is valid when $L > 1$. On the other hand, for deep cracks ($a \approx W$), the length of the plate should be a multiple of its width ($L > nW$ with $n > 1$). The (empirical) F_2 function is also listed by Tada et al.,[430] and reads as follows:

$$F_2 = Y\left(\frac{a}{W}\right) = \sqrt{\frac{2W}{\pi a} \tan \frac{\pi a}{2W}} \left(\frac{0.923 + 0.199 \left(1 - \sin \frac{\pi a}{2W}\right)^4}{\cos \frac{\pi a}{2W}} \right) \quad (5.25)$$

again with an accuracy better than 0.5% for any a/W. There are several stress intensity handbooks such as the book by Tada et al.[430] that was mentioned before, and the one by Murakami.[432] For many different two- and three-dimensional geometries, stress intensity factors are listed. For several geometries more than one solution exists and in general the range of applicability of a certain solution, as well as the way in which it was determined are mentioned in these handbooks.

5.3.2.2 Griffith theory

As mentioned, Griffith[1] made the important observation that the strength of glass rods is limited due to the presence of imperfections in the material. He then tried to derive a crack propagation criterion based on a balance between the elastic energy release due to crack growth and the amount of energy needed to create the new crack surface.

The amount of elastic energy released was computed from elastic unloading taking place above and below a slit. The area where energy release would occur

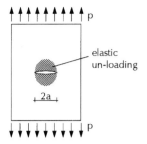

Figure 5.24 Energy release in a uniformly stretched plate with a slit of length 2a.

was assumed to be circular with radius a (semi-crack length), and the energy release was equal to

$$W_e = \pi a^2 \frac{\sigma^2}{E}$$ (5.26)

where σ is the far field stress and E is the Young's modulus of the material. The amount of energy needed to create a crack of length $2a$ is equal to

$$W_S = 2a\gamma$$ (5.27)

where γ is the energy needed to create a unit crack surface. The situation is clarified in Figure 5.24, where the slit-like crack in a uniformly stretched plate between fixed grips is shown. The shaded area is the area where the elastic energy release occurs due to the presence of the crack with length $2a$. Crack propagation would occur when the rate of energy release dW_e/da would surpass the increase of surface energy dW_s/da following

$$\frac{dW_e}{da} \geq \frac{dW_s}{da} = 2\gamma$$ (5.28)

Solving this equation leads to

$$\sigma = \left[\frac{E\gamma}{\pi a} \right]^{1/2}$$ (5.29)

In other words, a crack will propagate in tension when the external stress σ exceeds the square root of the Young's modulus times the surface energy divided by the crack length a. The difficulty is in assessing the surface energy γ. In literature one often finds G_{Ic} instead of γ. G_{Ic} is the critical energy release rate. It is equal to the surface energy consumed in the fracture process.

There is a simple relation between Irwin's critical stress intensity factor and Griffith's energy release rate: from a comparison of Equation 5.19 for the critical stress intensity factor K_{Ic} and the above equation for G_{Ic}, one can observe that

$$K_{Ic} = \sqrt{E G_{Ic}}$$ (5.30)

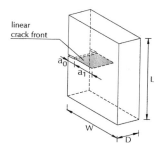

Figure 5.25 Single-Edge-Notched tensile specimen, definition of dimensions.

This relation is only valid in plane crack growth, i.e., for mixed-mode cracks this relation is, in general, not valid.

5.3.2.3 Example of crack growth analysis

A simple, but quite illuminating, example of the use of LEFM in crack growth studies is the effect of boundary rotations on crack propagation in a SEN tensile bar. In Section 7.2 this example is worked out by means of numerical simulations. It is of importance for understanding displacement-controlled uniaxial tensile tests. In displacement-controlled tensile fracture experiments often double-edge-notched specimens are selected to trigger crack growth at a known location. This was discussed in Chapters 3 and 4, where it also was shown that even in a double-edge-notched specimen, crack growth starts from one of the two notches. The main reason is the heterogeneity of the material. Thus, an asymmetric situation develops. Let us therefore assume a SEN tensile specimen as shown in Figure 5.25. In this situation the location of crack growth is uniquely defined. The specimen has length L, width W and thickness D. For D we assume unity throughout the following analysis. A single notch has formed in the specimen due to earlier load histories and extends with a linear front over a depth a_0. Furthermore, it is assumed that the crack front remains straight during crack propagation. In concrete, the initial notch a_0 could for example be explained from micro-crack processes in the pre-peak regime of the stress-strain diagram,[184] see also Sections 3.2 and 5.3.2.4. The stress intensity factor for a specimen with a crack of length a_0 is equal to

$$K_{I,a_0} = Y\left(\frac{a_0}{W}\right)\sigma_0\sqrt{a_0} \tag{5.31}$$

where σ_0 is the nominal, externally applied stress. Now let us assume that K_I has reached the critical value K_{Ic} when the crack length is equal to σ_0.

At this moment the maximum stress is carried by the specimen, and we can write

$$\sigma_p = \sigma_0 = \frac{K_{I,a_0}}{Y\left(\frac{a_0}{W}\right)\sqrt{a_0}} = \frac{K_{Ic}}{Y\left(\frac{a_0}{W}\right)\sqrt{a_0}} \tag{5.32}$$

We assume that stable, controlled crack propagation may occur, for example, by performing a displacement-controlled test as described in Chapter 4, and that K_{Ic} is constant during crack propagation. This is a very simplified and ad-hoc assumption, but for the purpose of this example it will do.

When the crack has extended to length a_1, the stress σ_1 that can be carried by the specimen is equal to

$$\sigma_1 = \frac{K_{Ic}}{Y\left(\dfrac{a_1}{W}\right)\sqrt{a_1}} \tag{5.33}$$

By using Equations 5.32 and 5.33, we can calculate the ratio σ_1/σ_p without explicitly knowing the value of K_{Ic}, following

$$\frac{\sigma_1}{\sigma_p} = \frac{Y\left(\dfrac{a_0}{W}\right)}{Y\left(\dfrac{a_1}{W}\right)} \cdot \left[\frac{\left(\dfrac{a_0}{W}\right)}{\left(\dfrac{a_1}{W}\right)}\right]^{1/2} \tag{5.34}$$

This equation represents the residual carrying capacity of a prismatic specimen containing a single crack, growing from one side of the specimen to the other side. According to the assumptions, Equation 5.34 can be regarded as the softening branch of the specimen. As can be seen from Figure 4.63, it is quite realistic to assume crack propagation through the specimen's cross-section during softening. The only things that seem to be neglected are the micro-crack processes leading to the initial notch a_0, and the bridging effects over the macro-crack as demonstrated in Section 3.3.1. The main issue that is elucidated here is the effect of boundary rotations on the shape of the tensile softening curve. By using Tada's geometrical factors for a SEN specimen loaded in tension between freely rotating platens (Equation 5.23), and the solution of Marchand et al. for the case with non-rotating platens (Equation 5.24), and substituting these formula in Equation 5.34, we can compute the effect of boundary rotations. In Figure 5.26 the results are shown.

In Figure 5.26a, the residual carrying capacity σ_1/σ_p is shown for a SEN plate with rotating boundaries; in Figure 5.26b, the same plate is shown but now with fixed boundaries. The analysis for freely rotating boundaries was carried out for different initial notch sizes a_0. The results show that an increasingly steep descending branch is found for decreasing initial notch sizes. This can be interpreted as follows. Under the assumption that the initial notch is formed by pre-peak micro-crack processes (as will be discussed in Section 5.3.2.4), larger or smaller micro-cracks may develop depending on the structure of the material of which the specimen is made. Clearly, larger micro-cracks must develop in coarsely grained materials before they can overcome the increased resistance caused by the presence of the larger aggregate par-

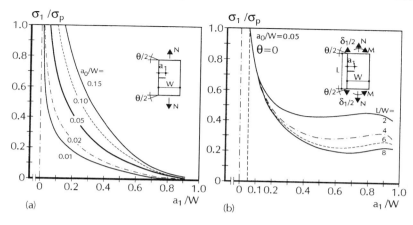

Figure 5.26 Relative stress σ_l/σ_p versus crack length a/W diagram for a SEN plate loaded in tension between freely rotating boundaries (a) and between fixed boundaries (b).[184]

ticles. For finely grained materials a smaller initial notch would be needed to trigger fatal macroscopic crack growth. As we have seen in Section 3.3.2 and as will be elaborated in Section 7.2, a significant influence of material structure on tensile softening exists. More specifically, decreasing the heterogeneity of the specimen leads to a more steep softening curve, which is an agreement with the above reasoning. The extreme case is softening of hardened cement paste. LEFM seems applicable to this material as we will see below.

In Figure 5.26b, the σ_l/σ_p-a/W diagrams are shown for a prism loaded between non-rotating boundaries, i.e., $\theta = 0$. The situation is now more complicated. Because the specimen boundaries are forced to remain parallel during crack propagation, the neutral axis of the intact cross-section shifts away from the centre of the specimen, and a bending moment develops at the specimen ends. Initially, the bending moment opposes further crack propagation, viz. it tends to close the crack. As a consequence, stable crack growth may occur during some interval and a plateau is found in the diagram of Figure 5.26b. The plateau increases during a short interval, indicating that a higher external load must be applied to propagate the crack. For hardened cement paste (hcp) a temporary increase of the softening branch has been observed as shown in Figure 3.58. This means that LEFM is applicable for hcp. Note, however, that crack initiation is not included in the present analysis. For mortar and concrete an increase of stress in the softening branch has never been observed. A bump is found, but never has the external stress been found to increase.[184]

An important observation is that the rotation not only depends on the boundary condition $\theta = 0$, but also on the flexural stiffness of the prism. From Figure 5.26b this can be seen. The shorter the specimen (low L/W ratios), the higher the plateau stress becomes. The results presented in Figure 4.41 confirm this tendency. Of course, the present analysis is a gross simplification of the real situation. As mentioned before, micro-crack processes and crack face bridging, caused by the heterogeneity of the material, have been neglected in the analysis. Using a non-linear fracture model for concrete, some of these simplifications are omitted and, as we will see in Section 7.2, numerical micromechanics can be used to further elucidate the various phenomena. Another approach, which is not worked out here, is the effective crack model. In such a model, an average crack and an effective fracture toughness K_{Ic} are defined for analyzing fracture behaviour of structures. Thus, all non-linear effects are lumped into these parameters. The effective crack length is a weighted average, and all effects from curvi-linear crack fronts, bridging and micro-crack growth are included.[314,315]

5.3.2.4 Application of LEFM at the meso-level

As may have been clear from the above, linear elastic fracture mechanics is very well suited for analyzing specimens containing a single discrete flaw. Mechanisms can be elucidated without having to know the most important material parameter K_{Ic}. The determination of K_{Ic} for concrete is not straightforward, mainly because we are not dealing with single isolated cracks, but rather with distributed micro-cracking, branching, bridging, pore toughening and frictional interlock, which all seem to play a more or less important role. If in spite of all this, one would like to apply LEFM to concrete, either the length and fracture toughness must be adjusted as in the effective crack models, or the level of observation must be changed as argued in Section 5.1. In other words, a detailed description of the material structure and micro-crack events must be included in the model. The task becomes tedious as we also have to describe crack interactions.

Many people have worked on analytical micromechanics models for describing the macroscopic stress-strain behaviour of concrete and rock.[265,433-440] In a numerical environment, some of the interactions are easy to handle, as shown by Wang and Huet[164] and Wang,[417] who used a particle model with randomly distributed micro-cracks to compute the stress-deformation response and cracking behaviour of concrete. The crack propagation criterion they used was a K_{Ic}-criterion.

Without entering into too much detail, we can "explain" the curvatures in the compressive stress-strain curve from micromechanics considerations. The same line of reasoning can be used for explaining pre-peak curvatures in tensile experiments. As discussed in some depth in Section 3.2.2, concrete contains distributed micro-cracks, even before any mechanical load has been applied. These micro-cracks are small, and the amount of energy dissipated by a single micro-crack can be computed using the following approach. First we must assume that LEFM is valid at the particle level of concrete (the meso-level). A slice of unit thickness is considered containing N micro-cracks with semi-major axis a_i ($i = 1,...,N$). Under a local tensile stress $\sigma_i(p)$ (which depends on the external applied nominal stress p), the cracks grow from a_i to $a_i+\Delta a_i$ (Δa_i

$<< a_i$), and assuming a local geometrical factor Y_i for each micro-crack, we can compute the amount of energy released as follows,

$$\Delta W_{lc} = \sum_{i=1}^{N} \left(2a_i \cdot \Delta a_1 + \left(\Delta a_1\right)^2\right) \cdot \frac{\pi \sigma_1 \left(p\right)^2}{E} \cdot Y_i \qquad (5.35)$$

where ΔW_{lc} denotes the energy released upon reaching the external stress level p, and E is the average Young's modulus of the composite material. The geometrical factor Y_i depends on the local crack geometry, interactions with neighbouring cracks, and on the boundary conditions and specimen size. In the analysis we assume that the average Young's modulus of the material remains constant during cracking. Thus, the only cause for stiffness degradation in the material is crack nucleation and growth. When a material with the same elastic stiffness as the heterogeneous material is considered, but now without cracks, the amount of energy stored in a specimen with volume $V_s = H.W.1$, is equal to

$$W_s = \frac{1}{2} E \varepsilon_0^2 V_s = \frac{1}{2} \frac{\left(p'\right)^2}{E} \cdot V_s \qquad (5.36)$$

In Figure 5.27, ε_0, p and p' are explained. It should be mentioned that no distinction has been made between different crack types in the present analysis. In concrete technology, cracks are often distinguished according to the material or interface where they appear.[34,441] Here, the cracks are considered merely as an efficient energy-release mechanism, and we are concerned about the deviation from linearity of the stress-strain curve when cracks appear and/or propagate. The energy released due to extension of individual micro-cracks is small compared to the energy contained in the entire specimen. The geometrical factor in Equation 5.35 depends on the crack-length increment Δa_i, and will be

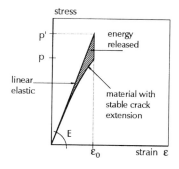

Figure 5.27 Energy release due to stable propagation of individual micro-cracks.[253]

several orders of magnitude smaller than the geometrical factor in Equation 5.36, viz. the specimen volume V_s. The total number of cracks should increase enormously to obtain an energy release in the same order of magnitude as the elastic energy stored in the specimen. Experiments have shown a very moderate increase of micro-cracks at low levels of external stress only (i.e., for $p <$ $0.3\sigma_p$).[34,441-443] Also fluorescent spray tests reveal only larger quantities of micro-cracks in the post-peak regime,[80] and not at very early stages of loading. However, the resolution might not have been sufficient in this technique. Nevertheless, an additional mechanism seems essential to explain the large curvatures of the pre-peak stress-strain curve of concrete in compression, not to speak of the post-peak behaviour. If the analysis of the previous section is valid, post-peak softening behaviour can be explained from the growth of large localised fractures. An intermediate mechanism is needed to explain the transition from moderate energy release due to isolated micro-cracking, to energy release caused by localised macro-cracks. An explanation for the transition might be a process called crack joining. Following a similar line of reasoning as above for the amount of energy released from the growth of individual micro-cracks, it is possible to investigate what would change when two micro-cracks join to form a larger combined crack. First, the amount of energy released from the growth of two identical isolated micro-cracks can be computed by setting $N = 2$, $\sigma_i(p) = \sigma(p)$ and $Y_i = Y$ in Equation 5.35:

$$\Delta W_{I,c} = 2 \cdot \left(2a \cdot \Delta a + (\Delta a)^2\right) \cdot \frac{\pi \sigma(p)^2}{E} \cdot Y \qquad (5.37)$$

When the crack tips are close to one another, joining of the cracks becomes possible (Figure 5.28), and the amount of energy released in the joining event can be expressed as

$$\Delta W_{I,c}^j = (2a + 2\Delta a)^2 \cdot \frac{\pi \sigma(p)^2}{E} \cdot Y^j - 2a^2 \cdot \frac{\pi \sigma(p)^2}{E} \cdot Y \qquad (5.38)$$

The geometrical factors Y and Y^j (the superscript j indicates joining) depend on local conditions and are unknown, but it seems reasonable that they differ only slightly as long as the crack length remains smaller than a characteristic size of the specimen. Under the assumption $Y = Y^j$, the surplus of energy release from the joining event can be computed by subtracting Equation 5.37 from Equation 5.38:

$$\Delta W_{I,c}^j - \Delta W_{I,c} = \left(2a^2 + 4a \cdot \Delta a + 2(\Delta a)^2\right) \cdot \frac{\pi \sigma(p)^2}{E} \cdot Y \qquad (5.39)$$

Because $a \gg \Delta a$, the energy release is now governed by the crack length a, instead of the crack length increment Δa. The crack length increment was the

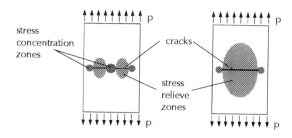

Figure 5.28 Increase of size of stress-relieved zones caused by joining of two separate micro-cracks.[253]

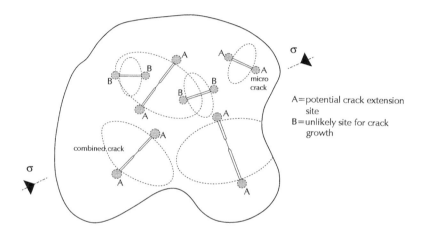

Figure 5.29 Development of preferential crack propagation sites caused by the increased size of stress-relieved zones.[253]

governing factor in the case of isolated micro-crack propagation. Thus, in terms of macroscopic stress-strain behaviour, crack joining could be responsible for a significant increase of curvature in the pre-peak regime. Joining events may occur over the complete specimen's volume. However, if at a certain location crack joining occurred, it is not very likely that a second event will occur nearby. The large increase of stress-relieved zones will undoubtedly affect the loading conditions of neighbouring cracks as shown in Figure 5.29. Thus, when crack joining events take place in various parts of the specimen volume, the number of sites where active crack growth might occur decreases rapidly, until a single location remains from which a localised macro-crack propagates.

The crack process sketched here might indeed be true. However, the crack-joining process is in contradiction to the observation that two neighbouring crack tips repel one another rather than coalesce.[154] Moreover, crack coales-

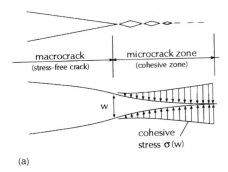

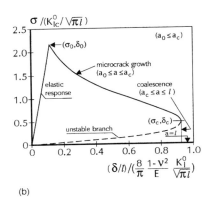

Figure 5.30 Effect of micro-crack growth and coalescence. Regular array of micro-cracks considered (a), and dimensionless stress-deformation response (b), after Ortiz.[438]

cence might lead to local snap-back instabilities.[438,444] In analyzing the response of a cohesive zone of the type "micro-crack cloud in front of a stress-free macro-crack", in analogy with the Dugdale-Barenblatt model for plastic metals (see also the following sections in this chapter), Ortiz[438] derived the following expression between the cohesive stress σ and the opening displacement w in the plane of the crack (Figure 5.30a):

$$w = \frac{\partial U}{\partial \sigma} = -\sigma \frac{8l}{\pi} \cdot \log \cos \frac{\pi a}{2l} = \frac{\sigma}{k} \tag{5.40}$$

E and v are the Young's modulus and Poisson's ratio of the material which is considered as an isotropic continuum, and l and a are defined in Figure 5.30a. The stress intensity factor for an array of collinear cracks is given[445] by

$$K_I = \sigma \sqrt{\pi l} \cdot \sqrt{\frac{2}{\pi} \tan \frac{\pi a}{2l}} \tag{5.41}$$

Micro-crack growth will occur only when the crack opening has increased to $w_0 = k \cdot \sigma_0$, at which point the critical stress intensity factor K_{Ic}^0 of the material is exceeded. Up to this point the σ-w relation is linear as shown in Figure 5.30b. Upon further increasing of the crack-opening, the traction decreases following the equilibrium path $K_I = K_{Ic}^0$, i.e., under the same assumption as made in the example of Section 5.3.2.3. Ortiz noticed a striking feature of the equilibrium path, namely at a micro-crack density $a_c/l \approx 0.91$ stable micro-crack growth cannot be sustained anymore. Opening beyond the critical value w_c leads to cleavage of the ligament between the cracks, which coalesce to form a continuous crack. The dynamics of the crack growth process were ignored by Ortiz for simplicity, and the load dropped to zero at the critical point as indicated in Figure 5.30b. Similar solutions were obtained by others.[444,446] Thus, crack coalescence would lead to snap-back behaviour. If this occurs at a local scale, i.e., when the size of the cracks and the amount of energy release are small with respect to the total specimen volume, global snap-back would probably not be observed.

In tension it was found that crack face bridging mechanisms play an important role in softening. In fact, the tail of the softening diagram can be explained from bridging, although some other mechanisms might be active as well (Section 3.3.1). Bridges in concrete and other disordered materials develop because neighbouring crack tips repel one another rather than coalesce. The material heterogeneity is considered as an important additional factor that may drive two approaching crack tips apart. Therefore, we should be careful accepting the above theoretical line of reasoning, simply because the mechanics are accepted. The slogan given at the beginning of this chapter seems very valid indeed. If we understand the mechanisms, the mechanics become simple. The LEFM solution in the example of boundary rotation effects in uniaxial tension in Section 5.3.2.3 was done in conjunction with the photo-elastic experiments presented in Figure 4.63. The numerical lattice analyses of Section 7.2 were carried out in conjunction with the impregnation experiments of Section 3.3.1, and other tests as well (four-point-shear and anchor pull-out). Also, the Hillerborg model which will be presented in Section 5.3.4 was developed in combination with displacement-controlled, uniaxial tension tests. This may indicate that a close interaction between experiment and theory development is essential to make progress in understanding the phenomena that are studied.

Different micromechanical models can be derived that may account for the increased energy release as well, without having to search for higher crack densities. One of the other possibilities is that at a given moment crack growth will accelerate at certain locations where weaker spots in the material structure appear. At these locations, larger cracks might grow more rapidly than at other sites, leading to increased energy release as well. Moreover, the relative crack size might become sufficiently large to trigger growth of a localised macro-crack. In Section 7.2 we will discuss this alternative possibility. Experimental verification of the pre-peak crack processes is extremely difficult. As shown in Section 4.3, detecting narrow short cracks is very complicated (the results may also be open to various interpretations), which implies that experimental validation of the pre-peak crack processes becomes a very tricky affair.

5.3.3 THE DUGDALE-BARENBLATT PLASTIC CRACK TIP

So far, only LEFM was discussed. In purely brittle materials, like hardened cement-paste, LEFM seems applicable, as for example indicated by the crack growth example between fixed end platens in Section 5.3.2.3. As shown in the beginning of this chapter, stresses become infinitely large at the tip of a slit-like crack. This was the reason for replacing the stress criterion for failure in the LEFM approach. Experiments and theoretical considerations have led to the conclusion that LEFM is not applicable to materials having a gross heterogeneity like mortar, concrete, rocks and ceramics. Of course, the heterogeneity should always be considered in conjunction with the size of the structure under consideration, as well as with the level at which we are interested to apply a given model. The structural size could be very large, such that the effects from individual grains become negligibly small. Moreover, when modelling at the macro-level is preferred, one should aim at describing the phenomena as realistically as possible at that larger scale. However, in all attempts one should remember that it would be interesting to develop a model that can be applied at a variety of size scales as was discussed in Chapter 1. So let us now return to the macroscopic level.

As mentioned, in real materials, stresses cannot increase to infinity. For example in ductile metals, yielding will take place, and as will be obvious from the contents of this book, softening may occur in softening materials. In macroscopic models these phenomena should be incorporated in the theory. For metals, the well-known Dugdale-Barenblatt model[2,3] was developed. In this model crack-tip plasticity was accounted for by replacing the crack and the length of the plastic zones by a longer crack with closing pressure at the tips as shown in Figure 5.31. The crack edges carry a load that is equal to the yield stress σ_y of the material. The wedge forces are selected such, that the stress intensity K_σ due to the externally applied stress σ is compensated for by the stress intensity $K_{\sigma y}$ from the wedge stresses. The two stress intensities are given by[426]

$$K_\sigma = \sigma \sqrt{\pi(a+t)} \tag{5.42}$$

for the far stress, and

$$K_{\sigma y} = 2\sigma_y \sqrt{\frac{a+t}{\pi}} \ arccos \frac{a}{a+t} \tag{5.43}$$

for a constant wedge stress σ_y. The size t of the plastic crack tip zone can then be computed, and is equal to

$$t = \frac{\pi^2 \sigma^2 a}{8\sigma_y^2} = \frac{\pi K^2}{8\sigma_y^2} \tag{5.44}$$

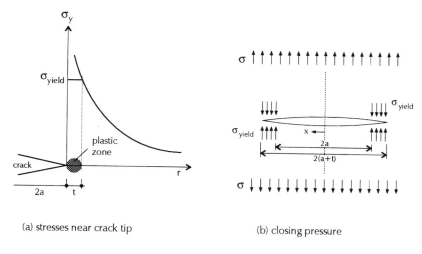

(a) stresses near crack tip   (b) closing pressure

Figure 5.31 Plastic crack-tip model.

Based on these formulas, some researchers have tried to derive a plastic-zone correction for stress intensity factors, but when the plastic zone becomes larger than the actual crack, the application of such corrections becomes rather doubtful because of the limited validity of the expressions for K, which are based on elastic solutions. Much can be said about the shape and size of the plastic-zone, but we will leave it to the principle here, as it forms the basis for the Fictitious Crack Model.[5] The FCM has been developed for describing crack growth in concrete, and will be presented in the following section. For an arbitrary wedge-stress distribution near the crack tips, Equation 5.43 changes to[447]

$$K_{\sigma y} = 2\sqrt{(a+t)/\pi} \cdot \int_{a}^{a+t} \frac{\sigma(x)dx}{\sqrt{(a+t)^2 - x^2}} \tag{5.45}$$

where $\sigma(x)$ is the stress distribution near the crack tips, and the x-coordinate runs from the origin in the centre of the crack towards the tip as indicated in Figure 5.31.

5.3.4 FICTITIOUS CRACK MODEL

Based on the ideas behind the Dugdale-Barenblatt plastic crack-tip zone, Hillerborg and co-workers proposed in 1976 the Fictitious Crack Model (FCM) for analyzing crack growth in cementitious composites. Of course, due to the singularity at a sharp notch, concrete would crack immediately. The tensile strength is considerably lower than in plastic metals for which Dugdale and Barenblatt developed the model presented in the previous section. In fact, the idea is quite simple. The tensile stress-deformation diagram of concrete is

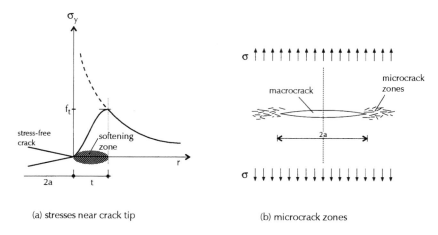

(a) stresses near crack tip   (b) microcrack zones

Figure 5.32 Stress transfer in a process zone (or cohesive zone) in front of a stress-free macro-crack, as assumed in the FCM[5] and Crack Band Model.[140]

characterised by a long descending branch after the strength peak has been exceeded. This means that when it is assumed that a crack is formed as soon as the tensile strength of the material is exceeded, this crack is not immediately stress-free, but some carrying capacity remains perpendicular to the crack face. The carrying capacity of the crack depends on the crack width. With increasing crack width the stress transfer over the crack gradually decreases. A model, quite similar to the FCM was worked out by Bažant and Oh.[140] They proposed that a micro-crack zone, identical to the Dugdale-Barenblatt plastic zone for metals, extends in front of a stress-free crack in concrete. In the Fictitious Crack Model, the crack zone is assumed to be a line, in the Bažant and Oh model a band of certain width is assumed. The crack band width in this latter model is assumed to be a material constant which can be directly related to the microstructure of the concrete. In many cases it is assumed to be a multiple of the largest aggregate particles in the material. Note, however, that Bažant quotes different values in various papers.

5.3.4.1 Principle of the model

The FCM is shown in Figure 5.32. The crack tip zone is drawn as it was originally proposed, viz. as a "cloud of distributed micro-cracks". In Section 3.3.1 it was demonstrated that a major role was played by bridging, but as will be argued in Section 7.2, distributed cracking probably plays an important role as well. Next to this stress-transfer zone, or process zone in front of the stress-free crack, the stress profile over the process zone is shown as well. Sometimes the process zone is referred to as cohesive zone. In the ideal situation, Hillerborg argued, the necessary model parameters could be determined from a stable uniaxial tensile test.[5,142] The loading platens should remain parallel to one another during the complete experiment to ensure uniformity of loading.

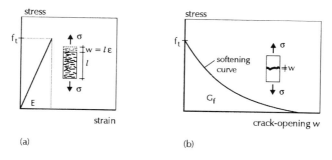

Figure 5.33 Fracturing in a tensile bar. In the pre-peak regime distributed cracking is observed, and a stress-strain diagram can be used (a); in the post-peak regime a stress-deformation diagram is used because the crack localises in a narrow zone (b).

However, as was already discussed in Section 4.1.3.2, assuring uniformity over the process zone is the major difficulty in heterogeneous materials, and — more seriously — this seems in contradiction to the nature of the material. The other important idea behind the FCM is that the tensile stress-deformation diagram should be split into two parts. This can best be explained with Figure 5.33.

Before cracking, deformations are distributed in a concrete bar which is subjected to uniform uniaxial tension. Based on the micro-mechanisms described in Chapter 3, micro-cracking may take place in the pre-peak regime, but is necessarily distributed over the complete specimen volume. At peak, a macroscopic fracture zone is assumed to initiate, which localizes at a certain weak spot in the tensile bar. If a measuring device (A) is located over the crack zone (or process zone) a stable descending branch is measured, whereas a measuring device (B) located over an "intact" part of the bar will show unloading. It is exactly this type of behaviour that we have seen in Figure 3.34. Because the deformations in the post-peak regime are localized in a narrow zone, Hillerborg decided that it was not logical to express the response in terms of strain. The post-peak behaviour should be modelled as a stress-crack-opening relation, rather than a stress-strain diagram. This is depicted in Figure 5.33. Evidently, unloading also takes place in the part of the measuring length l which is situated over the process zone. Therefore, in order to obtain a stable result, the measuring length should not be too long, as was argued in Chapter 4 (Figure 4.4). Experiments presented in Section 3.3 indicate that the pre-peak stress-strain behaviour is more or less linear. However, depending on the size of the specimen, and the possible interactions with the loading device, non-linearities may appear in the pre-peak regime, as was the case with the 50 mm long specimen (type C) in Figure 4.41. Now let us assume that the pre-peak part is linear, and that unloading proceeds along a straight line, parallel to the initial loading curve. Following Figure 5.33, the area under the post-peak stress-crack opening diagram, corrected for the elastic deformation in the measuring length, is then defined as the fracture energy G_f.

$$G_f = \int_0^{w_{max}} \sigma(w)\,dw \qquad (5.46)$$

The function $\sigma(w)$ should be determined from a stable displacement-controlled uniaxial tensile experiment, or alternatively from a stable displacement-controlled three-point-bend test as described in Section 4.6. In the latter case, inverse modelling is essential, see Section 6.2. The stability is a point of main concern, as debated by Hillerborg on several occasions[448] (see next section). If it *assumed* that the stability conditions are fulfilled, and that indeed *a true property* of the material is measured in a displacement controlled uniaxial tensile test, one may attempt to fit the experimental results to simple functions. Such functions would have universal applicability, but only if the assumptions are true. An example of a function fitted to experimental data is[84]

$$\frac{\sigma(w)}{f_t} = \left[1 + \left(c_1 \frac{w}{w_c}\right)^3\right] exp\left(-c_2 \frac{w}{w_c}\right) - \frac{w}{w_c}\left(1 + c_1^3\right) exp(-c_2) \qquad (5.47)$$

The best fit was obtained with $f_t = 3.2$ MPa, $c_1 = 3$, $c_2 = 6.93$ and $w_c = 160\ \mu$m for normal weight concrete, which is shown in Figure 5.34. c_1 and c_2 are empirical constants, and depend on the available data set. Other functions were derived earlier, for example by Reinhardt,[187] directly based on the Dugdale plastic crack tip model,

$$\frac{p(t)}{f_t} = 1 - \gamma\left(\frac{(a+t)-x}{t}\right)^n \qquad (5.48)$$

where γ is a factor between 0 and 1. $\gamma = 1$ implies total stress release at the tip $(x = a)$, whereas $\gamma = 0$ means no reduction of stress at all. The factor γ is important as it determines the shape of the softening function at the crack tip.

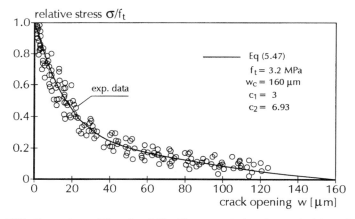

Figure 5.34 Comparison of Equation 5.47 with post peak data from uniaxial tensile tests.[84] (Reprinted by kind permission of Dr. Hordijk.)

Reinhardt presents a lengthy discussion on the matter. However, in the end a power softening law is proposed,

$$\frac{\sigma_t}{f_t} = 1 - \left(\frac{w}{w_c}\right)^k \tag{5.49}$$

where $k = 0.31$ and $w_c = 175$ μm gave the best fit of experimental results. Foote et al.[449] propose a power law as well, viz.,

$$\frac{\sigma_t}{f_t} = \left[1 - \frac{w}{w_c}\right]^n \tag{5.50}$$

where f_t is the peak tensile strength and w_c is the crack opening where $\sigma = 0$. For different values of the power n, different softening curves are found. The case $n = 0$ means no softening but perfectly plastic behaviour is retrieved. For $n = \infty$, $\sigma = 0$ an ideal brittle response is obtained. For all intermediate values of n some type of softening curve is found.

Basically, these curves all lead to similar results as has been demonstrated over and over, e.g., by Elfgren.[450] For computational purposes, often a simple function is selected, for example a linear softening law, a bi-linear diagram (Chapter 6) or any of the above functions.

5.3.4.2 Stability considerations

Basic to the application of the FCM is the tensile stress-crack opening diagram. As mentioned before, Hillerborg has argued at several occasions that the test should remain stable during the complete loading, i.e., no snap-backs are allowed (see also Chapter 4, Figure 4.4). One of the assumptions made is that the stresses and deformations should be distributed uniformly in the whole experiment. Or, stated differently, only translations are allowed in the localization zone, and rotations are prohibited. We have touched upon these matters already in Chapter 4 and Section 5.3.2.3, where experimental results and analyses with a simple LEFM model were shown successively. Rotation of the crack surfaces is almost unavoidable, because the material is heterogeneous. The heterogeneous nature of the material may cause a deviation of the neutral axis from the axis of loading. Moreover, the specimen is not always correctly aligned in the testing machine, and small load-eccentricities will happen as well. Hillerborg[448] and Fanping Zhou[451] argue that the rotational stiffness of the specimen/machine system should be large enough to counteract the rotations taking place in the fracture zone. In their analyses, it is assumed that all sections remain planar and that a damage zone has developed over the entire cross-section of a notched specimen (see Figure 5.35). Inside the damage zone, the softening curve is assumed to be valid; outside the softening zone, the material is considered to be linearly elastic.

The load P is applied with an eccentricity e as shown. The rotational stiffness of the machine is K_r. The moment equilibrium yields a stability

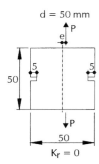

Figure 5.35 Double-edge-notched tensile specimen for stability analysis.[451] (Reprinted by kind permission of Dr. Fanping Zhou.)

criterion for the experiment, and the rotational stiffness of the machine should satisfy

$$\frac{1}{K_r} < -\frac{6h}{Ebd^3} + \frac{6}{b_c d_c^3} \cdot \left(-\frac{dw}{d\sigma}\right)_{min} \tag{5.51}$$

where E is the Young's modulus of the uncracked material, b_c and d_c refer to the width and the thickness of the specimen between the notches (Figure 5.35), and b, d and h are the overall specimen dimensions. The factor $-(dw/d\sigma)_{min}$ is the maximum (steepest) slope of the softening curve. The right side of the inequality, Equation 5.51, must always be larger than 0 since $K_r > 0$. This means that for a material with softening behaviour $-(dw/d\sigma)_{min}$, the specimen dimensions should be selected such that

$$\frac{hb_c d_c^3}{bd^3} < E\left(-\frac{dw}{d\sigma}\right)_{min} \tag{5.52}$$

This case can also be interpreted as the case where $K_r = \infty$, i.e., when truly non-rotating specimen boundaries are achieved. In Section 4.1.3.2 the rotational stiffness of the tensile set-up in the Stevin Laboratory was discussed. In this case, the rotational stiffness was quite large, but the installation of internal load-cells below the lower loading platen affected the rotational stiffness considerably. Equation 5.52 indicates that when the specimen size is selected too large, or when the notch depth is not chosen with care, unstable behaviour will occur, irrespective of the magnitude of the rotational stiffness of the testing machine. Note, that when $-(dw/d\sigma)_{min} = 0$, i.e., when a truly brittle material is considered, Equation 5.52 can never be fulfilled. This implies that for a completely brittle, homogeneous material no softening can be measured. Thus, it might be concluded that a certain deformation gradient is needed in the specimen in order to achieve stable softening. In practice, deformation gradients may be caused by eccentricities in the specimen and/or the loading

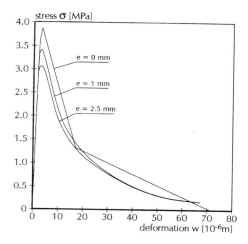

Figure 5.36 Influence of eccentricity on σ-w curves.[451] (Reprinted by kind permission of Dr. Fanping Zhou.)

arrangement. Next to this, differential temperatures during hardening of the concrete, and/or non-uniform drying may lead to deformation gradients in the specimen as well.[182] Note that these factors are not included in the present analysis.

As an example, we consider the small 50 mm specimen (type C) of Figure 4.41. The notches were 5 mm deep, leaving an effective area of 50 × 50 mm². Using Equation 5.52, it is possible to compute which materials will remain stable, and which materials will lead to an unstable test result. Stability is guaranteed for this specimen size when

$$\left(-\frac{dw}{d\sigma} \right)_{min} > \frac{125}{3E} \qquad (5.53)$$

assuming, of course, that the machine stiffness is infinite $(K_r = \infty)$. Because $-(dw/d\sigma)_{min}$ is not known in advance, always some preliminary experimentation is needed.

An example of a computational result of Fanping Zhou[451] is given in Figure 5.36. The response of the above type C specimen was computed using a bi-linear softening diagram. The original bi-linear diagram is plotted in Figure 5.36, together with two computational results with different eccentricity, i.e., $e = 1$ and 2.5 mm. In the analysis it was assumed that the specimens could rotate freely $(K_r = 0)$.

From this figure it can be concluded that load-eccentricity has a significant effect on the softening behaviour. Actually, it is very difficult indeed to avoid eccentricities in experiments, even if the specimens are prepared with the utmost care. In view of the heterogeneity of the material it is rather doubtful if a uniform distribution can be achieved at all. And, more importantly, is the effort realistic? Carpinteri and Ferro[292] constructed a three-jack system and performed uniaxial tensile experiments in which uniformity of deformations in a central zone of the specimen was achieved through careful adjustment of the loading during the softening regime (see also Section 4.1.3.2). In Section 7.2.2, the situation in their experiment will be analyzed with a brittle lattice model at the meso-level.

A similar analysis as shown above for the stability of a uniaxial tensile fracture test can be carried out for a compressive experiment as well.[89] The outcome is similar, except that for compression the complete specimen softens rather than of only a narrow band.

5.3.4.3 Crack-band width

In the fictitious crack model, the crack is assumed to be a line crack with zero width. A similar model was developed by Bažant and Oh,[140] the so-called crack-band model. In this model, a process zone is also assumed (see Figure 3.33b) which can transmit some load. The crack-band has a certain width, and a value of $3d_{max}$ is generally assumed. The crack-band width was computed from an inverse analysis. Slightly deviating values are therefore sometimes found. The impregnation experiments presented in Section 3.3.1 showed that, at least in the long tail of the softening diagram, the behaviour is governed by crack overlaps near stiff particles in the material. The size of these overlaps is never more than the largest aggregate size d_{max}. However, when computations are made at the meso-level, distributed micro-cracking is found in uniaxial tensile test simulations, in particular around the peak of the stress-crack opening diagram (see Section 7.2). This would imply that the crack-band width should be much larger than a multiple of the largest aggregate size, and that it probably would include the complete specimen. The crack-band width specified by Bažant and Oh should therefore be considered as a weighted average of the complete fracture process.

The crack-band model is relatively easy to implement into smeared finite element programs. For example, Rots[452] has implemented the model into the DIANA finite element program. The advantage is that the state variables stress and strain can still be used. The alternative is to use the discrete FCM, but this would be most favourable in combination with a remeshing technique. The smeared model is mesh dependent, i.e., different meshes will give a different result. Recently, it was proposed to overcome this mesh sensitivity by using higher order continuum theories, as was done by De Borst and Mühlhaus.[383] Different types of higher order continua are distinguished, such as Cosserat models, gradient plasticity models and non-local theories. We will not delve into all the details associated with such models. More information can be found in relevant books.[376] The mesh sensitivity is overcome because an internal length scale is introduced in the constitutive equation. Pamin[453] describes these matters for gradient plasticity models. The constitutive law for the uniaxial case is written as follows

$$\sigma = f(\varepsilon) - c\frac{d^2\varepsilon}{dx^2} \tag{5.54}$$

where c is a positive phenomenological contant and $f(\varepsilon)$ defines the non-linear softening function. Following this approach the constitutive law has been separated in a computational softening curve $f(\varepsilon)$ and a length-dependent factor. Consequently, the result of an analysis does not depend anymore on the finite element mesh. The problem now is to determine the internal length, which (in view of the foregoing discussion) is most likely a variable. An inverse procedure seems the only way to validate the model. The definition of higher order continua makes sense from micromechanical considerations. However, it will take further research to solve the internal length scale problem. Moreover, at present many different higher order theories have been developed and it is not straightforward which model should be applied for a given problem.

5.3.4.4 Mixed fracture modes

So far the discussion on the FCM was limited to mode I fracture (tensile opening) only. As was shown in Figure 5.20, two other fracture modes are distinguished in Linear Elastic Fracture Mechanics. These modes were referred to as mode II (in-plane shear) and mode III (out-of-plane shear). Next to these pure crack modes, mixed-mode situations of mode I, II and III can occur. Mode II and III are normally considered of secondary importance. However, for example in interface fracture, they may play a more prominent role. For some time, there has been an enormous activity in the concrete fracture community to find out more about the mode II, III and various mixed-mode loadings.[211,454] In Section 3.4 a variety of specimen geometries for studying these other fracture modes was presented, and the discussion on interpretation of the various experiments was initiated. In Section 7.3, we will return to this discussion of mixed-mode I and II and mode II experiments by means of numerical simulations. Here, we will discuss the extension of the FCM to mixed-mode I and II loadings.

In view of all non-uniform opening problems in tensile testing (and which have been central to many of the discussions so far), some doubt exists concerning the validity of the Fictitious Crack Model. The main objective of the model was to obtain a better insight in fracture of concrete. The recognition of all the non-linear fracture processes and the localization of deformations in tensile cracking has been of utmost importance for progress in the field. The loading on a process zone is never purely tensile, but in plane-stress the process zone will be subjected to shear as well. Following this line of reasoning, mixed mode I and II should be studied as well. The process zone (or cohesive zone) is subjected to simultaneous shear and tension, and both opening and sliding will occur, as shown in Figure 5.37.

Thus, the behaviour in the process zone should be described by means of a more complex constitutive equation than the simple laws given in Section 5.3.4.1. The relation is given, in incremental form, by

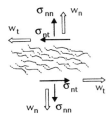

Figure 5.37 Process zone subjected to combined tension and shear.

$$\begin{vmatrix} \Delta\sigma_{nn} \\ \Delta\sigma_{nt} \end{vmatrix} = \begin{vmatrix} A_{11} & A_{12} \\ A_{21} & A_{22} \end{vmatrix} \cdot \begin{vmatrix} \Delta w_n \\ \Delta w_t \end{vmatrix} \tag{5.55}$$

The stresses and displacements are defined in Figure 5.37. The index n denotes normal to the crack, index t indicates tangential. In the crack-band model, Equation 5.55 must be written in terms of strains, and the element size h should be incorporated. With $\varepsilon_i = \delta_i/h$, this leads to

$$\begin{vmatrix} \Delta\sigma_{nn} \\ \Delta\sigma_{nt} \end{vmatrix} = \begin{vmatrix} B_{11} & B_{12} \\ B_{21} & B_{22} \end{vmatrix} \cdot \begin{vmatrix} \Delta\varepsilon_{nn}^{cr} \\ \Delta\varepsilon_{nt}^{cr} \end{vmatrix} \tag{5.56}$$

where the superscript cr denotes cracking. The total shear strain increment with contributions from the uncracked (elastic) concrete and the crack sliding displacement can now be written as

$$\Delta\gamma_{nt} = \Delta\gamma_{nt}^e + \Delta\gamma_{nt}^{cr} = \left[\frac{2(1+v)}{E} + \frac{1}{B_{22}} \right] \cdot \Delta\sigma_{nt} = \left[\frac{1}{G} + \frac{1}{B_{22}} \right] \cdot \Delta\sigma_{nt} \tag{5.57}$$

In the early days of smeared crack analysis, the term between straight brackets was combined to a single quantity, which is called the shear retention factor β. Multiplied by the shear modulus G it accounts for the decrease of shear stiffness due to cracking:

$$\Delta\gamma_{nt} = \frac{\Delta\sigma_{nt}}{\beta G} \tag{5.58}$$

where β can also be written as

$$\beta^{-1} = 1 + \frac{G}{B_{22}} \tag{5.59}$$

Different expressions for the shear retention factor β have been proposed by various researchers.[454] In some cases, the B_{12} and B_{21} factors are non-zero, and the matrix becomes even asymmetric. In Section 3.4 it was shown that secondary cracking occurs when the normal crack opening w_n is very small, i.e., smaller than 200 to 250 mm.[220] However, this limiting crack opening may depend on the structure of the material under consideration, as well as on possible confinement perpendicular to the crack. The above formulation was in part developed based on conventional aggregate interlock theories that were mentioned in Section 3.4. Such theories seem valid for very large crack-openings (i.e., beyond w_c), but certainly not for narrow softening cracks as mentioned above. As will be shown in Section 7.3, a simple brittle numerical lattice model employing mode I fracture only is capable of simulating "mixed-mode" fracturing at small crack openings.

5.3.5 SUMMARY OF THE BASIC MODELS

In the previous sections an overview of possible fracture initiation and propagation criteria for continuum based models was given. Some of the models can be applied at different scales, like for example the linear elastic fracture models. Thus, although a certain order has been followed, it remains difficult to completely discriminate between continuum models for modelling at the macro-level, and the discrete lattice models which have been quite successful at the meso-level.

In summary, we have touched upon classical strength of materials where complex failure surfaces in three dimensions were formulated, the linear elastic fracture mechanics, and non-linear fracture models. The linear elastic fracture mechanics can be applied at various dimensional levels. The strength of materials seems to be dependent on the scale of observation (Section 4.2.3), whereas scale effects can be computed using the non-linear fracture models like the Fictitious Crack Model. Several examples were included, showing the application of a given model to a specific practical problem. The non-uniform crack problem in tensile testing was used at two occasions. It was shown that the models can in principle be quite helpful in giving a better understanding of experiments, even when the models are quite simple. In the next section examples of fracture laws that can be used in conjunction with lattice type models is presented.

5.4 FRACTURE LAWS FOR LATTICE TYPE MODELS

In Section 5.2 lattice-type fracture models were presented. Basically, we stepped down the dimensional hierarchy, and the behaviour of the material was modelled at the meso-level. At the same moment the continuum was discretized into a lattice of linear elements. The basic element in the lattice is a spring (bar) or a beam. Both types of elements are drawn in Figure 5.38. In a truss element (bar or spring), only normal forces are transmitted as shown in Figure 5.38a; the nodal forces and moments acting on a single beam element are shown in Figure 5.38b. For a spring (or bar) element the axial stiffness must be specified (K or EA), whereas for a beam element a flexural stiffness (EI) must be prescribed. For a beam element as shown here, failure can be modelled in various manners.

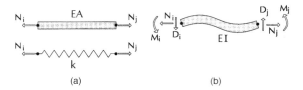

Figure 5.38 Forces and bending moments acting on a beam.

Basically, all the fracture criteria presented in Section 5.3 can be used when slight modifications are made. Moreover, lattice models can be used at different dimensional levels, and the philosophy behind the model can be completely different as well. Three examples are presented. In the first example, the failure of a regular lattice of atoms is analyzed by means of a spring model. An energy criterion is used for fracturing the springs. Such analyses can be quite helpful in determining the theoretical strength of materials. However, for simulating fracture spring models are of less interest, as the coherence in a lattice may disappear as too many elements are fractured. This may lead to undesirable rotational stiffness of parts of a lattice. In the second example, the most simple fracture law for a beam model for meso-level fracture analyses is presented. In this particular example it is assumed that the lattice has been projected on top of the material structure. The third example comprises a particle composite where the centres of the particles are connected by means of beam elements. From a principal stress analysis a normal-load/shear-load criterion is derived to simulate failure in the centre of the beams. Thus, the last two examples are basically based on classical strength of materials theory, whereas the first example is an energy approach. Note that softening models or any other energy approach could be applied in lattice models as well. This has for example been done by Jirásek and Bažant.[402] Note that plasticity can be included in lattice models as well, even in a quite simple manner.[461] The results obtained from the simple lattice model are surprisingly realistic, in particular the fracture mechanisms as will be demonstrated in Chapters 7 and 8.

5.4.1 AN ATOMIC VIEW ON FRACTURE

Purely crystalline solids can be considered as built from a regular array or lattice of atoms. Depending on the nature of the crystal, the atoms can be all the same, or alternatively they may be different. In the latter case, the bonds between the atoms will be different as well. In crystallography, twofold and threefold symmetry is quite common, as are multiples of these numbers, but for a long time fivefold symmetry was assumed to be non-existent. In 1984, however, fivefold symmetry was observed as well, although the patterns are not periodic anymore. Therefore, crystals displaying fivefold symmetry are reffered to as quasi-crystals.[455] Before these quasi-crystals were found, fivefold plane tilings were constructed by Penrose in 1973.[456] In Figure 1.3, a drawing of a Penrose tiling was shown under the header "crystal structure". In this

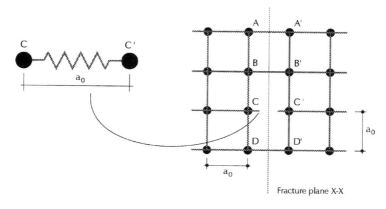

Figure 5.39 Regular square lattice of atomic bonds.

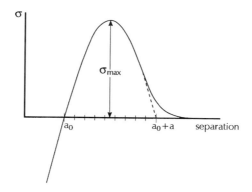

Figure 5.40 Bond stress to distance relation for the atomic bonds in Figure 5.39.

chapter, we do not go deeper into crystallography. Instead, fracturing in a regular square lattice of atomic bonds is analyzed as shown in Figure 5.39.

In this book we will consider purely brittle fracture for lattices only. The atoms, and the bonds between them are all identical. At equilibrium the spacing between the atoms is a_0. The bond between two atoms depends on their mutual distance. When, for example, the atomic lattice is stretched by applying a mechanical tensile load, the distance between the atoms increases and the bond stress increases as well. A maximum spacing exists at which the bond strength reaches a peak as shown in Figure 5.40. After the maximum, when the atoms are further separated, the bond stress decreases. At a given maximum separation, fracture occurs. In contrast, when the atomic lattice is compressed, the spacing decreases, and the atoms are forced to move from their equilibrium state. A repulsive force develops and energy must be supplied to decrease the spacing. The shape of the bond stress-spacing relation of Figure 5.40 is approximate. On basis of the example of Figure 5.39 an estimate of the theoretical or expected tensile strength of materials can be made.[39] The atomic lattice is in fact a central force lattice. The bond springs can transmit axial forces only. The analysis shown below can be found in many different text-books, and is included here to show the most basic type of lattice.

The bond stress-spacing relation of Figure 5.40 can be approximated by a sine function as shown. The tail of the diagram is cut-off, and tensile loading is considered only. The bond stress-separation function can be written as

$$\sigma = \sigma_{max} \cdot sin\left[\frac{\pi}{a}(x - a_0)\right] \qquad (5.60)$$

where σ_{max} is the peak in the bond stress-spacing diagram, and the coordinate x is defined in Figure 5.40. The maximum bond stress can be computed by relating the slope of the bond stress-displacement diagram in the equilibrium state (spacing a_0), with the Young's modulus E of the material. For small values of $(x-a_0)$,

$$sin\left[\frac{\pi}{a}(x - a_0)\right] = \frac{\pi}{a}(x - a_0),$$

and σ_{max} can be calculated from

$$\frac{d\sigma}{dx} = \frac{\pi}{a} \cdot \sigma_{max} \qquad (5.61)$$

where, following conventions of linear elasticity, $\sigma = E.\varepsilon$ and $\varepsilon = x/a_0$, i.e., the relative stretch of the lattice. Consequently, we can write

$$\sigma = E\varepsilon = E \cdot \frac{x}{a_0} \qquad (5.62)$$

The derivative to x follows immediately and is equal to

$$\frac{d\sigma}{dx} = \frac{E}{a_0} \qquad (5.63)$$

Combining Equations 5.61 and 5.63 yields

$$\sigma_{max} = \frac{E}{\pi} \cdot \frac{a}{a_0} \qquad (5.64)$$

Generally it is assumed that $a = a_0$, and the peak bond stress becomes

$$\sigma_{max} = \frac{E}{\pi} \qquad (5.65)$$

Thus, the theoretical strength of a crystalline solid can be very high indeed, viz. up to one third of the Young's modulus of the material. For example for diamond, which is most frequently quoted in such examples, the Young's modulus is approximately 1200.10^3 MPa, and the theoretical strength would be around 400.10^3 MPa. For alkali resistant glass fibre, which is sometimes used in concrete, with $E = 70.10^3$ MPa, the theoretical strength would be $\sigma_{max} = 23.10^3$ MPa. In reality only a value of 70 MPa is measured on a single fibre.[457] Many other examples can be found in textbooks.[39]

The basis for the above computation is the bond stress-spacing diagram of Figure 5.40. It is in fact the constitutive law in the model. The maximum theoretical cleavage strength of a crystalline solid can also be expressed in terms of surface energy, i.e., the amount of energy needed to create a unit fracture surface. The idea is in principle similar to the Griffith crack that was treated in Section 5.3.2.2. The basic idea is that the area under the bond stress-displacement diagram of Figure 5.40 is equal to twice the surface energy γ needed to create the fracture surface (note: two times the surface energy must be counted because two crack surfaces are formed). It follows that

$$\int_{a_0}^{a_0+a} \sigma_{max} \cdot sin\left\{\frac{\pi}{a}(x - a_0)\right\} = 2\sigma_{max} \cdot \frac{a}{\pi} = 2\gamma \qquad (5.66)$$

The earlier expression for the maximum bond stress (Equation 5.64) can be rewritten as

$$a = \frac{\sigma_{max}\pi a_0}{E}$$

and with Equation 5.66 we obtain

$$\sigma_{max} = \sqrt{\frac{\gamma E}{a_0}} \qquad (5.67)$$

The theoretical cleavage stress can now be expressed in terms of the surface energy γ and the spacing of the atoms in the lattice at equilibrium. Note that from combining Equations 5.67 and 5.65, the surface energy γ can be expressed in terms of the Young's modulus and the lattice spacing at equilibrium a_0.

In practice, the theoretical strength can of course never be obtained because impurities or voids in the atomic lattice (see Figure 5.41) will cause local stress concentrations. As a result a lower strength will be measured. In fact, the discussion follows similar lines as before when linear elastic fracture mechanics was introduced in Section 5.3.2. The type of lattice model is easily accessible to numerical computation. Equation 5.60 is but one simple way of describing the interactions between the atoms. Nowadays computations involving up to two million atom bonds can be performed on super-computers,[458] and this development is expected to continue as more powerful computers are being developed. In principle, many of the fracture criteria that were presented in Section 5.3 of this chapter could be used in conjunction with a lattice of springs or beams. In the following section a number of different fracture laws that were used in the beam model of Figure 5.3 are presented.

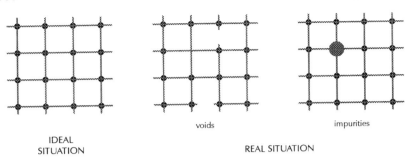

Figure 5.41 Voids and impurities in crystal lattices.

5.4.2 FRACTURE LAWS FOR BEAM MODELS

Spring (or truss) models (sometimes also referred to as central force lattices) are of course the most simple, as only effects from normal force are considered. More complicated are beam models where normal forces, shear forces and bending moments can occur in the lattice elements. Surprisingly, in the past few years parallel developments have occurred in continuum mechanics and in lattice modelling, which seem to point in a direction where flexure must be included in the constitutive law for correctly simulating localization of deformations in disordered solids. So, let us examine different ways of modelling fracture in a lattice. Either an energy approach can be used, resembling fracture mechanics, or a strength-based model can be employed. Here, only the last type of fracture law is presented.

As shown in Figure 5.38b, normal forces, shear forces and bending moments can be transmitted in a beam model. From a simple, linear elastic analysis, for example based on the finite element method, the various loadings on each beam in a lattice can be determined. The criterion to fracture a beam can be based on either of the three load components, or alternatively an effective stress can be computed that can be used in the analysis. The simplest way would be to compute the loads in a beam model, but then to use the normal load only to decide whether a beam should fracture. Thus, the criterion becomes

$$\sigma_{eff} = \frac{N}{A} < f_t \qquad (5.68)$$

The criterion resembles the Rankine criterion (Equation 5.3) that was mentioned earlier. A simple model like this was used by Schorn and Rode.[407] Their model was a statically indeterminate truss model, and both a tensile and compressive strength was specified for each of the bar elements.

The model of Equation 5.68 is probably too simple, at least when used in a beam model. The increase of stress in the outer fibres of the beams caused by the bending moments is neglected. If the contribution from flexural stresses is included, an effective stress based on normal force and bending moment could be used following

$$\sigma_{eff} = \frac{N}{A} \pm \alpha \frac{\left(|M_i|, |M_j|\right)_{max}}{W} < f_t \tag{5.69}$$

where $W = bh^2/6$ is the section modulus, and the subscripts i and j refer to the two nodes of the beam.[396] The flexural component is multiplied by a constant α, which decides how much bending is taken into acount. For tensile fracture simulations, the last criterion has proven to be quite effective; for compression, however, problems are encountered. The major problem is the determination of the α-factor, which cannot be measured directly from experiments. An inverse procedure is needed to determine this parameter. In fact the problem is quite similar to the internal length problem in the higher order continuum models that was touched upon in Section 5.3.4.3.

A fracture law similar to Equation 5.69 was used earlier by Herrmann et al.[393] They relate their fracture law to a Von Mises yield criterion, and from the observations in Sections 3.5 and 5.3.1 it may be clear why the fracture law will not lead to realistic simulations of compressive fracture. It should be mentioned here that tensile stress concentrations will still play an important role in concrete subjected to external compression. As argued in Chapter 3, much of the non-linear stress-strain behaviour of concrete in compression seems to be caused by cracking at the meso-level. Thus, perhaps a tensile criterion would still suffice, although it might be argued too that a shear component becomes active as soon as the cracks have nucleated.

It is worthwhile noting that a beam is always fractured instantaneously when the maximum tensile stress in the outermost fibres of a beam reaches the strength assigned to that particular beam. Some researchers use a hierarchic lattice, in which a beam consists of a number of parallel bars at a lower level. Fracturing is then simulated by successively fracturing of the lower level bars. Only after all the lower level bars are fractured, the global bar is completely stress-free.[459] This parallel bar arrangement means that softening is introduced at the bar level. Note that in this chapter only brittle models are considered, i.e., a strength criterion is used, and it is assumed that the fractured beams are immediately stress free.

The above two fracture criteria are not very much different from the strength-based fracture laws presented in Section 5.3.1. As a matter of fact, the link can be made quite directly by computing the principal stresses in a lattice beam, and subsequently translating the principal stress combination to a shear-load/normal-load criterion. This was originally proposed by Beranek and Hobbelman,[413,460] who derived a particle model for concrete and brick. This model follows the principle shown in Figure 5.6b, except that all particles have the same size. When a particle composite is loaded in compression, high stress concentrations appear in the interfaces between the particles, as shown in Figure 3.7. At present, quite some debate is going on regarding the validity of the fracture law.[413,414,460] In Figure 5.42 the basic assumptions in the model formulated by Beranek and Hobbelman are shown. The contact layer with thickness d and area A between two spherical aggregates is shown. Failure will

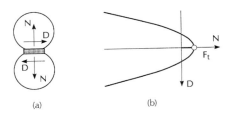

(a)   (b)

Figure 5.42 Contact layer between rigid particles (a), and failure contour for combined normal and shear load (b), after Beranek and Hobbelman.[413]

occur as soon as the largest positive principal stress exceeds the tensile strength of the material in the contact zone. In the contact layer, only normal stresses σ_x are transmitted; lateral stresses σ_y and σ_z depend on the Poisson's ratio of the contact layer material only. Following conventional stress analysis,

$$\sigma_y = \sigma_z = \frac{v}{1-v}\sigma_x = \eta\sigma_x,$$

where v is the Poisson's ratio of the contact layer material. The y- and z-axes are selected such that only shear stresses are transmitted. Mohr's circles can be drawn in the yz-coordinate system, and the principal stresses follow from

$$\sigma_{1,2} = \frac{1}{2}\left(\sigma_x + \sigma_y\right) \pm \left[\frac{1}{2}\left(\sigma_x - \sigma_y\right)^2 + \sigma_{xy}^2\right]^{1/2} \tag{5.70}$$

After dividing by σ_x and replacing the quotient σ_y/σ_x by η, Equation 5.70 can be written as

$$\sigma_{1,2} = \frac{1}{2}(1+\eta)\sigma_x \pm \frac{1}{2}\sqrt{(1-\eta)^2\sigma_x^2 + 4\sigma_{xy}^2} \tag{5.71}$$

σ_1 is the major principal stress and is tensile, whereas σ_2 is compressive. Equation 5.71 can be rewritten as a function of σ_{xy} in terms of f_t, σ_x and η,

$$\sigma_{xy} = \sqrt{f_t^2 - f_t(\eta+1)\sigma_x + \eta\sigma_x^2} \tag{5.72}$$

A graphical representation of this function is given in Figure 5.43 for values of $0 < \eta < 1$. It is a hyperbolic function with parameter η. For a certain value of η the graph shows the boundary of the area representing combinations of σ_x and σ_{xy} which are below the critical stress. The influence of non-zero values of η is obvious: the failure contour expands and the general state of stress passes from plane-stress for $\eta = 0$ to plane-strain for $\eta = 1$.

In a lattice analysis, the fracture law shows for each beam element whether the critical stress (σ_{crit}) is exceeded under a certain prescribed force or displacement. The element with the highest critical stress relative to the tensile strength can be identified. In general, a test load of unit force or displacement is applied. Because the computations are linearly elastic, the real force or displacement

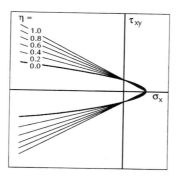

Figure 5.43 Influence of η on shape of failure criterion.[414]

causing the critical element to collapse is given by the quotient $\sigma_{act,max}/\sigma_{crit}$, where σ_{act} is the governing stress in the critical lattice beam. At failure, the largest principal stress σ_1 is equal to f_t, and we obtain,

$$
\sigma_{x,crit} = \frac{1}{2} f_t \cdot \left[\frac{(1+\eta) \pm \sqrt{(1-\eta)^2 + 4\sigma_{xy}^2/\sigma_{x,act}^2}}{\eta - \sigma_{xy}^2/\sigma_{x,act}^2} \right] \tag{5.73}
$$

It has already been mentioned that a value for η has to be selected. It has been decided to take $\eta = 0$ as a starting value in the computations, i.e., plane stress is assumed in the contact layer between the particles. For this value of η the hyperbolic failure surface becomes a parabola. The shape of the applied fracture law, in terms of forces is then:

$$
N_{crit} = \frac{1}{2} F_t \cdot \left[\frac{-1 \pm \sqrt{1 + 4 \cdot (D/N_{act})^2}}{(D/N_{act})^2} \right] \tag{5.74}
$$

In Figure 5.44 a graphical representation is given of the complete procedure. For the combination of the actual normal force (N_{act}) and shear force (D) computed in a beam element, the percentage of the critical combination (N_{crit}, D_{crit}) is calculated. The beam with the highest percentage is removed, and the computation starts all over again, similar as in the original approach.[396] As mentioned, the fracture law (Equation 5.74) was originally derived by Beranek and Hobbelman.[413] A comparison of the fracture laws (Equations 5.69 and 5.74) is made in Section 7.5. There biaxial failure contours are computed by means of a numerical lattice simulation.

Either of the effective stress criteria (Equations 5.68, 5.69 and 5.74) may be used to decide whether a beam in the lattice must be removed. Depending on the heterogeneity implemented in the lattice, the local stress-distributions, and the external load and boundary conditions, the most critical element is deter-

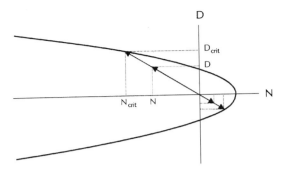

Figure 5.44 Increase of failure load with factor N/N_{crit} after Beranek and Hobbelman.[413]

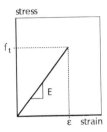

Figure 5.45 Fracture law for a single lattice beam.

mined. As mentioned, fractures are created by simply removing the critical lattice beam from the mesh. Effectively, this means that an elastic-purely-brittle stress-strain law is used as shown in Figure 5.45. As soon as the effective stress is reached, a vertical stress-drop occurs when the beam is removed. The energy consumed in the process is given by the area under the curve of Figure 5.45.

Plastic behaviour of materials can in principle also be taken into account. In that case a beam is not immediately removed from the mesh, but instead a step-wise reduction of the Young's modulus of the material of the lattice beams is simulated. In Figure 5.46 an example of such a fracture model is shown.[461] The criterion to reduce the Young's modulus is based on exceeding a maximum effective stress in the relevant lattice beam. Either one of the strength criteria mentioned in this section might be used. Through a step-wise decrease of the Young's modulus of a lattice beam, 3D effects are included in 2D lattice analyses. In the next chapter an example is shown.

It may be obvious that many possible fracture laws can be used in lattice type models. In continuum-based models, as presented in Section 5.3, there is still some (small) hope that the parameters for the fracture laws can be mea-

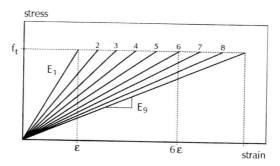

Figure 5.46 Stepwise reduction of the Young's modulus to simulate plastic fracturing in the lattice model.[461]

sured directly in an experiment, although such efforts are hampered by boundary and size effects. In lattice models, the derivation of the fracture law parameters is not straightforward. Similar as in the higher order continuum models, parameters are introduced that seem to diminish computational problems. However, these parameters, like the internal length scale in the gradient models (Section 5.3.4.3) and the flexural coefficient α in the beam model fracture law (Equation 5.69), can be determined only indirectly through inverse analysis. In the next chapter we will discuss parameter validation procedures both for models operating at the macro-level and for lattice type models operating at the meso-level.

PART 3

SYNTHESIS

Chapter 6

VALIDATION OF MODEL PARAMETERS

6.1 INTRODUCTION

As shown in the previous chapter, different types of fracture laws should be used in models depending on the scale of representation of the material. In the lattice model, and other meso-level models, the material structure is included directly in the model in the hope that a simple fracture law suffices. Preferably, a simple brittle fracture law would suffice. In macroscopic models, concrete is treated as a continuum, and the non-linearities caused by the heterogeneous material structure must be included in the constitutive law. In addition to the material parameters, most models require the definition of model-related parameters that are a direct consequence of the element discretization. In most cases the model parameters lack a sound physical basis. In this chapter the determination of the fracture parameters in both the lattice model and the Fictitious Crack Model are described. In the Fictitious Crack Model the softening law must be determined, and this is the topic of Section 6.2. The softening parameters cannot be measured directly in an experiment, as may have become clear from Chapter 4. Note that only the procedure for estimating the material related parameters in the Fictitious Crack Model are described here. The additional parameters needed in, for example, enhanced continuum models, i.e., the internal length scale, will not be presented here, as up to this moment no reliable procedure has been developed to estimate such parameters.

As may be clear from the previous chapter, for the lattice model somewhat more work must be done before the model can be used. More specifically, the elastic stiffness and Poisson's ratio of the complete lattice should resemble corresponding values for an isotropic continuum. The parameters in the lattice model affecting the stiffness are all geometrical parameters and can be found in a simple and straightforward mapping procedure that is described below as well. The lattice model parameters are elucidated in Section 6.3.

Although the term "validation" is used in the title of this chapter, this does not really cover the contents. The validation includes a great number of model analyses that are presented in Chapters 7 and 8. The success of such analyses may show the range of applicability of a given model. Chapter 6 should be considered as an introduction to the final two chapters of the book.

6.2 FICTITIOUS CRACK MODEL

For the Fictitious Crack Model of Hillerborg and co-workers, the tension softening diagram is the main material property that must be determined. In Chapter 4, more specifically in Sections 4.1.3.2, 4.2.3 and 4.6, it was mentioned that direct measurement is not possible. Boundary condition and size effects have a significant influence on the fracture parameters. Because of the indeterminate nature of the fracture parameters, only an inverse procedure can be used to tune the model. The Hillerborg model is a so-called line-crack model, i.e., originally it was proposed to model crack growth between finite elements (Figure 5.1b). The crack-band model and the enhanced continuum models smear the fracture properties over the element size (Figure 5.1a), and for each model an additional parameter is needed, viz. the crack-band width and the internal length scale, respectively.

Using the original Fictitious Crack Model, the softening diagram must be estimated through an inverse analysis because of the fundamental problems of fracture models mentioned above. Hillerborg and Petersson proposed to use a uniaxial tension test between non-rotating loading platens as the best way to determine the softening diagram (Section 4.1.3.2). Later a displacement-controlled three-point-bend test was proposed, from which the fracture energy can be computed directly, as was discussed earlier in Section 4.6. However, if this procedure is followed, the tensile strength and the shape of the softening diagram, should be determined from an inverse procedure, or additional testing must be done. Inverse modelling has become an important topic in engineering, in particular in those areas where numerical models are used, see for example Bui et al.[462] Inverse procedures for the determination of the softening parameters have also been developed in the past few years.[171,315,372,463,464] The procedure is quite straightforward. A softening diagram is assumed, and in a finite element analysis of the fracture geometry under consideration, the deviation between the computational and experimental outcomes is determined. Depending on the difference between computation and experiment, the analysis is repeated using a modified softening diagram. This procedure is repeated over and over until the error has become sufficiently small, i.e., small enough to fall within predetermined bounds. As the shape of the softening diagram is quite complex, it is often approximated by a bi-linear relation as shown in Figure 6.1.

Both the SOFTFIT program developed by Roelfstra and the General Bilinear Fitting developed by Planas et al. are based on a bi-linear approximation of the real softening diagram. The procedure used in the SOFTFIT approach is shown in Figure 6.2. Initial values of E, f_t, s_1, w_1 and w_c are given, and the fracture response is computed using the finite element method. The tensile strength can also be estimated from other experiments like the Brazilian test or a uniaxial tensile test, but also from the following relation:[463]

$$f_t = (G_f\text{-}52)/8.5 \qquad (6.1)$$

It is suggested that G_f is related to the compressive strength following

$$G_f = 0.97f_c + 41.8 \qquad (6.2)$$

where f_c is given in [MPa], which results in G_f in [N/m]. If the error between computed and experimental outcome is too large, the input parameters are adjusted in a systematic manner until the required accuracy has been obtained.

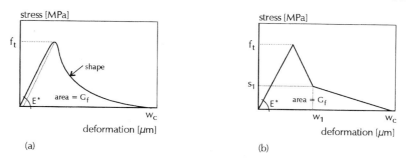

Figure 6.1 Curvilinear softening diagram (a) and bi-linear approximation with definition of the relevant parameters.

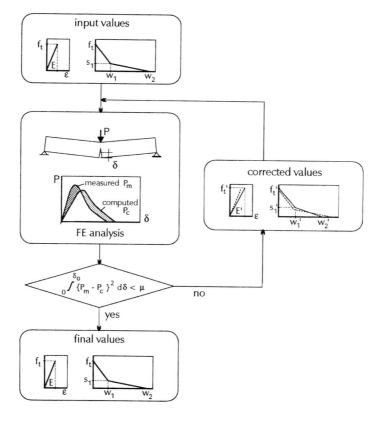

Figure 6.2 Procedure for estimating softening parameters with SOFTFIT, after Wittmann et al.[372,463] (Reprinted with kind permission from RILEM, Paris.)

The computational diagram should then be used for analyzing the fracture response of other structures.

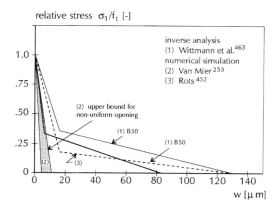

Figure 6.3 Comparison of softening parameters from a SOFTFIT analysis with computational diagrams obtained from simulations of the non-uniform crack opening problem in uniaxial tension.

The parameters for the "inverse" softening diagram determined by Wittmann and co-workers for concretes of different compressive strength are shown in Figure 6.3 and Table 6.1. The limitations of the inverse modelling procedures may be evident. The parameters can only be used for analyzing structures in which cracks nucleate and grow under identical circumstances. Again, extrapolation is dangerous, and in applying the results to situations that have not been explored before by experiment, serious attention must be given to the selection of the softening parameters. Fortunately, because of the abundant amount of research done to date, some faith in at least some of the fracture parameters has

TABLE 6.1
Estimated Parameters from SOFTFIT Analysis
for Concretes of Different Strength and
Composition.

f_c [MPa]	d_{max} [mm]	G_f [N/m]	f_t [MPa]	s_1 [MPa]	w_1 [μm]	w_c [μm]
Wittmann et al.[463] *three-point-bend test*						
30	16	71	2.2	0.8	17	130
60	16	90	4.5	0.8	17	130
Brühwiler[171] *wedge splitting test*						
55	12	76	4.4[a]	1.1	16	67
36	32	146	2.4	0.8	31	230
54[b]	120	235	2.0[c]	0.8	89	366

[a] determined from a Brazilian test
[b] determined from uniaxial tension test
[c] larger specimens were used for dam concrete

developed. The tensile strength and the initial slope of the softening curve are known to be the most important parameters. This is also clear when the results in Table 6.1 are compared. Wittmann found that the results were most sensitive to variations in tensile strength and the break point (s_1, w_1) in the softening diagram. The tail was relatively unimportant for different types of concrete. Moreover, numerical analyses of, for example, the boundary rotation effect in uniaxial tension reveal that the initial part of the descending branch is an essential characteristic of the softening relation for concrete. The slope can be estimated using Equation 5.52 for a specimen of given size and shape. The specimen should be tested in a machine with infinite stiffness, $K_r = \infty$. For the tensile machine in the Stevin Laboratory it can be computed that the initial slope for a mortar mix should be $(-dw/d\sigma)_{min} > -5.36 \; 10^{-3}$. Finite element analyses using the smeared crack model in the package DIANA, which is based on Bažant and Oh's crack-band model, reveal that the non-uniform opening in uniaxial tensile testing (which was addressed in Sections 4.1.3.2 and 5.3.2.3) can be simulated when a very steep part of the softening curve is assumed. In Figure 6.3 different "computational softening diagrams"[253,452] are compared to the relations from the inverse modelling approach used by Wittmann. Both computational approaches could capture the non-uniform opening, but the initial steep branch in the "computational softening diagrams" were much steeper than the results from the inverse analyses. The deformed specimen shape is plotted in Figure 6.4, which was similar for both numerical analyses. The typical waggling effect for a tensile specimen loaded between non-rotating end-platens is revealed and indeed, as was shown in Figure 4.63, two fracture zones develop. The tail of the softening diagram seems the result of stress transfer between the overlapping fracture zones, as was suggested in Section 3.3.1. This point will be worked out in more detail in Section 7.2.

One final remark regarding the inverse modelling is made here. It seems that the same degree of accuracy can be reached using softening diagrams with different combinations of f_t, w_1, s_1 and w_c. In other words, it is questionable whether a *unique* softening diagram can be found when the inverse analysis is limited to a single geometry. Figure 6.3 would indicate that the non-uniform opening in uniaxial tension cannot be simulated using the "inverse" diagrams determined by Wittmann and co-workers. In this figure the upper bound for the first steep part of the softening diagram needed to compute the "waggling" effect is shown. The result from the inverse analysis approach is clearly out-of-bounds. This would plead for an approach where the inverse analysis is not based on observations of softening in a single specimen geometry. Rather, many different specimen shapes and sizes should be explored, like the three-point-bend test *and* the splitting tensile test *and* the uniaxial tensile test *and* the Brazilian test *and* every other conceivable geometry where *mode I fracture prevails*.

Finally, it should be mentioned that softening models are not only used at the macro-level. Meso-level models developed by Roelfstra,[465] Stankowski,[466]

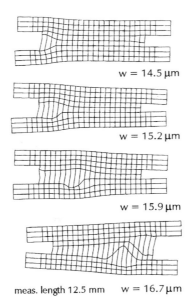

w = 14.5 μm

w = 15.2 μm

w = 15.9 μm

meas. length 12.5 mm   w = 16.7 μm

Figure 6.4 Growth of two interacting fracture zones in a numerical simulation of a uniaxial tensile test loaded between non-rotating end-platens.[253]

and Vonk[89] all use a softening relation for mode I crack growth at the particle level, whereas sometimes a softening model for shear is implemented as well. The model used by Vonk is elucidated in Section 7.4.2. The reason for applying softening at the meso-level stems from the fact that the matrix material cannot be regarded as a purely brittle material, as small sand particles are still contained in the matrix. As shown in Figure 3.49, cement mortar containing up to 2 mm sand still displays a significant softening behaviour. No experimental information is available to tune such models, however, and the only way left is to explore the sensitivity of an analysis to parameter variations.

6.3 LATTICE MODEL

In the lattice model the parameters can be divided into two categories. The first category concerns those parameters that are related to the elastic stiffness of a complete lattice, whereas the second set of parameters comprises those related to the fracture strength of the lattice beams. Important in the first category are the (constant) cross-sectional area of the beams and the Young's moduli of the material of which the beams are made. The most attractive property of lattices is that the lateral expansion is a natural part of the chosen discretization. Lateral deformations appear as a consequence of the triangular lattice structure itself. The choice of the Poisson's ratio of the lattice beams themselves seems therefore of less importance. Thus, the global Poisson's ratio depends on the geometry of the total lattice and the cross-sectional properties of the beams, as will be shown below. For the specific case where the structure of the material is mapped on a lattice, different Young's moduli are assigned to beam elements appearing in different parts of the material structure. For example, for the particle composite described in Section 5.2.2, the lattice beams falling inside the aggregates will get the Young's modulus of the aggregate material. The matrix beams and the bond beams (i.e., those beams that intersect the interface between aggregate and matrix) are given the same Young's modulus. The beam length and cross-sectional area must be selected such that the material structure can be described in sufficient detail, but also in such a manner that the overall Poisson's ratio of the lattice resembles the initial Poisson's ratio of the concrete that is modelled. In this context "initial" Poisson's ratio refers to the value for the material without cracks, thereby neglecting the fact that initial cracks may be present due to temperature and humidity effects in early stages of hardening of the concrete. In addition to the particle structure of the concrete, it may also be decided to include other material structural elements in the analysis, such as the larger pores in the cement matrix and bond zone.[461]

6.3.1 PARAMETERS RELATED TO THE ELASTIC LATTICE PROPERTIES

The first step in the identification procedure is the determination of the length of the lattice beams. The length of the beams depends directly on the size of the smallest characteristic material dimension that is to be included in the model. For example, for a particle composite such as concrete and sandstone, the size of the smallest particle included in the model determines the size of the (lattice) beams. In most of the analyses that are presented further on in this book, a lower cut-off for the particle size is made at 2 or 3 mm. This is necessary because reducing the particle size would require inclusion of many more elements in the model, and in the end an enormous computational effort is needed. The length of the lattice beams should be smaller that one third of the diameter of the smallest aggregate particle. In that case, not much difference is detected anymore in tensile stress-crack opening diagrams. In Figure 6.5, three examples of crack patterns of meshes with different d_{min}/l-ratios are shown. The length $l = 2.5$ mm, 1.667 mm and 1.0 mm. The minimum aggregate size was 3 mm.

As can be seen from the crack patterns in Figure 6.5, much of the detail is lost when longer lattice beams are selected, viz. $l = 5/2$ mm. In that case many of the small particles are lost in the model. The situation improves when the beam size is reduced, and for the case $l = 5/3$ and $5/5$ mm not much difference between stress-crack opening behaviour is found as shown in Figure 6.6, although the crack patterns are still slightly different. Similar results were obtained in compressive simulations.[468] In Section 6.3.3 the effect of the number of small particles included in the mesh and the effect of introducing porosity is discussed.

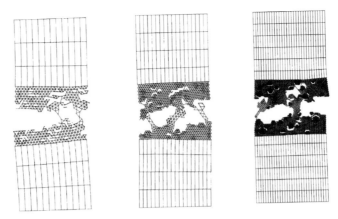

Figure 6.5 Deformed meshes of cracked specimens for three simulations with different beam lengths: (a) $l = 2.5$ mm, (b) $l = 1.667$ mm and (c) $l = 1.0$ mm.[467]

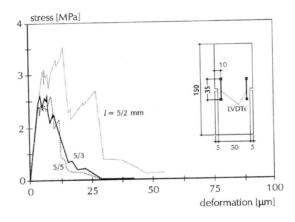

Figure 6.6 Tensile stress-deformation curves for three simulations with different beam lengths: (a) $l = 5/2$ mm, (b) 5/3 mm and (c) 5/5 mm.[467]

   Recently, it was shown that for a regular triangular lattice without particle overlay, where 25% of the beams was given a reduced strength, the fineness of the lattice has a significant effect on ductility.[469] The strength of the lattice was not affected. Thus, in addition to pure geometrical effects, the element discretization has a direct influence on the numerical outcome. This suggests that it is not sufficient to describe the material geometry in detail, but that other (unknown) aspects are important too. It should be mentioned here that in lattice analyses the "continuum assumption" is made on the level of the individual beams. This means that a beam is considered to be made of an isotropic

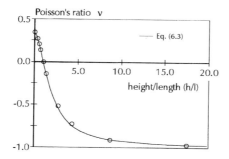

Figure 6.7 Relation between the Poisson's ratio of a triangular random lattice (50 * 50 nodes, *s* = 5 mm) and the ratio *h/l*.[400]

material that can be modelled as an elastic continuum. Given the mesh-dependency of the original smeared-crack models, which might be resolved by using enhanced continuum models (Section 5.3.4.3), such an approach should perhaps also be considered for the lattice type models at the level of the individual beams, at least when a softening law is used.

Now that the beam length has been "decided", the cross-sectional area of the elements can be determined. For a 2D analysis, in general rectangular cross-sections are used. Other shapes are possible too and, for example, for 3D analysis it seems more appropriate to use circular cross-sections.[295] Let us confine the discussion to 2D lattices. The height and thickness of the beams must be defined. Values for beam height *h* and thickness *b* must be selected such that the Young's modulus and Poisson's ratio of the complete lattice resemble the values measured in the elastic stage for the real material that is modelled. For a random triangular lattice the following relation between the height over length ratio (*h/l*) and the Poisson's ratio of a lattice of 50 * 50 nodes with maximum randomness on a grid of *s* = 5 mm (see Figure 5.4) can be obtained[400]

$$v = \left[\frac{4}{\left(3 + (h/l)\sqrt{3}\right)} \right] - 1 \tag{6.3}$$

This relation is shown graphically in Figure 6.7. As can be seen, v varies between 1/3 for *h/l* = 0 and -1 for large *h/l*. For *h/l* > 1/1, v becomes negative. This implies that the specimen will shrink in lateral direction under uniaxial compression, which is of course physically not acceptable for concrete. Thus, one should be careful to select the correct value of *h/l* which fits to the material under consideration. For different lattices the same procedure must be followed. Using Equation 6.3 and for a known length *l* of the beams (which was related to the particles in the material structure), the height *h* can be computed for a material with a given Poisson's ratio. The Poisson's ratio can also be determined analytically. For example, for a regular triangular lattice, Schlangen and Garboczi[401] derived a relation between the Poisson's ratio, the cross-sectional area A and moment of Inertia I, which can be simplified to

$$v = \frac{1-\left(h/l\right)^2}{3+\left(h/l\right)^2} \tag{6.4}$$

for rectangular cross-sections $A = b \times h$. For concrete with $v = 0.2$ this means that $h/l = 1/\sqrt{3}$ must be selected.

The last geometrical adjustment concerns the beam thickness b. The overall Young's modulus of the complete lattice should match to the elastic stiffness of the material under consideration. Obviously, the beam properties affecting the overall stiffness of a lattice are the beam cross-section ($b \times h$) and the Young's moduli of the beams themselves. The Young's moduli of the individual beams are adjusted to resemble realistic ratios of the (macroscopic) Young's moduli of aggregate, matrix and bond zone. Thus, the ratios E_a/E_m and E_a/E_b should be specified, where the subscripts a, b and m stand for aggregate, matrix and bond zone, respectively. For convenience, the stiffness of the matrix and bond zones are set to be equal, i.e., $E_a = E_m$. In fact, this is an approximation because, for example, in normal weight gravel concrete, the bond zone has a larger porosity than the bulk cement matrix (see Figure 2.19). Therefore, in general, the stiffness of the bond zone will be much smaller than the stiffness of the bulk matrix. In the analyses that are presented in Chapters 7 and 8, the reduced stiffness is neglected, but a lower fracture strength is assumed for the bond zone. Typical ratios of Young's moduli for the lattice beams are $E_a/E_m = 70/25$, with $E_m/E_b = 25/25$. Note that the overall stiffness is also affected when the larger pores are included in the mesh.[461] The thickness of the beams is finally adjusted such that the overall stiffness of the lattice resembles the macroscopic Young's modulus of the concrete. Note that a composite stiffness is computed which undoubtedly will depend on the aggregate content as discussed in Section 3.2.1. In fact, the aggregate volume could be varied to compute the changes in overall stiffness. This has not been done to date. The total procedure for mapping the elastic properties of a 2D-lattice is summarised in Table 6.2.

6.3.2 FRACTURE PARAMETERS

The second group of parameters of the lattice model contains the fracture parameters. For fracturing a beam, a law must be derived. The most simple law is a purely brittle mode I fracture law. As soon as the strength of a beam is reached it fractures completely following Figure 5.45. This is the only case that will be considered here. In Section 5.4.2 a number of approaches to compute the effective stress in a lattice beam were described, but this does not have much relation to the selection of the strength of the individual beams. If a particle overlay is used, following Figure 5.8, beams falling in the matrix,

TABLE 6.2
Overview of the Mapping Procedure of the Elastic Properties in a 2D Lattice with Particle Overlay

Step 1   *Beam length l*

From smallest feature in the material structure, for example minimum aggregate size $l < d_{min}/3$

Step 2   *Beam height h*

In combination with l related to the overall Poisson's ratio of the lattice, for example Equations 6.3 and 6.4

Step 3   *Beam thickness b*

Together with the stiffness of the individual beams related to the overall Young's modulus of the complete lattice

Step 4   *Elastic stiffness of the lattice beams*

Ratio E_a/E_m is defined according to realistic macroscopic data from uniaxial tension tests on aggregate and matrix materials. The bond zone stiffness is taken to be equal to the matrix stiffness: $E_m = E_b$.

Note: The lattice consists of beams with rectangular cross-section $A = bh$.

interface and aggregate material should all have their own strength. As was discussed in Section 2.2.3, the interfacial zone between aggregate and matrix is very weak, and most results indicate that this value is one of the key-parameters in a lattice analysis. The determination of the three strength parameters ($f_{t,m}$, $f_{t,b}$ and $f_{t,a}$, where the subscripts m, b and a stand for matrix, bond and aggregate phase respectively, and t indicates tensile fracture) is as difficult as that of the softening parameters described in the previous section. The only advantage seems that now single valued parameters are needed instead of complex softening functions.

When the lattice model was used first, representative macroscopic strength values were assumed.[396] Moreover, it should be realized — and this is a point where much confusion exists — that the lattice analyses are needed to better understand crack growth mechanisms in laboratory scale fracture experiments. In this sense, the model must be regarded as a truly experimental tool! Many people are tempted, because of a number of successes with the model, to see the model as a competitive model to macroscopic non-linear fracture models. This has, however, never been the intention. The goal is much less far-reaching.

As was discussed in Section 4.2.3, the tensile strength of concrete is dependent on the size of the specimen used. Therefore, it was considered that the ratios between the various strength parameters should be correctly specified, rather than the absolute values. As mentioned, the model is mainly used in a qualitative manner. The ratios that have been mostly used to date are based on characteristic macroscopic strength values for matrix, interface and aggregate.

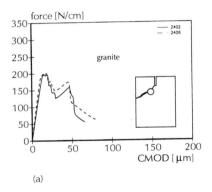
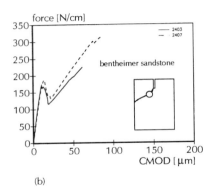

(a)   (b)

Figure 6.8 Load-crack-mouth-opening displacement diagrams in direct splitting tests on idealized concrete. Cylindrical aggregates are embedded in a homogeneous cement matrix. Different types of aggregates are used, like sandstone and granite as shown in the Figures (a) and (b).[52]

The ratios are

$$f_{t,m}/f_{t,a} = 5/10 \qquad f_{t,b}/f_{t,a} = 1.25/10 \qquad (6.5)$$

These ratios are for normal weight concrete containing stiff river gravel particles. For lightweight concrete, or high strength concrete, or even sandstone, different ratios must be assumed. In the above approach all beams falling in each of the three distinct material phases that are distinguished at the meso-level have the same strength and stiffness. In reality, it seems more appropriate to use a distribution of strength and stiffness for elements falling in the cement matrix. The same should be done for the bond and aggregate zones. Different strength and stiffness distributions can be selected, such as a normal distribution or a Weibull distribution, but such refinement of the model has not been carried through to date.

At present, development of a more rational inverse procedure for the determination of the strength parameters has been attempted. By studying crack growth in an idealized particle composite, as shown in Figure 6.8, we try to determine the strength of matrix and bond zone. This is done through inverse analysis, where the typical strength peaks in the stress-crack-mouth-opening displacement diagrams and the fractal geometry of the crack pattern are used as response parameters to fit the parameters as good as possible.[53]

One final remark should be made. The fracture law (Equation 5.69), has been most deeply studied to date. An effective stress is computed from the normal forces and bending moments. In this equation, a factor α is introduced, which regulates how much bending is taken into account. The factor α seems to lack a sound physical basis. Variation of the parameter has shown that for a long tail in the tensile softening diagram a low value (i.e., $\alpha = 0.005$) is preferential. The compressive response changes from brittle to ductile when α decreases from 1.5 to 0.1 as shown in Figure 6.9. More about this is given in

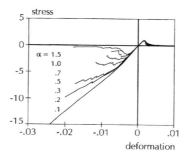

Figure 6.9 Effect of α in the fracture law on the stress-deformation behaviour of a $5 \times 5 \times 5$ noded, random triangular lattice.[400]

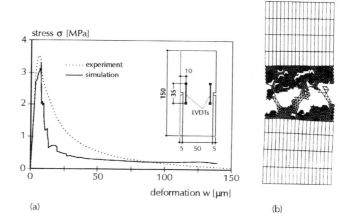

(a) (b)

Figure 6.10 Calibration of the model to a uniaxial tensile test on a concrete prism.[467]

Section 7.4. Here it is mentioned that the present lattice model seems to lack an essential ingredient to capture compressive fracture satisfactorily.

A comparison between a numerical lattice simulation and a uniaxial tension test is shown in Figure 6.10. The test is carried out in displacement control. The specimen is a small prism ($50 \times 60 \times 150$ mm³), with two 5 mm deep notches at half height. The length of the lattice beams is 1.667 mm, and only particles between 3 and 8 mm are included in the mesh. In Figure 6.10a a comparison is made between the experimental and numerical stress-deformation diagrams, whereas in Figure 6.10b the computational crack pattern at the end of the analysis is shown. The computational stress-deformation diagram indicates too brittle behaviour but, as will be further elucidated in Section 7.2, the crack growth mechanisms seem quite well captured by this version of the lattice model. In the analysis of Figure 6.10 the parameters as mentioned in the above sections have been used.

6.3.3 SMALL PARTICLE EFFECT, POROSITY AND 3D EFFECT

In the remainder of this book, attention will be given to the outcome of numerical simulations, both from macro-level fracture models as from lattice type models and other meso-level models. The results will be used mainly to explain failure mechanisms and to underscore experimental results that were presented in earlier chapters. As was mentioned before, the title of this chapter is a bit misleading and does not really cover its contents up till now. Validation of the parameters in the numerical models includes a critical comparison of numerical results with experimental observations. All parameters should be varied extensively for different specimen geometries and loading conditions. In advance of the coming chapters, the effect of material structural elements, such as neglecting the small aggregate particles in the mesh, and/or neglecting the large scale porosity in the matrix and bond-zones are discussed here. Both issues seem to directly affect the post-peak brittleness in the lattice computations, which is so clearly visible from Figure 6.10. Another issue directly affecting the brittleness is the fact that the three-dimensional fracture process is approximated in a two-dimensional analysis. This so-called 3D-effect can be analyzed following two distinct approaches. In the first approach, a full three-dimensional lattice is used. The model parameters must be tuned again for the 3D-case. A disadvantage of this method is the huge increase of computational effort, which makes it virtually impossible to make sufficiently detailed 3D-analyses with present-day hardware and software. Another approach is to adjust the fracture law, as pointed out in Figure 5.46. In this approach a beam is not completely removed as soon as the strength of the element is reached, but rather, its stiffness is reduced stepwise. Thus, it is assumed that a crack does not extend over the full specimen thickness. This is quite a realistic approach, as crack fronts are non-linear as pointed out in Figures 3.39 and 4.72. Essentially, this quasi-3D approach resembles the effective crack models that are based on linear elastic fracture mechanics in which an effective crack length is used, rather than the crack length visible at the specimen's surface. Summarizing in the following paragraphs the small particle effect, the influence of porosity and two approaches to include three-dimensional effects in the analysis are described. All these efforts are directed towards increasing the confidence in the lattice approach.

6.3.3.1 Omitting small particles

In every lattice analysis presented in Chapter 7 and 8, not all small particles in the concrete mixture are included in the mesh. The reason for this is of a purely practical nature. The computational effort becomes too large for an analysis at a workstation when the lattice contains more than 50.000 elements. The size of a lattice is directly coupled to the size of the smallest material feature in the mesh. Of course, larger lattices could be computed using super-computers, but the experience is that even than the total elapsed time is simply too large at present. Perhaps more detail can be included in the analyses when

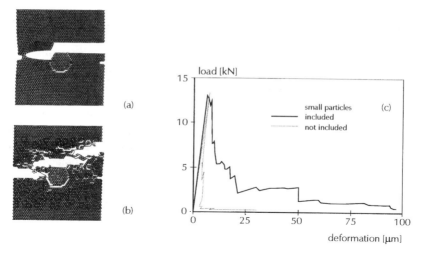

Figure 6.11 Small particle effect: lattice containing a single large aggregate (a), lattice containing a single large particle and randomly distributed small particles (b) and computed stress-crack opening diagrams (c).[470]

faster computers become available in the future, or when more advanced and faster solvers are developed. However, at the same time it is felt that *fundamental shortcomings of all numerical approaches cannot be solved by continuously including finer detail in the analyses.* The simple fact that, irrespective of the level of representation, selected (meso- or macro-) parameters must be determined through inverse analysis means that numerical models will always lack predictive power. Or, in other words, the models will only be capable of reproducing results of situations that have been explored before by experiment. Drawing conclusions on situations that have not been studied experimentally is a haphazard undertaking. In Chapter 8 a very fine example of the "predictive capabilities" (or rather the inablitity of many models to predict fracture mechanisms) of numerical models will be presented. In spite of this fundamental short-coming of numerical models it is still possible to apply the computations to improve the insight in fracture mechanisms. One important aspect is the small particle effect. Because of computational limitations, this effect is difficult to demonstrate on a full-scale analysis of a laboratory scale fracture specimen. However, a model analysis will suffice. In Figure 6.11 the simulation of the fracture response of a square element containing a single, large aggregate particle is compared to a situation where many small particles are included in the otherwise homogeneous matrix.

The size of the total lattice is 80 × 80 mm, the diameter of the large aggregate particle is 20 mm, whereas the small particles have a diameter of 4 mm. The fractured meshes are shown in Figure 6.11a and b, respectively, whereas the load-deformation diagrams of the two analyses are shown in

Figure 6.11c. The difference in behaviour will be evident. The amount of side-cracking, bridging and branching increases substantially when the small particles are included in the mesh. The ductility beyond peak is also enormously affected as can be seen from Figure 6.11c. The fine detail in the crack patterns has also been observed in impregnation experiments and SEM crack-detection tests (see Chapters 3 and 4). The increase in crack density depends of course on the level of observation in such experiments, but as shown here, the same is true for the computations. Related to this is the determination of the fractal dimension of crack patterns. Comparison between experimental and computational crack patterns only makes sense when the same resolution is obtained.[471,474] The small particle effect was also shown in computations of the same small prism that was used in the analysis of Figure 6.10. The conclusion supports the result of Figure 6.11.

6.3.3.2 Including porosity

Porosity has a large effect on strength as well. Notably, the water-cement ratio is one of the major controlling factors. It was perhaps most convincingly shown in the simple tests of Birchall et al.[26] that were discussed in Section 2.1.3. Crack growth may also be affected by the larger pores. They may either cause local stress concentrations in the cement matrix and the interfacial zone, leading to crack nucleation and growth. On the other hand, the larger pores might also act as crack-arrest mechanism, see for example Huang and Karihaloo.[472] Porosity can also be included in a meso-level model, provided of course that the number of elements remains limited (for computational reasons!). This either sets a limit on the number of pores that is included, or on the pore-size. Using a lattice with $l = 5/3$ mm long beams, Arslan et al.[473] analyzed the effect of porosity on the stress-crack opening relation and crack-growth mechanisms under uniaxial tension. The pores were included in the mesh by randomly removing beams from the regular triangular lattice that was used. Pores were therefore all of the same size, viz. equal to the beam size. Because of this size limitation it was felt that the total porosity should not be taken too high. Three cases were compared, namely 0, 6.35 and 12.77% total porosity in the matrix and bond zones. The pores were simply placed in the material by means of randomly generated numbers over element numbers. The specimen geometry was identical to that used in the analysis of Figure 6.10. An alternative fracture law was used, in which the beam was not immediately removed after the maximum strength was reached. Rather, the stiffness was reduced in six steps using the fracture law of Figure 5.46. Thus, complete fracture of a beam was achieved if a total strain of 6ε was exceeded.

In Figure 6.12 the computed relative load-deformation curves are shown for the three different porosities, whereas in Figure 6.13 the final crack patterns are shown. Note that the load-deformation diagrams were all normalized with respect to the maximum load obtained for the 0% porosity case. A number of observations can be made. First of all, because of the application of the

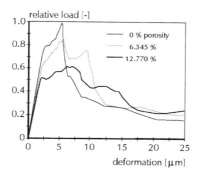

Figure 6.12 Effect of porosity on the relative load-deformation behaviour under uniaxial tension.[461]

alternative fracture law, the computed load-deformation diagrams indicate a decreasing brittleness as compared to analyses where the simple brittle fracture model Equation 5.69 and Figure 5.45 were used. Figure 6.12 clearly shows a decrease of peak strength with increasing porosity. An increase in ductility is observed too. This suggests that crack initiation occurs at an earlier stage as compared to the zero-porosity case, whereas crack arrest and multiple cracking must be held responsible for the increased ductility. The crack patterns of Figure 6.13 were taken at the end of the analyses and indeed show larger crack densities with increasing porosity. In particular, in the 12.77% porosity case, crack growth is not confined to the area between the notches, but rather cracks nucleate and grow at many different places in the specimen. Finally, localization of cracking occurs at a single location, in the example of Figure 6.13c

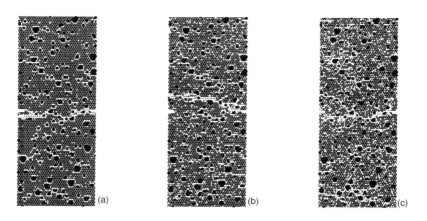

Figure 6.13 Computed crack patterns for different total porosities: (a) 0%, (b) 6.35% and (c) 12.77%.[461] Note, that the porosity is drawn in the same way as cracks. Isolated white spots in the matrix and bond zones are pores of single beam size.

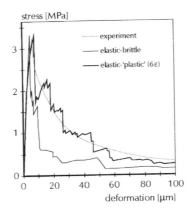

Figure 6.14 Effect of fracture law on the computed stress-deformation diagram in uniaxial tension.[461]

between the notches. The total porosity in hardened cement paste is of course much higher than the values used here. As mentioned before, computational limitations set a bound on the porosity that can be included at the level of discretization. It is expected that with further increasing porosity (increase of the number of pores of smaller size), a further decrease of load will be found as well as an increase of ductility. However, at some stage, the size of the pores will be reduced so much that their effect on mechanical properties will gradually disappear.

6.3.3.3 3D-effects

Three-dimensional effects can be included in the analyses in different manners. The simplest is to use an effective crack model as shown in Figure 5.45. By stepwise reduction of the stiffness of a beam, instead of removing it completely, partial crack growth over the thickness of the specimen is simulated. In Figures 3.39 and 4.72 curvilinear crack fronts were shown, which suggest that the above assumption is useful. The effective fracture law following Figure 5.46 was used by Arslan.[461,473] In Figure 6.14, the effect of the number of steps needed to achieve complete fracture of a beam on the stress-deformation diagram in uniaxial tension is shown. For simplicity only the 6ε case is compared to an analysis in which the brittle fracture law of Figure 5.45 is used. Also, a representative experimental result is included in the figure. Note that no porosity was included in these analyses. The increase in ductility is quite substantial, and the 6ε is shown here because it came closest to the experimental data. The number of steps was varied from 3ε, 6ε to 9ε.[473] The 3ε results lay between those of the brittle simulations and the 6ε case, but the 9ε analysis was almost identical to the 6ε analysis.[473] These lattice analyses show that an effective 2D model can be used to improve the stress-deformation

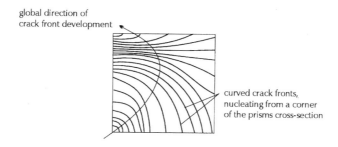

Figure 6.15 Non-uniform crack growth in a prism subjected to uniaxial tension (schematic, based on experimental observations from Chapters 3 and 4).

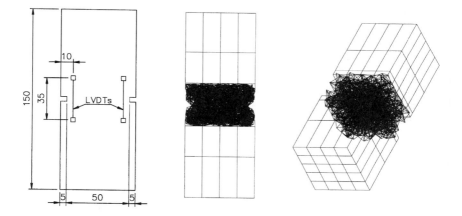

Figure 6.16 Calibration test geometry (a), front view of the random lattice (b), and isometry of the 3D mesh (c).[400]

response, but in order to simulate three-dimensional crack-growth processes realistically, a full 3D-analysis must be carried out. Note that in the uniaxial tension tests on cylinders that were presented in Figure 4.39, but also in the results of Figure 4.59, non-uniform crack growth from a single point along the circumference of a specimen was observed. The crack front would then slowly extend over the full cross-section as shown schematically in Figure 6.15.

Fully 3D-analyses were carried out by Schlangen.[295,400] Part of a small tensile specimen (again the same geometry as was used in Figure 6.10) was modelled using a three-dimensional random lattice, which was constructed along similar lines as described before for two dimensions (Figure 5.4). The specimen geometry, the frontal view of the finite element mesh, and an isometry of the 3D mesh are shown in Figure 6.16. Note that only the part of the specimen where cracks were expected to grow was modelled as a lattice; the remainder of the specimen was modelled using volume elements available

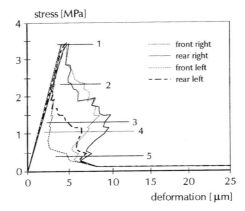

Figure 6.17 Stress-displacement diagrams at the four corners of the specimen of Figure 6.16. Stages numbered 1 through 5 refer to the crack graphs of Figure 6.18.[400]

in the finite element package that was used to perform the necessary linear elastic computations. Instead of using beams with rectangular cross-sections, circular cross-sections were chosen. For the fracture law, Equation 5.69, two bending moments had to be considered. Results of the analysis are shown in Figure 6.17 and 6.18. Because the mesh was much coarser than that of the 2D-lattice used in Figure 6.10, no direct comparison of the stress-deformation curves can be made. Therefore, the stress-deformation diagrams at the four corners of the prism are shown in Figure 6.17. The picture that emerges is quite similar to the outcome from the cylinder tests between non-rotating end-platens (Figure 4.39b). Highly non-uniform deformations are computed. The crack-growth process visualised in Figure 6.18 reveals the mechanism depicted schematically in Figure 6.15. In Figure 6.18 five stages of crack growth are shown that are also identified in the stress-deformation diagrams of Figure 6.18. Because the cracks are difficult to visualise in a three-dimensional plot, three projections are shown, namely on the planes ABCD (figure in column a), the back plane marked by line CD (column b) and the side plane marked by line AC (column c). Most clear is the non-uniform crack growth in the specimen's cross-section (column a in Figure 6.18). The crack nucleates at the right front corner, propagates diagonally through the cross-section (stage 2), but then rotates such that the crack front runs parallel to AC in stage 3. The crack-front rotation continues in stage 4, and the last part to fracture is the front left corner of the specimen. Thus, the fracture mechanism becomes quite realistic in a fully 3D-analysis. However, improvements of the stress-deformation behaviour were not detected because, for computational reasons, only a relatively coarse mesh could be used. It is expected that future developments in computer software and hardware will allow for fully fledged three-dimensional analyses. However, it should be emphasized that 3D-analyses are only part of the

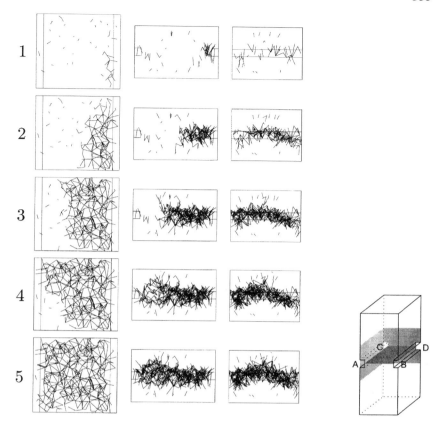

Figure 6.18 Five stages of cracking in the analysis of Figure 6.17. For each stage of cracking a section through the notches, a front view and a side view of the specimen are shown. In each section the beams that were removed from the lattice are projected onto the specific sectional plane.[400] The lettering scheme is explained in the text.

solution. Inclusion of small particles, as well as porosity, has a large effect on the stress-deformation behaviour also.

Further developments of crack-detection techniques will undoubtedly lead to a much improved assessment of crack densities and crack distributions. However, it is felt that such refinements will not substantially change the view that has emerged from the previous chapters.

The analyses presented in the following two chapters are all two-dimensional representations. In many cases this is justified because we attempted to grow forced cracks into two-dimensions. As will be shown, fracture processes can be better understood with the aid of numerical simulations, even if the models are not perfect. Discrepancies between experimental and numerical outcomes can, in most cases, be explained from issues discussed in this chapter, but sometimes additional reasons can — and will — be given.

Chapter 7

NUMERICAL SIMULATION OF PLAIN CONCRETE FRACTURE EXPERIMENTS

Nature must be explained by Nature and not by our own views

Abraham Trembley (1710-1784)

7.1 INTRODUCTION

Up till now, we have discussed the structure of cement and concrete (Chapter 2), mechanical properties as they are usually presented in text books on concrete technology (Chapter 3), experimental techniques and their problems (Chapter 4), and numerical techniques (Chapters 5 and 6). In the final two chapters of the book, an effort will be made to integrate the results from numerical simulations, both at the meso-level and the macro-level. Experiments and numerical simulations both serve two goals. Experiments are used to further explore the behaviour of materials and structures, but experiments are also needed to quantify parameters needed in numerical (or analytical) models. On the other hand, numerical models are developed for structural and materials engineering. In particular in numerical models, it is possible to include the boundary conditions of the problem in a very detailed and realistic manner, which makes numerical models suitable for designing experiments as well. The optimum situation develops when a very deep interaction between experiment and numerical simulation can be achieved.

Over the years it has been recognised that material behaviour cannot be separated from the conditions under which the material is used or studied. For example, in laboratory tests for measuring fracture parameters for concrete, the interrelation between material properties and structural environment can be seen from boundary conditions and size effects. Thus, the presentation of material behaviour independent of the conditions under which the properties were measured does not seem to make much sense. For example, with the aid of knowledge of slenderness effects and boundary restraint effects on the uniaxial compressive strength, it would be no problem at all to help someone to prove that he has achieved the requisite compressive strength in his concrete mixture. With the aid of numerical tools, effects from the structural environment can be studied in great detail. Even if the models are not correct, they may give a much increased insight in material behaviour and structural behaviour, which was unthinkable in times when only analytical tools were available.

In the past years, I have enjoyed a very fruitful collaboration with two of my Ph.D. students, Vonk and Schlangen, who both have great expertise in the field of numerical modelling. Each of them developed a meso-level fracture model of concrete.[89,156] Schlangen's work was related to tensile fracture, whereas more emphasis was on compressive fracture in Vonk's thesis. In this chapter, it will not be attempted to present all the results of the numerical simulations in an exhaustive manner. Instead, a number of examples will be shown. Important is the principle of the approach, rather than the individual examples, which are bound to become out of date when new, even more detailed, results become available in the future as computer technology advances at a very rapid rate. Similar as in Chapters 5 and 6, it is tried to draw a comparison between results obtained with macroscopic models and mesoscopic models.

In the ubiquitously used macro-models, no particle structure is included in the discretization. Non-linear material models are implemented, such as the softening models presented in Section 6.2, which are of course in many cases an idealized representation of "real" material behaviour. In the meso-level models, internal structure of concrete is included, and the material laws are defined for the different materials found at the meso-level, i.e., bond strength and stiffness, matrix strength and stiffness, and aggregate strength and stiffness. Fibres might also be included at this scale, but this has not been attempted to date. In fibre-concrete analysis, similar as in reinforced concrete analysis, the bond between steel (fibre or reinforcement) and concrete must be included as well.

In this chapter, some issues regarding plain concrete fracture are discussed. First, in Section 7.2, crack growth and crack face bridging at the meso-level are explored to explain tensile softening at the macro-level. Results from different particle models used by different researchers are compared. The impression is that the bridging effect may be dependent to some extent on the boundary conditions under which a fracture zone grows. Therefore, the effect of boundary rotations in a uniaxial tensile test is studied (again) by using the lattice model. Here a comparison is made with macroscopic approaches. Moreover, it is shown how differences in material structure (e.g., different types of aggregates) affect the crack growth processes at the particle level of the concrete. Second, in Section 7.3, the size and boundary restraint effects (see Section 3.2.4 for the experimental results) in uniaxial compression are addressed. At present, lattice models seem not very well suited for modelling compressive fracture, but by means of the UDEC model,[89] or other models such as the rigid particle model by Liu et al.,[475] better results are obtained, although at the cost of more complex meso-level material laws. Third, in Section 7.4, the four-point-shear beam of Section 3.4 is analyzed using a meso-level lattice model. The results are compared with simulations using continuum based finite element analysis. In Section 7.5, finally, an example is given where the biaxial failure contour of concrete is computed by means of a meso-level analysis. In this way the phenomenological models that were presented in Section 5.3.1, might be based on more rational physical grounds, which will eventually lead to much improved numerical simulation models. In Chapter 8 some examples of structural analysis of reinforced concrete are presented.

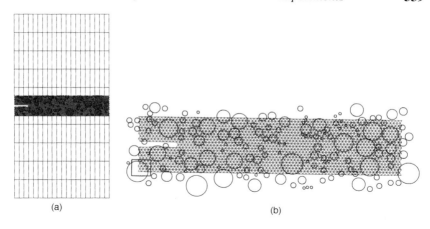

Figure 7.1 Element mesh for the tensile SEN specimen (a). The lattice area is shown enlarged in (b).[396]

7.2 UNIAXIAL TENSION

Uniaxial tension has been discussed in almost every chapter thus far, both from an experimental and analytical point of view. Here, for the last time the issue of non-uniform opening in tension is discussed. Numerical techniques are used to elucidate the problem. Analyzing the same example with many different tools clearly shows the advantages and disadvantages of each of these approaches. In parallel to the non-uniform opening in tension, the relevant micro-crack processes are analyzed.

7.2.1 CRACK FACE BRIDGING

The impregnation experiments presented in Section 3.3.1.1 have been analyzed by means of the lattice model.[396] The same boundary value problem was analyzed by means of an LEFM-based model.[417] A comparison between results obtained from the two approaches is made here. In the lattice analyses the parameters as given in Section 6.3, Equation 6.5 were used. The element mesh is shown in Figure 7.1a, and, as can be seen, only the area where cracks were expected to grow is modelled as a lattice. The lattice area is shown enlarged in Figure 7.1b. The remainder of the specimen is modelled using four-noded plane-stress elements that are available in the finite element package (DIANA) that was used for solving the set of equations. Modelling part of the specimen as lattice has the advantage that the (required) computational effort is very much reduced. On the other hand, micro-crack processes outside this area might be substantial (see Figure 6.14), but are now missed. For the present example this is no problem.

In the model aggregates, between 2 and 8 mm are included in the mesh. The length of the lattice elements is 1.25 mm, which results in omitting some of the

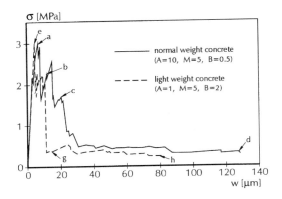

Figure 7.2 Stress-crack opening diagrams for simulations of normal weight and lightweight concrete.[396]

2 mm particles. In fact, this value is at the brink of the allowable (normally a ratio of three between minimum particle size and beam length should be used[467]). The boundary conditions are identical to those prescribed in the experiments: the top and bottom edges of the specimen are prevented to rotate during crack propagation. This is quite essential in such simulations: the boundary conditions should correspond to those used in the experiment. Tensile fracture propagation in concrete is highly sensitive to variations in the boundary conditions.

In Figure 7.2 two load-deflection curves obtained with this geometry are shown. Curve 1 concerns a simulation of normal weight concrete using the parameters specified before (Section 6.3), Curve 2 is a simulation of lightweight concrete. In this second analysis the ratios $f_{t,m}/f_{t,b}$ and $f_{t,a}/f_{t,b}$ were changed and set to 2.5 and 0.5, respectively, corresponding to strength values of 1, 5 and 2 MPa for the aggregate, matrix and bond zone beams. Corresponding to the discussion in Section 2.2.2, a higher bond strength should be adopted for lightweight concrete simulations.

Both curves of Figure 7.2 have a more or less identical shape. After a rising branch, a peak is reached, followed by a steep descending branch with a shallow tail at larger crack openings. A slight difference between the two curves is the somewhat increased brittleness of the lightweight concrete, indicated by the steeper falling branch of the initial part of the softening curve. This corresponds to the experimental observations presented in Section 3.3.2.1 (Figure 3.49). For each of the simulations cracking is plotted at four stages of crack-opening. These stages are identified in the load-crack opening diagrams with the letters (a), (b), (c) and (d) for the normal weight concrete, for which crack patterns are shown in Figure 7.3. Cracking at stages (e), (f), (g) and (h) for the lightweight concrete simulation are shown in Figure 7.4.

Let us first consider the normal weight concrete results. At peak (Figure 7.3a), distributed microcracking appears in the notched region of the specimen.

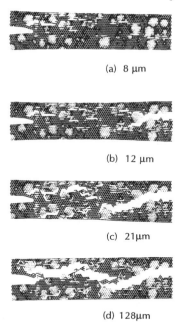

(a) 8 μm

(b) 12 μm

(c) 21μm

(d) 128μm

Figure 7.3 Crack growth in the normal weight concrete simulation of Figure 7.2. The stages are identified with the letters (a) through (d).[396]

The micro-cracks develop primarily at the interface between the aggregate and matrix, which is no surprise in view of the low interfacial strength that was specified. Into the steep part of the descending branch, larger cracks start to appear. In this particular analysis, a larger crack developed near a concentration of large aggregate particles at the "un-notched" side of the specimen as shown in Figure 7.3b. At the end of the steep part of the softening curve [stage (c)], more large crack branches appear, and the specimen is almost fully cracked. In the tail of the softening diagram, a full crack traverses the specimen, but stress transfer is still possible at isolated spots called crack face bridges, see Figure 7.3d. The development of the crack face bridges seems in agreement with the impregnation experiments (see Plates 4 and 5) as can be seen from a comparison of the numerical results with the experiments presented in Figures 3.42 through 3.47. Note that crack growth from the "un-notched" side of the specimen was also observed in some of the experiments, see for example the crack fronts in Figure 3.39.

In the lightweight concrete simulation a relatively low aggregate strength was specified, which causes the initial micro-cracks to appear in the aggregate particles rather than at the interface between aggregate and matrix as shown in Figure 7.4e. Subsequently, the distributed micro-cracks expand into the matrix (Figure 7.4f and g), and finally, at full separation, crack face bridges remain, quite similar to the normal weight concrete analysis. Note, however, that the

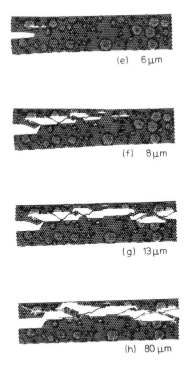

(e) 6 µm

(f) 8 µm

(g) 13 µm

(h) 80 µm

Figure 7.4 Crack growth in the lightweight concrete simulation of Figure 7.2. The stages are identified with the letters (e) through (h).[396]

crack faces are now more smooth than in the analysis of the normal weight concrete. Crack growth through the lightweight particles is in agreement with the observations from the impregnation experiments, see Figure 3.42d and Plates 2 and 3. In the experiments, small crack face bridges were observed in the lytag particles themselves, indicating small scale heterogeneity in the aggregates as well. This was recently confirmed in tensile splitting tests.[54] The details of the analyses require further improvement, but at present this cannot be achieved without an enormous increasing computational effort.

As was already mentioned in Section 6.3.2, the lattice analyses give too brittle results, which can be explained from the fact that too much detail at the microstructural level has not been taken into account. For example, in the analysis of Figure 7.1, all particles below 2 mm were excluded, although in Figure 3.36 it was shown that 2 mm mortar has a distinct softening behaviour. The brittle fracture law assumed in the lattice model should therefore be replaced by a softening law if no more detail is included in the analysis. But then, of course, one might as well return to the continuum-based particle models like the numerical concrete model.[465]

Other particle models, like Vonk's UDEC model and the LEFM based model of Wang and Huet[164] and Wang,[417] are capable of representing the

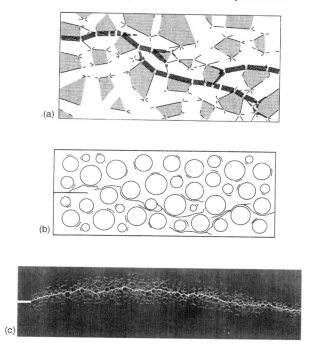

Figure 7.5 Crack face bridging in tension. Results from numerical simulations with UDEC of Vonk[89] (a), the LEFM based particle model of Wang[417] (b), and the lattice model of Bolander and Kobayashi[476] (c). (Figures (a) and (b) are reprinted by permission of the authors. Figure (c) is reprinted by kind permission of Aedifcatio Verlag GmbH.)

overlapping crack mechanism as well. Lattice analyses by Bolander and Kobayashi,[476] using a model that was identical to the above lattice model, also show the same mechanism in the tail of the diagram. Some of the relevant results are reproduced in Figure 7.5. Note that Vonk[89] used a tensile softening model at the meso-level. In his model, the softening branch was linear, and the maximum crack opening where the crack would become stress-free was — rather arbitrarily — set at 20 μm, see also Section 6.2.

The result of Wang (Figure 7.5b) was obtained for the same boundary value problem of Figure 7.1. In his model, Wang allowed for crack growth based on the analysis of the mixed mode I and II stress intensity factor of *pre-defined* interfacial cracks. As soon as the computed stress intensity factor exceeded a critical value, the cracks would propagate. Crack growth was limited to the aggregate-matrix interface and to the matrix material, but was never allowed to enter the aggregate particles. In comparison to the lattice model this is a rather limiting factor. Moreover, the LEFM-based fracture law is more complicated than the brittle, strength-based fracture law in the lattice model. In both cases, however, the meso-level material parameters are equally difficult to determine.

Other particle models, like the model developed by Zubelewicz and Bažant[404] and the more recent random particle model of Jirásek and Bažant,[402] do not reveal the bridging mechanism as clearly as the results shown in Figure 7.5. In the random particle model developed by Jirásek, the interaction between particles was modelled as a lattice of bar elements, which does not seem to be the best choice for fracture problems, as was also indicated in Herrmann and Roux.[394]

From the current experience with lattice models, it is suggested that tuning models to crack mechanisms, like crack fractality,[54] rather than to overall stress-deformation behaviour, might lead to substantially more reliable numerical models. We return to these matters later on. One other remark is in place here. Numerical models are capable of simulating bridging only *after* the mechanism is observed in experiments. Vonk and Schlangen were closeby when the mechanism was found, and their models were the first to include the bridging mechanism. Others followed later. The earlier models do not even recognize the mechanism, which might indicate the importance of testing. Unfortunately, most researchers prefer to work with computer models and to simulate material and structural behaviour. The role of experiment is usually very much underestimated. This is a serious problem. *A model can tell us anything, but only an experiment can discriminate!*

7.2.2 EFFECT OF BOUNDARY ROTATIONS ON TENSILE SOFTENING

An interesting effect, several times alluded to before, is the non-uniform opening in uniaxial tension. Figure 6.4 showed that two fracture zones might develop, suggesting that the bridging mechanism might appear not only at the level of the material structure (particle structure), but at a larger scale as well. Also, the fact that two fracture zones are interacting in a uniaxial tensile test — a test that has long time be considered as being the "best" experiment to determine the softening properties for the Fictitious Crack Model — gives some room to doubt whether the macroscopic fracture models are really based on secure physical observations. Carpinteri and Ferro[292,300] tried to overcome the non-uniform opening problem by designing complicated tensile tests where uniform deformations were forced on the specimen by means of an electronic control system in combination with three hydraulic actuators, see also Figure 4.45. The quotation at the beginning of this chapter seems more in place here than anywhere else in this book (although another example will be given in section 7.4). The advantage of forcing a specimen to behave as assumed in a model is not clear, and it seems that in this way we are rather fooling ourselves instead of making progress in science.

The situation created by Carpinteri and Ferro in their experiment was simulated with the lattice model. For computational reasons the experiment was modelled in two dimensions. Here, only the smallest specimen is considered. Thus, $b = 50$ mm and the specimen thickness $t = 100$ mm. The boundary

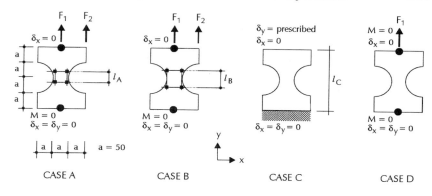

Figure 7.6 Specimen geometry tested by Carpinteri and Ferro,[300] and the four different boundary conditions analyzed with the lattice model.[477]

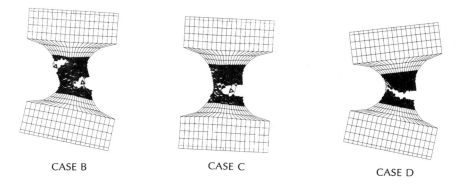

CASE B   CASE C   CASE D

Figure 7.7 Cracking for three simulations with particle structure 2 for boundary condition B, C and D. For each analysis cracking after removal of 250 beams.[294]

conditions in their tests are shown as case B in Figure 7.6. Uniform deformations are kept in a 50 mm wide zone in the central section of the specimen. F_2 indicates the location of the hydraulic actuator that is used to generate the counteracting moment (see Section 4.1.3.2 for a more detailed description of the experimental technique).

For comparison three additional boundary conditions were analyzed with the lattice model, namely uniform deformations in a 35 mm wide zone (case A), parallel end-platen displacement (case C) and freely rotating end platen (case D). Because the response of case A and B did not differ too much, only the results for simulations B, C and D are shown in Figure 7.7. For each case two different analyses with a different generated particle structure were carried out.[477]

Here only the crack patterns pertaining to the second particle distribution are shown. In Table 7.1, the computed fracture energies up to an average crack

TABLE 7.1
Fracture Energies for Simulations of
Dog-Bone Shaped Specimen Geometry
Used by Carpinteri and Ferro[300] for
Four Different Boundary Conditions[477]

Simulation	Grain structure	G_f (N/m)[a]
CASE A (ℓ_{meas} = 35 mm)	1	40.0
	2	44.0
CASE B (ℓ_{meas} = 50 mm)	1	40.6
	2	43.8
CASE C (fixed)	1	35.3
	2	32.5
CASE D (rotating)	1	29.8
	2	11.6

[a] Fracture energy values are based on 35 mm gauge length, and cross-sectional area of $b \times t = 50 \times 100$ mm².

opening of 100 µm are shown for all cases and for each of the two particle distributions. Clearly visible are the differences in failure mode depending on the selected boundary condition. Most significant is perhaps the difference between freely rotating and fixed platens. When fixed platens are used — and it seems that the cases B and C can be compared to one another — two sets of cracks develop, namely at the left and right side of the specimen. However, at the end of the simulation only a single crack has traversed the specimen, and the second branch closes again. This is identical to the crack face bridging mechanism reported in Section 3.3.3.1. In contrast, the simulation with freely rotating boundaries shows the development of a single crack zone, which eventually follows a kinking path (case D). Figure 7.8 shows that cracking proceeds much faster in the simulation with freely rotating boundaries (case D). In this figure the deformations at the left (w_1) and right side (w_2) are shown. For boundary condition B the deformations at the left and right are equal, i.e., $w_{1,B} = w_{2,B}$, and a gradual process of cracking can be seen. The stage "200 beams removed" is well beyond peak (viz. at the kink between the steep falling branch and the tail of the softening diagram), but deformations are relatively small. This indicates the brittleness of the lattice model. On the other hand, when the specimen boundaries are free to rotate, large differences between $w_{1,D}$ and $w_{2,D}$ are observed. These differences appear from the beginning of the simulation and become very significant after 100 beams have been removed. Around 170 beams removed, $w_{2,D}$ increases fast, indicating that the final stage of cracking has been reached. At this point, the specimen is almost separated into two halves and only a few beams are still connecting the two parts. Consequently relatively large deformations occur and the result of the analysis is not very trustworthy anymore. The differences in crack density also appear

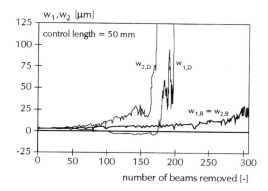

Figure 7.8 Deformations at the left (w_1) and the right (w_2) side of the specimen, plotted against the number of beams removed for the analyses with boundary condition B (uniform displacements over 50 mm gauge length) and D (freely rotating end platens).[478]

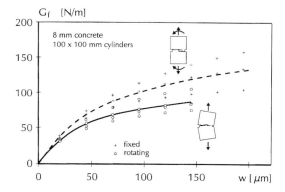

Figure 7.9 Effect of boundary rotations on the mode I fracture energy of concrete.[295]

in the computed fracture energies in Table 7.1. The fracture energy for the case D with freely rotating platens is much lower than the three other simulations. However, from Table 7.1 the significant effect of the generated particle distribution can be seen too, in particular for the simulation D. This indicates that a larger scatter in test results is to be expected in tests between hinges. The results of the experiments presented in Section 4.1.3.2 support this observation.

From the cylinder tests between fixed and freely rotating boundary conditions, that were presented in Section 4.1.3.2, the diagram of Figure 7.9 has been derived. The area under the stress-crack opening curves represents the work of fracture. This quantity has been determined up to different values of axial crack-opening. The results are gathered in Figure 7.9, which shows that the energy needed to reach a certain average crack-opening is consistently higher for the tests between fixed loading platens. The differences are even as high as

30 to 40%. In fact, this figure demonstrates that at least two interacting fracture zones develop in the test between fixed loading platens. Obviously, this is caused by the stress redistributions due to the bending moments that can develop in the fixed test. In the freely rotating experiment, the specimen simply opens from one side, and no stress-redistributions can occur. Because the two interacting fracture zones can develop in the fixed specimen, a large bridging zone between the overlapping crack zones can form, where substantial stress-transfer can occur. Note, however, that due to the heterogeneity of the concrete, smaller sized bridges appear in the main crack zones as well. This is perhaps best illustrated in Figure 3.45.

The large interacting fracture zones were observed earlier in photo-elastic coating experiments, see Figure 4.63. In addition to the large difference in fracture energy (Figure 7.9), it was also found that the tensile strength of a concrete or sandstone specimen loaded between freely rotating loading platens was substantially lower than when the test was carried out between fixed platens.[295] The scatter was significantly larger in the rotating tests as well, exemplifying that a larger freedom for crack nucleation and propagation exists under those boundary conditions.

7.2.3 NON-UNIFORM OPENING: COMPARING DIFFERENT APPROACHES

The non-uniform opening problem in uniaxial tensile tests runs as a red ribbon through the book. The phenomenon has been treated at a variety of scales, and by using different experimental and numerical techniques. It is a direct consequence of the heterogeneous nature of the concrete (or rock) that is studied. It is simply impossible to manufacture symmetric specimens made of heterogeneous materials. Again, the quotation of Abraham Trembeley applies. Why do we want to force the specimen to break in a symmetric manner if Nature is against us? This only leads to misconceptions of the true working of Nature. The better attitude seems to observe the behaviour of concrete as unbiased as possible. An ignorant student can probably obtain more refreshing new insights than the greatest authority in the field.

The non-uniform opening was first shown in experiments, in Section 4.1.3.2. Following that, in presenting different crack detection techniques, the mechanism was observed as well. In Chapter 5, a simple LEFM-based analysis sufficed to demonstrate the difference in the softening curves for a tensile test loaded between fixed or rotating loading platens (Section 5.3.2.3). In Chapter 6, the non-uniform opening phenomenon was used to determine the slope of the steep part of a bi-linear softening diagram for macroscopic fracture analyses of concrete. And finally in this chapter, the phenomenon has been analyzed with a meso-level fracture law.

Now what can we learn from all this? In the first place it shows that the role of experiments and an unbiased view on matters being studied are essential ingredients to come to new fundamental insights. This implies that people should not be absorbed too long in a certain field of research. Second, it shows that a problem should first be analyzed using the simplest available analytical techniques. Here LEFM was used, which is known to be incorrect for modelling macroscopic fracture of concrete, but by some simplifying assumptions it was possible to explain why a bump should appear in the softening diagram of a tensile specimen loaded between non-rotating end-platens. However, the tail of the softening diagram cannot be explained by means of the LEFM model. The macroscopic bi-linear softening diagrams, which were presented in Chapter 6, suggest that two different mechanisms must occur in the descending branch. However, such a phenomenological approach is not capable of elucidating the problem further, and a physical explanation cannot be found. The meso-level models are an outcome in this respect, as they show in great detail the propagation of fracture zones, and the interaction of cracks with the structure of the material. At least qualitatively, such models suggest various mechanisms. The validity of the analyses can be tested against experimentally determined crack geometries, which seem a more sensitive criterion for validating fracture experiments than the stress-crack-opening diagram. Validation of fracture models on the basis of stress-crack-opening diagrams might perhaps lead to correct energetic results; however, phenomenology will never lead to correct physical insight.

Perhaps the best illustration here is the analysis of the non-uniform opening problem by means of a smeared crack model.[479] In the paper, the authors correctly conclude that an interaction between experimentalists and computational specialists is essential to come to good results. They also conclude that a steep drop is essential in the first part of a bi-linear computational softening diagram as described in Section 6.2 in order to simulate the bump in the softening curve of a test carried out between fixed loading platens. However, they tried to "predict" the response of a tensile test between rotating platens, and the conclusion is drawn that no essential difference exists for (numerical) experiments between fixed or rotating platens. Obviously, this is in contradiction to all the results presented in the various chapters up till now. This is remarkable, even more so because early experiments on granite by Labuz et al.[345] already showed that a bump does not appear in a test between rotating end platens. From the computational point of view the above mentioned analyses[479] are of a very high standard indeed, but here the point is made that computation alone is not sufficient. The experimental and numerical tools are in great need of one another, which is the message of this example.

One last remark should be made. The numerical models seem quite capable of simulating mechanisms *after* they have been observed in experiments. However, this is no constraint to use the models also for designing experiments, although one should of course be careful. Not only is it possible to analyze the state of stress in a specimen with complicated shape (see Figure 4.55), but it also might be attempted to use the models as a means to hint at a possible control parameter in fracture experiments. This can be tried, even if the model is not perfect. An example of this approach is given in the next section on combined tension and shear.

7.3 COMBINED TENSION AND SHEAR

In Section 3.4 different views on shear failure were presented. It was suggested that the scale of observation is important when shear mechanisms are studied. Following a lively discussion with Dr. Bažant of Northwestern University in Evanston, we attempted to simulate (the supposed) shear failure in a four-point-shear beam by means of the lattice model. In this section the results of the analyses are presented and compared to findings obtained with different approaches. The lattice model is again a meso-level approach, and comparisons are made with macroscopic analyses performed by other researchers. Following this, in Section 7.3.2 an example is given of the application of numerical modelling for the design of experiments.

7.3.1 "SHEAR" IN A FOUR-POINT-SHEAR BEAM?

The four-point-shear beam which was used by Bažant and Pfeiffer[198] (see Section 3.4.1.1) was analyzed by means of the lattice model. In addition, the experiments carried out by Bažant and Pfeiffer were reproduced and tested under varying boundary restraint, see also Section 4.1.2.2. The geometry of the beam is shown in Figure 3.68b, whereas the variation of the support rotations is elucidated in Figure 4.7. Analyses were carried out for the beam with $d = 150$ mm. The computed crack patterns obtained from the analyses with a regular triangular lattice with particle overlay are shown in Figure 7.10. The computed load-displacement curves are shown in Figure 7.11 for fixed and freely rotating supports, respectively. In addition to the computed curves, results from experiments on concrete and Felser sandstone have been included in these diagrams.

When freely rotating supports were selected, two small curved cracks would grow from the notches. Failure was always through the growth of one of these two curved cracks. In the example of Figure 7.10a the top crack propagated until failure occurred. The rotating supports allow for free horizontal movement of the specimen halves separated by the crack. In earlier analyses,[156] it was shown that the bond strength between the aggregates and matrix has a substantial effect on the amount of side cracking observed in the analyses. For a low bond strength between $f_{t,b} = 0.5$ to 2 MPa, the amount of side cracking was found to be substantial (as can also be seen from Figure 7.10a where $f_{t,b}$ = 1.25 MPa), but for increasing bond strength up to $f_{t,b} = 5$ MPa, a single sharp crack was found. Surprisingly the variation in global maximum loads was only between 14.6 kN and 19.4 kN for a variation of local bond strength of an order of magnitude (i.e., between 0.5 and 5 MPa). This confirms observations made in Chapter 2, where it was concluded that the local bond strength between aggregate and matrix has a relatively small effect on the global strength of concrete only.

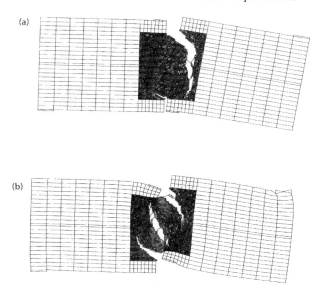

Figure 7.10 Effect of rotational freedom of supports and load-points in a DEN four-point-shear beam test. Analyses with the lattice model with regular triangular lattice with particle overlay. Cracking in a beam with freely rotating supports and free lateral deformations of the two outermost supports (a), and fully fixed supports against lateral deformations and rotations (b).[274]

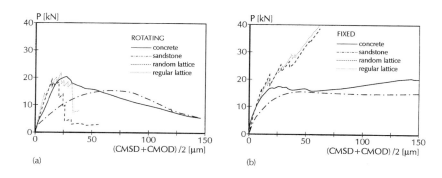

Figure 7.11 Lattice analysis: load-displacement curves for the beams of Figure 7.10.[274]

Returning to the four-point-shear analysis between fixed supports, it was found that two curved cracks would nucleate and grow simultaneously from the two notches, and final failure would occur through the growth of a third splitting crack between the two curved cracks. This is in complete agreement with experimental observations (Figure 3.66b). The load-displacement curves were found to be highly sensitive to variations in the rotational freedom of the supports of the beam as can be seen in Figure 7.11. Note, however, that the

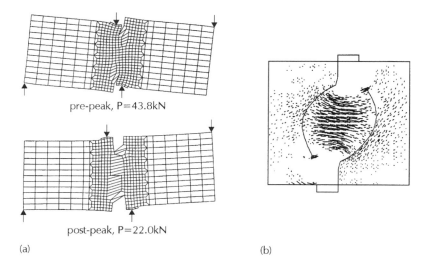

pre-peak, P=43.8kN

post-peak, P=22.0kN

(a)   (b)

Figure 7.12 Crack growth from simulations of Bažant and Pfeiffer's DEN four-point-shear beam: (a) smeared crack analysis by Rots et al.[480] and (b) discrete crack approach based on LEFM by Swartz and Taha.[208] (Figure (a) is reprinted by kind permission of Springer Verlag, New York. Figure (b) is reprinted by kind permission from Elsevier Science, Ltd., Kidlington, U.K., from *Engineering Fracture Mechanics.*)

analysis between rotating supports is generally too brittle, whereas the fixation of the supports seems overestimated in the analyses (Figure 7.11b). Recently, it was shown that results from analyses with regular triangular lattices are dependent on the orientation of the lattice.[416] In particular, the orientation was important when a homogeneous lattice was analyzed, thus without random distribution of strength or particle overlay. The analysis between fixed supports seems to suffer from this orientation problem to some extent. It seems that the effect is somewhat reduced when heterogeneity is added, although the effect cannot be circumvented. In analyses with random triangular lattices (on the basis of Figure 5.4) the effect decreases substantially. The fracture mechanism suggested in Figure 7.10 is confirmed in analyses based on a random lattice.[274]

The four-point-shear problem has received attention from many researchers. The problem has also been analyzed by means of a smeared crack model by Rots and De Borst,[480] and by means of a discrete fracture model by Ingraffea and Panthaki.[210] Using the program CRACKER developed by Ingraffea, Swartz and Taha[208] confirmed the findings of Ingraffea and Panthaki. The outcome of the analysis of Ingraffea and Panthaki was shown earlier in Figure 3.66a. The results from Rots and De Borst are shown in Figure 7.12, along with the discrete crack patterns computed by Swartz and Taha. Both the smeared model and the discrete crack model confirm the growth of two curved cracks for the top and bottom notch. Note that the curvature is more pronounced in the discrete crack analysis, as compared to the smeared crack analysis.

Now the main question should be answered. What is the real failure mechanism in the four-point-shear beam: Mode I or Mode II? The crack patterns observed in experiments can very well be simulated by means of a mode I criterion as shown by means of the lattice analyses. Also, the brittle LEFM criterion of Swartz and Taha leads to realistic crack patterns. In these analyses it is assumed that fracture proceeds as soon as $K_I > K_{Ic}$, whereas the direction of crack propagation follows directly from the direction of the maximum circumferential stress in the crack tip element. Note that in Ingraffea's discrete crack model always an initial crack must be specified. This is probably the reason why the central splitting crack is not found in this analysis.

Rots and de Borst[480] used a smeared crack model based on the crack-band model.[140] In their analyses they assumed that crack nucleation was in the direction of the major principal tensile stress. Only after nucleation, and after rotation of principal stresses, shear could be transferred in the plane of the crack. Basically their analyses showed two cracks propagating from the notches, but these cracks joined. The nucleation and growth of the third crack was never observed. Thus, all these analyses have in common the fact that crack nucleation is governed by exceeding the tensile strength in the material, and shear seems to have a minor effect only. In Section 3.4.2 it was suggested that instable crack growth in the four-point-shear test of Bažant and Pfeiffer may have led them to conclude that shear failure occurred. This becomes very likely in view of all these different analyses. Once more this is a good example that shows that testing should not be taken too light-heartedly.

The behaviour of a SEN four-point-shear beam is closely related to the above DEN four-point-shear beam. The case was worked out in detail in Schlangen.[156] Generally, it appears that continuum-based models like the smeared crack model of Rots[452] (in all its variants), and the higher order continuum models (for example Pamin and De Borst[482]) are not capable of simulating curved crack growth. Boundary element analyses, for example Ohtsu,[483] or the discrete crack model of Ingraffea, seem to give much better results as far as crack patterns are concerned. In Figure 7.13 a comparison is made of a number of different approaches to the SEN four-point-shear beam, which was tested by, among others, Arrea and Ingraffea[484] (see also Section 4.1.2.2, Figure 4.12). The results suggest that a detailed description of the stress concentrations in the area of the crack-tip is of prime importance. In smeared crack analyses the stress gradients are distributed over larger areas. On the other hand if the size of the finite elements is reduced, a strong mesh-dependent behaviour is found.[485] To overcome the mesh dependency, higher order continuum models are proposed, but even then the problem is not solved satisfactorily, and the crack refuses to grow towards the far end of the opposite loading platen.

More research is certainly needed to elucidate these matters. The lattice model seems quite suitable for describing curved crack growth. However, mesh dependency is of concern here as well, and problems are still encountered in simulating correct stress-displacement curves.[411] Generally the lattice analy-

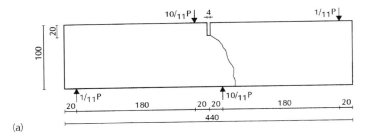

(a)

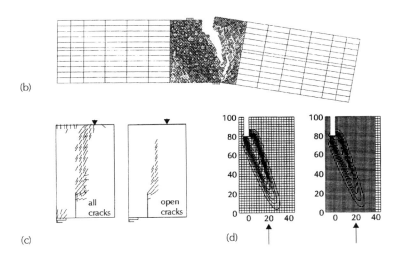

(b)

(c)

(d)

Figure 7.13 Comparison of experimentally observed crack growth in a SEN four-point-shear beam[162] (a), lattice analysis[162] (b), smeared crack analysis[486] (c), and gradient plasticity analysis by Pamin and De Borst[482] (d). Characteristic in the experiment is that the curved crack grows towards the far end of the opposite loading platen. (Figure (c) is reprinted by kind permission of Dr. De Borst.)

ses are too brittle, but for the interpretation and design of experiments this is no objection. The point where the simplicity of an elastic/perfectly brittle fracture law should be abandoned in favour of, for example, tensile softening and/or frictional sliding in the model at the level of the lattice beams has not been reached yet.

7.3.2 BOUNDARY EFFECTS IN BIAXIAL TENSION/SHEAR EXPERIMENTS

The numerical models are very suitable for the design and improvement of fracture experiments. In Figure 4.55 an example was shown of the application of a simple, linear elastic, plane-stress analysis for optimising the shape of a dog-bone tensile specimen. The fracture models can be used too, but one should then be very careful in judging the obtained results. An example of the application of the lattice model for the determination of the control parameter in the four-point-shear beam was given in Section 4.1.2.2., Figure 4.11. Here another example will be given of the development of the biaxial tension-shear experiment.[214]

The test set-up for the biaxial tension-shear experiments was described in Section 4.3, Figure 4.57. It is a quite complicated set-up, and in order to maintain parallel end-platen displacement throughout the complete fracture experiment, bending moments develop because asymmetric crack growth cannot be avoided in concrete. The mechanism is identical to what was presented in Section 7.2.2 for uniaxial tension. In a series of SIFCON experiments in this test set-up, it was found that often glue failure would occur along the glue platens. In the experiments this was prevented by glueing an additional steel platen along the side of the specimen where the (horizontal) steel platen was normally torn loose.

Some insight in the stress distributions around the loading platen can be obtained from a lattice analysis. The analysis was performed by Nooru-Mohamed et al.[487] and indicated the effectiveness of the additional steel platen. Moreover, it was found that the stiffness of these additional steel platens had a significant effect on crack growth in the specimen itself. Results of the analyses are shown for a load-path where first shear was applied to $P_s = -5$ kN (Figure 7.14a), -10 kN (Figure 7.14b) and $P_{s,max}$ (Figure 7.14c and d). The shear-load was kept constant, while the axial deformation was increased to complete failure. A random lattice was used.

For comparison, the experimental crack patterns are gathered in Figure 7.15. There is not only a difference between the various analyses in the level at which the lateral shear load is applied, but also differences appear due to the additional steel platens along the sides A and B (see Figure 7.14a). In the analyses of Figure 7.14b and c the additional platen was simulated by adding an additional row of plane stress elements, whereas the stiffness was doubled in analysis Figure 7.14d. Without the additional row of elements in the analysis of Figure 7.14b, a crack would run along the lower left loading platen. The additional layer of elements prevented this crack to develop, and a curved crack pattern was found in the analysis, which corresponds very well to the outcome of a successful experiment (Figure 7.15b).

For the analysis at the highest shear-load level $P_{s,max}$, an additional layer of elements did not suffice, and cracking in the glue-layer could still occur as can be seen in Figure 7.14c. Doubling the layer of elements was tried, and now the curved cracks would develop as was observed in the experiment (compare Figure 7.14d and Figure 7.15c). Clearly, the (structural) environment of the concrete specimen has an important effect on crack growth processes. Therefore, in numerical analyses, it is essential to include the environment as realistically as possible. In the days of analytic modelling this was almost

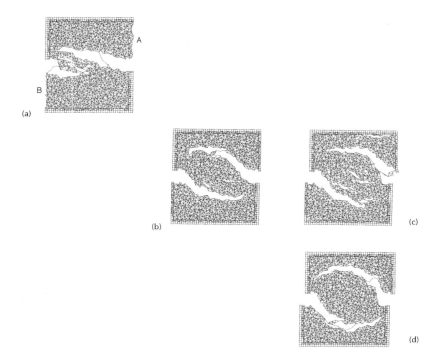

Figure 7.14 Effect of additional steel loading platens on the behaviour of concrete specimens subjected to biaxial tension/shear.[487] $P_s = -5$ kN without additional steel platen (a), $P_s = -10$ kN with a single row of elements added at locations A and B (b), $P_s = P_{s,max}$ with a single row (c) and with a double row of elements added (d).

always impossible, but with the introduction of numerical tools these boundary condition effects can be captured quite easily. Obviously, a strong interdependence exists between experiments and numerical simulations. It should be mentioned that the other load-paths tested by Nooru-Mohamed can all be simulated by means of a lattice model equipped with a simple, purely brittle, tensile failure criterion. Clearly, a comparison between the four-point-shear beam experiments and analyses of the previous section, and the results of the biaxial tensile/shear experiments and simulations in this section, suggests an important correspondence between the two different fracture geometries (see also Figure 3.77). Finally, it is mentioned that in Nooru-Mohamed's experiments it was decided to monitor crack growth in the glue layers by means of a number of additional extensometers that were placed over the glue layer between the steel loading platen and the concrete specimen (Figure 4.58). If cracking was registered along the glue layer, the experimental result was discarded.

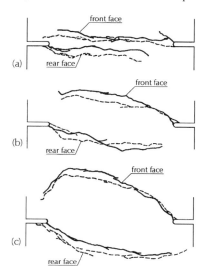

Figure 7.15 Cracking in concrete under biaxial tension/shear.[487] The constant shear load levels were (a) P_s = -5 kN, (b) P_s = -10 kN and (c) $P_{s,max}$.

7.4 UNIAXIAL COMPRESSION

So far, experiments were analyzed where mode I (tensile) fracture played a dominant role. In uniaxial compression, but also under multiaxial states of stress and in fibre concrete, the effect of shear may be more substantial. In the last few years it has been attempted to use the simple brittle beam lattice model also for simulating compressive fracture. However, until now this has been without success. The model in its present form seems to be too simple to capture the more complicated failure mode in compression. Perhaps one should enhance the lattice model to incorporate frictional slip as soon as tensile cracks have developed. However, such features can be incorporated only at the cost of more complexity, which is not in the interest of the experimentalist. So, let us proceed and see what can be achieved with the "simple" lattice model (Section 7.4.1). After that, in Section 7.4.2, the boundary friction and size effects in uniaxial compression are analyzed using Vonk's extended UDEC model, which is a more advanced particle model for concrete.

7.4.1 LATTICE MODEL ANALYSES

The tensile and shear analyses with the lattice model presented so far gave quite realistic fracture patterns. Therefore, it is challenging to investigate whether compressive failure can be captured with the model too. Can compressive fracture of concrete be modelled by means of a simple meso-level mode

I fracture model only?[400,411,468] Compressive simulations were also carried out by Rode,[407] using a simple truss model. These analyses confirm the results that follow.

In the first analyses, small square specimens were subjected to uniform boundary displacement, and it was assumed that unrestrained lateral deformations could occur. Because of computational reasons, the problem was modelled in two dimensions. This is probably a more severe assumption in the case of compression than in uniaxial tension. The effective stress fracture law based on normal stress and bending (Equation 5.69) was used. The amount of bending taken into account depends on the constant α which should be specified. Earlier parameter studies showed that α should be small to give the longest tail in the tensile softening diagram. The default value for α is 0.005. In analyses with a regular triangular lattice with particle overlay, either axial splitting or diagonal cracking can be obtained with the model, depending on the value of α. If $\alpha = 0.5$, more or less vertical cracks form after many de-bonding cracks develop along the interfaces between aggregate and matrix (note: again the default values for bond, aggregate and matrix strength were used here). For higher values of α, i.e., up to 0.75 and 1.0, the major crack zones would develop under a larger inclination. The different cases are compared in Figure 7.16. For unrestrained lateral deformations, vertical splitting would have to be observed in the analyses. This suggests that lower values for α should be used in compressive simulations. Earlier it was suggested that inclined cracks can develop in normal concrete subjected to uniaxial compression.[433] It was proposed that individual micro-cracks near larger stiff aggregates would join to form larger macro-cracks. The inclination would be the result of the geometric positions of the stiff aggregates with respect to each other. Therefore, similar as for tensile simulations, the analyses should be accompanied by carefully conducted experiments. At present the experimental information concerning crack densities and orientations has not been developed to a point where a simple comparison with numerical simulations can be made. Therefore, many comparisons are based — most unfortunately, but unavoidably — on information regarding the overall stress-deformation characteristics only (see also the next section).

For lower values of α, the ratio between overall compressive strength and tensile strength of a lattice increases as was shown on the basis of a random triangular lattice, see Figure 6.9. This would suggest that a low value must be selected. However, even more than in tension, the stress-deformation curves display brittle behaviour. Clearly, an additional mechanism is missed.

7.4.2 CONTINUUM PARTICLE MODELS

Compressive failure can also be modelled using continuum particle models, where the interaction between the particles is modelled by means of interface elements.[89,466,475] In soil mechanics such an approach is also used.[488] Note, however, that in soils cohesion between the particles is either non-existent or

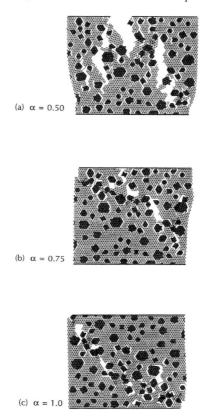

(a) α = 0.50

(b) α = 0.75

(c) α = 1.0

Figure 7.16 Effect of α on the failure of a lattice subjected to uniaxial compression.[468]

plays a much less pronounced role. Boundary restraint and slenderness effects were treated rather exhaustively by Vonk in his thesis. A limited number of numerical results obtained by Vonk will be presented here. First of all let us consider the effect of boundary restraint. In Figure 7.17 the basic characteristics of the interface model used by Vonk are shown. His particle model was shown before in Figure 5.9c. Aggregate and matrix are connected through two-dimensional interface elements. Moreover, interface elements were included in the matrix, which allows for simulating matrix-cracking too. As may be clear from Figures 3.6 through 3.8, a transition from interfacial cracking to matrix cracking seems an essential mechanism in the failure of concrete under compression. In the direction normal to the interface, tensile failure may occur according to a simple linear elastic/linear softening relation (Figure 7.17a). In the direction of the interface, shear is modelled through a linear elastic/linear softening model for frictional slip (Figure 7.17b), and for combined states of stress a simple Mohr-Coulomb criterion with tension cut-off has been used

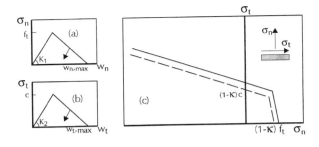

Figure 7.17 Interface model.[89] (Reprinted by kind permission of Dr. Vonk.)

(Figure 7.17c). In the softening regime isotropic shrinkage of the failure surface is assumed, as indicated with a dashed line in Figure 7.17c. Dilatancy can be included in the model as well. Obviously, quite a number of elements have been added in comparison to the "simple" lattice model. Tensile softening has been incorporated, although the effect is small because $w_c = w_{n,max} = 20 \ \mu m$. There is also a frictional component σ_t, which may cause failure as well.

In an extensive parameter study, Vonk found that both friction and the cohesive strength have a significant effect on the computed stress-deformation response in uniaxial compression. The parameter set used in the following two examples was partly guessed and partly based on this parameter study. In all, 14 parameters must be specified in the model of Figure 7.17. Most important are the tensile strength of the interface, which was given the relatively high value of 3 MPa. The cohesive strength of the aggregate-matrix interface was set to $c = 10$ MPa. Interface elements were also modelled in the matrix itself, which allowed for crack growth through the matrix. The tensile and cohesive strength of the matrix material were set to 6 and 20 MPa, respectively. The maximum crack opening in the tensile softening diagram was chosen 20 μm, and 1 mm was selected for the cohesive softening diagram (Figure 7.17b). These values were the same for interface elements between the aggregate and matrix, or in the matrix. In tension the result of Figure 7.5a was obtained with the same parameter setting.

In Figure 7.18 the effect of boundary restraint on the axial stress-deformation diagram in uniaxial compression is shown. The value for $tg \ \varphi$ indicates the amount of friction along the top and bottom edges of the specimen. When $tg \ \varphi = 0$, lateral deformations are not restrained, whereas full confinement can develop when $tg \ \varphi = \infty$. Clearly, a more ductile response is computed when the frictional restraint increases, and the peak strength increases as well. The crack patterns in the different analyses showed the development of almost vertical splitting cracks for the no-friction analysis, to the well known hour-glass failure mode for high frictional restraint, see Figure 7.19. The latter failure mode is normally observed in uniaxial compression tests on cubes (loaded between rigid steel platens), see also Figure 3.18. The results seem both qualitatively and quantitatively in agreement with experimental observations

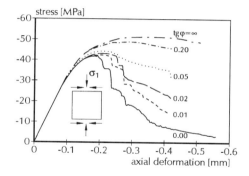

Figure 7.18 Computed influence of boundary restraint on the nominal stress-deformation behaviour in uniaxial compression.[89] (Reprinted by kind permission of Dr. Vonk.)

presented in Figures 3.19, 3.20 and Plates 6 through 9. However, one should realize that the model is not very simple anymore, and many parameters must be guessed. Further developments in this area are certainly needed.

The second example concerns the analysis of the slenderness effect, which confirms that localization of deformations occurs in uniaxial compression as well. Results of the numerical simulations carried out by Vonk are gathered in Figure 7.20. Note that slightly different values for the model parameters were used in these analyses. The cohesion at the matrix-aggregate interface was increased to 11 MPa, and in the matrix interfaces to 22 MPa. Moreover, dilatancy was introduced. For prisms with a width of 50 mm, and height varying between 50, 100 and 200 mm, almost perfect localization was found in the computations. This can be seen from Figure 7.20a. If these diagrams are translated to stress-post-peak deformation diagrams, they will all fall on top of each other. The crack patterns for the three analyses are shown in Figure 7.20b. All cracks are drawn. Preceding the development of stress-free, localised cracks, many distributed cracks develop along the interface between aggregate and matrix. This seems to be in agreement with experimental observations.[34]

The next test for the UDEC model would be to find out whether crack growth in other structures can be modelled equally well. For such analyses, the same parameter settings should be used as in the present analyses of the uniaxial compression tests. Together, the experiments and the analyses, give a better insight in compressive failure. More quantitative information on crack patterns and growth should be retrieved from future experiments in combination with (numerical) analyses.

7.5 FAILURE CONTOURS

In Sections 3.5 and 4.1 some attention was given to multiaxial testing. A complete understanding of the mechanical behaviour of materials should include knowledge on the behaviour under complex stress. In multiaxial com-

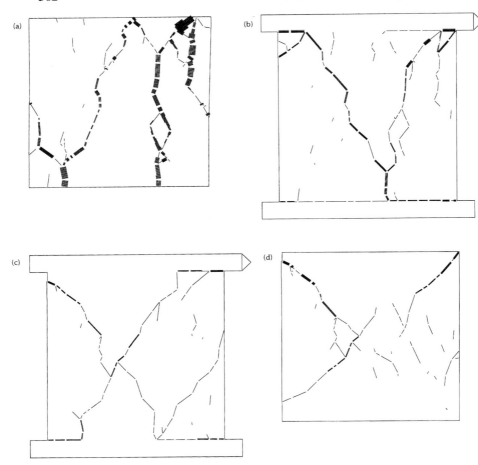

Figure 7.19 Computed crack patterns for the analyses of Figure 7.18.[89] Only the (open) stress-free cracks are shown. (Reprinted by kind permission of Dr. Vonk.)

pression the experimental difficulties increase substantially. For example, not only boundary shear between platen and specimen are important, but also the interaction between loading platen in different directions will have to be taken into account as was discussed in Chapter 4. In a multiaxial test on a cubical specimen, crack initiation normally occurs from the edges of the specimen.[311] The crack initiation in the corners of the specimen was shown by means of an impregnation technique (Section 4.4.3). Thus, more than for any other experiment, numerical support analyses seem (at least part of) the answer to solving experimental problems in multiaxial testing.

As may be obvious from the previous sections on uniaxial tensile crack simulations, combined tension and shear failure, and uniaxial compressive

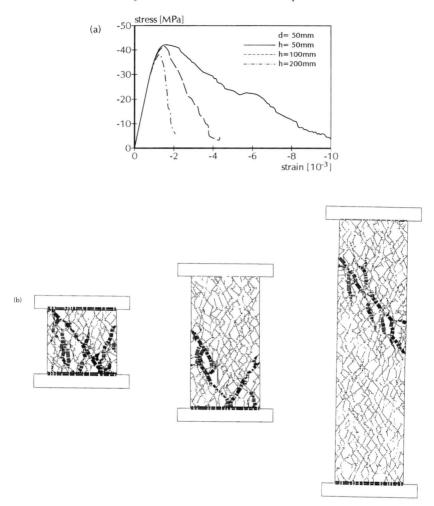

Figure 7.20 Effect of specimen slenderness on the nominal stress-deformation behaviour (a), and computed crack patterns (b).[89] (Reprinted by kind permission of Dr. Vonk.)

failure, we may still be far away from a complete meso-mechanical description of concrete under generalized stress. An interesting attempt to model the biaxial failure contour of concrete using a lattice type model was made by Beranek and Hobbelman.[413] The failure law Equation 5.74 which was derived for the inter-particle type lattice model shown in Figure 5.6b, was implemented in the lattice model of the Stevin Laboratory. A comparison was made with the fracture law based on normal force and bending moments (Equation 5.69). Earlier computations carried out by Beranek and Hobbelman showed a good match between the computed failure contour and experimental data (Figure

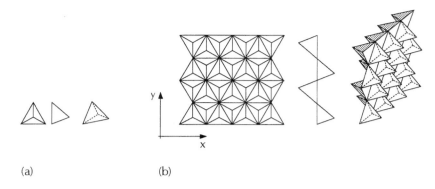

(a)   (b)

Figure 7.21 Front view, side view and isometry for the 1 × 1 mesh (a) and for the 4 × 4 mesh (b).[414]

3.83). Because of the problems encountered in lattice analyses of uniaxial compressive failure, it seemed a good idea to compare the two lattice models, neglecting for a moment the philosophical differences between the approach of Beranek and our approach in the Stevin Laboratory (Figure 5.6). Using three lattices of different size constructed from tetrahedrons, biaxial failure contours were computed. The mesh sizes were as follows: a mesh consisting of a single tetrahedron (Figure 7.21a), referred to as the 1 × 1 mesh, a 4 × 4 mesh containing 28 tetrahedrons (Figure 7.21b) and a 21 × 24 mesh containing 984 tetrahedrons.

In addition to the fracture law, the effect of the rotational freedom of the lattice nodes was studied too.[414] The effect of changing the fracture law is shown in Figure 7.22. The shear force criterion Equation 5.74 is compared to the normal load/bending moment criterion Equation 5.69. For the latter criterion two different values of α were taken, viz. 0.005 and 0.5. The computations were carried out on the 4 × 4 mesh. The use of tetrahedrons allows for crack growth in the out-of-plane direction, which is essential in equi-biaxial compression tests (see also Section 3.5.4). In the example of Figure 7.22, the rotational freedom of all the nodes in the plane $z = 0$ were suppressed.

Figure 7.22 is quite clear. Using the early normal force/bending moment criterion, the failure contour is found to be open-ended. In fact this is similar to the analysis of an inclined Griffith crack in a biaxial stress field.[10] A LEFM-based analysis will also lead to a failure contour which resembles the behaviour in biaxial tension and biaxial tension/compression quite satisfactorily, but which is open ended in the biaxial compression regime. The constant α has an important effect on the shape of the failure contour in the biaxial tensile and biaxial tensile/compressive regime. When the shear force criterion is used, the out-of-plane lattice beams can fail under in-plane biaxial compression, and the failure envelope is now closed. Moreover, the shape of the contour resembles the experimental envelope curve of Kupfer[236] quite well (Figure 3.83). The shear criterion gives the same result as the normal force/bending moment

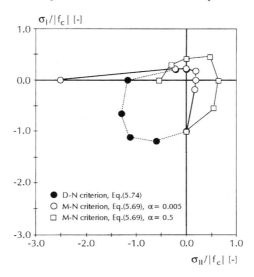

Figure 7.22 Effect of fracture law in the lattice model on the biaxial failure contour.[414]

criterion of Equation 5.69 in the biaxial tension and biaxial tension/compression regime when $\alpha = 0.005$.

The above results suggest that by selecting another failure law for the lattice beams, modelling of concrete fracture under generalized multiaxial stress is possible. However, the stress-deformation diagrams show too brittle behaviour. As far as this disadvantage of the lattice model is concerned, no improvement is obtained by changing the fracture law. Finally, it should be mentioned that by means of a meso-level fracture model for concrete under multiaxial stress, failure contours, stress-strain curves under triaxial stress and path-dependency can be studied much deeper than on the basis of experiments only. Also, experimental issues such as loading platen interactions and the effect of stress-concentrations in the corner of the specimens in truly triaxial tests can be studied, which will in the end improve the reliability of the multiaxial tests considerably. Phenomenological descriptions of failure surfaces (Section 5.3.1) will then become history. However, quite some research is needed before the above is accomplished.

Chapter 8

FRACTURE MECHANICS FOR STRUCTURAL ANALYSIS

8.1 INTRODUCTION

Fracture mechanics of concrete is still a relatively young area of research. The Hillerborg model, which set off all recent research activities in fracture mechanics, dates back to 1976. Since then, many numerical models have been developed, which can be applied for structural analysis. The book has thus far focused on the determination of fracture mechanics parameters from experiments. Testing and test-interpretation were central to the book. An approach was advocated where numerical models are developed for simulating laboratory scale fracture experiments. The goal is to obtain a better insight in experiments, and to assess the reliability of laboratory procedures. Of course, when a model is capable of simulating size and boundary condition variations in laboratory experiments, it should also be possible to explore the fracture behaviour of other structures.

In this chapter we will address two examples of the application of numerical tools in structural analysis. These are not simulations of real reinforced concrete structures, but rather structural details that are still small enough to be addressed by means of meso-level models. First the bond-slip between steel and concrete is studied. Detailed analyses can be carried out with the lattice model, or any other type of meso-level fracture model, to simulate the fracturing around the lugs on a steel reinforcing bar. Also it is possible to study the effect of adhesion between steel and concrete. In the end, such analyses may help to optimise the bond between steel and concrete, which is of prime importance in structural engineering. Also as mentioned in Chapter 1, meso-level analysis of plain concrete behaviour under generalized stress, bond-slip relations and aggregate interlock mechanisms may help to improve macroscopic constitutive models for structural analysis (Figure 1.6). The bond-slip analyses are presented in Section 8.2.

Next to this example, the pull-out of a steel anchor bolt will be analyzed. This example was part of a Round Robin analysis carried out by RILEM committee 90FMA "Fracture Mechanics Applications".[489] The Round Robin analysis was comprised of about 25 different contributions. The pull-out problem was simplified to a two-dimensional case, which was more easily accessible for numerical analysis than a full three-dimensional analysis. Note, however, that a three-dimensional case could also be analyzed, if at least the relevant three-dimensional models were available. In Section 8.3 we address the two-dimensional anchor pull-out problem. Tests were carried out in various laboratories, and a comparison will be made between different fracture mechanics approaches such as LEFM, various non-linear fracture mechanics models and meso-level analyses.

In Section 8.4, a possible application of fracture mechanics theories to an integrated design approach of materials in structures is presented. As may have become clear from the book, the fracture behaviour of concrete (but also of other brittle, disordered materials) depends to a large extent on the interaction between boundary conditions, structural size and material structure. An integrated approach where all these factors are considered, in order to optimise the fracture behaviour of full scale structures, seems the best and most direct type of application of the new knowledge and theories. In a very simple example of the development of a uni-directional fibre reinforced concrete truss-system, it is shown how simple ideas from fracture mechanics can be applied in the future.

8.2 ANALYSIS OF BOND-SLIP BETWEEN STEEL AND CONCRETE

The bond between steel and concrete can be looked at in different manners. The most common point of view is to consider the concrete and steel as two independent materials, connected through interface elements. In a sense no difference exists between the interface layer between stiff, dense aggregates and a porous cement matrix. Ideally, bond-slip analyses are carried out in 3D, by incorporating three-dimensional interface elements in the finite element programme. [380,490,491] The constitutive law for the interface elements comprises a non-linear, axial shear stress-slip relation plus, in the case of ribbed reinforcing bars, rules for radial stress-deformation law. Radial stresses and deformations are caused by wedging forces from the mechanical interlock of the ribs. For the determination of the various bond-slip relations, tension pull or pull-out experiments are carried out, as for example shown in Figure 8.1. The wedging forces caused by the ribs are balanced by tangential tensile stresses in the concrete, which may eventually lead to out-of plane cracking as shown in Figure 8.1c as well.[492]

A different approach is to study bond-problems by applying fracture mechanics tools. Crack nucleation and growth in the porous interface layer, the mechanical interlock caused by the ribs at the steel bar surface and frictional slip can all be incorporated into a fracture mechanics-based approach. Global analyses of bond-slip behaviour using fracture mechanics and finite elements have been carried out by, among others, Ingraffea et al.,[493] Rots[452,494] and Mazars et al.[495] In such analyses, usually the pull-out of a single reinforcing bar in concrete or the behaviour of a so-called tension-pull specimen is analyzed. More detailed analyses, particularly concentrating on the fracturing of the concrete around single ribs have been carried out in the past.[496,497]

In fracture mechanics-based analyses, normally the cracking of the concrete surrounding the rebar is the main source of the observed macroscopic bond-slip behaviour. For example, in impregnation experiments performed by Goto,[498] as well as in the more recent tests by Otsuka,[499] short, inclined cracks starting from the ribs at the rebar surface were detected, as shown schematically in

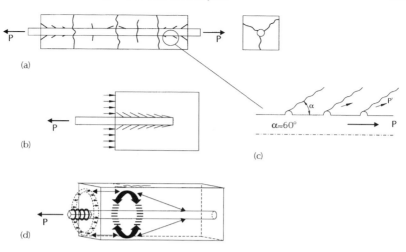

Figure 8.1 (a) Tensile-pull specimen and (b) pull-out experiment for studying bond-slip phenomena in reinforced concrete. Crack growth is shown schematically. In a tensile pull specimen, primary and secondary cracks may develop depending on the length of the specimen, the thickness of the concrete cover, concrete quality and the type of rebar used. Around the ribs at the rebar short inclined cracks develop as shown schematically in (c). In (d) the tangential tensile stresses caused by the wedging forces from the ribs at the rebar are shown schematically. The last figure is adapted from Tepfers,[492] who assumed that the radial stresses could be interpreted as an internal radial loading in a concrete ring.

Figure 8.1. Based on such observations, it can be attempted to separate the various local contributions to the bond-slip behaviour, such as adhesion between steel and concrete, crack growth in the weak interface layer between steel and concrete, crack growth from the ribs at the lugs and frictional slip between steel and concrete after the cracks have developed. Moreover, the large advantage of the numerical simulations is that the problem can be tackled taking into account realistic boundary conditions as they were used in the experiments to which the simulations are compared.

For the analysis of the global bond-slip behaviour a macroscopic model, for example the smeared crack model or the discrete crack model, can be used. Ingraffea and co-workers[493] were the first to apply fracture mechanics tools for studying bond-slip phenomena. They modelled the problem in two-dimensions. However, because of the wedging forces and the tangential tensile stresses, radial crack growth should also be allowed for. Rots[452] was one of the first to attempt this. He performed axi-symmetric analyses of the tension-pull specimen of Figure 8.2a by means of a smeared crack analysis. Some results from his analyses are shown in Figure 8.2. Crack growth occurred in a direction inclined from the axial load, whereas abundant radial crack growth was found too. In Figure 8.2b two stages of crack growth are shown, whereas the computed load-displacement diagrams are shown in Figure 8.2c. The radial cracks are smeared over the axi-symmetric ring elements that were used in the

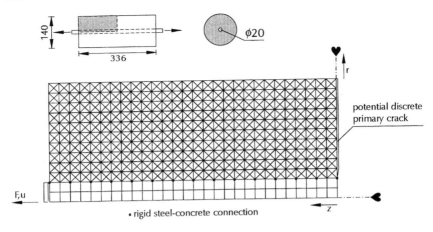

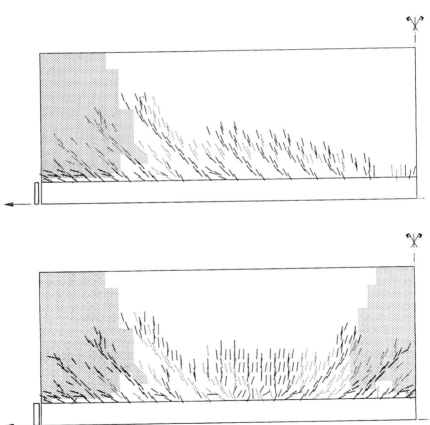

Figure 8.2 Axi-symmetric, finite element model of a tension pull specimen (a), two stages of cracking, i.e., just before primary cracking (b) and after primary cracking (c), and computed load-displacement diagrams (d), after Rots.[452] (Reprinted by kind permission of Dr. Rots.)

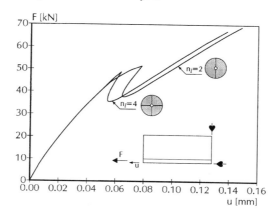

Figure 8.2 (continued).

analysis. Normally, however, in experiments cracks are localised, and in Figure 8.2c it is attempted to lump the tangential displacement in either two or four localised radial cracks. A small difference appears. The drop in axial load is caused by the development of a primary crack in the middle of the specimen. The numerical tool could be adopted in bond-slip experiments with the purpose of determining bond stress-slip relations for different types or rebars, concrete covers and different concrete qualities. However, one should always keep in mind that the macroscopic fracture mechanics models used in the analyses (in this case a conventional non-linear fracture mechanics model based on the crack-band model[140]) are a very rough approximation of the complete fracture process. The parameters used are determined from an inverse analysis. Therefore, a parameter study should form an integral part of the analysis. Engineering judgement is another factor that should be kept on board.

The bond-slip problem can also be studied at a more detailed level. Vos[496] was among the first to study the local fracture behaviour around a rib on a steel reinforcing bar. At that time the models were not really suited for studying such phenomena, but with the introduction of the meso-level models (or numerical concrete), the local behaviour can be studied in great detail. An example of a lattice analysis of the problem studied by Vos is shown in Figures 8.3 and 8.4. A small section of concrete near a rib on a rebar was modelled. Because of computational limitations, it is, unfortunately, not yet possible to carry out such detailed analyses in 3D. The model is part of a slip layer as shown in Figure 8.3a. On top of the finite element model a spring with stiffness k is attached to simulate the effect of the surrounding concrete. Because the stiffness of this concrete layer is difficult to estimate, two extremes were analyzed, namely $k = 0$ and $k = \infty$.

Crack growth near the rib is shown in Figure 8.4, for the two different spring stiffnesses analyzed. The lattice used is a random lattice with particle overlay. Particles between 0.2 and 2 mm have been included in the analysis. The

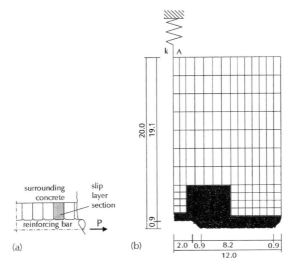

Figure 8.3 Representation of a slip-layer according to Vos[496] (a), and random lattice model with particle overlay (b). The part of the slip-layer where no cracks are expected to grow is modelled using 4-noded plane-stress elements.[497]

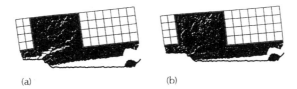

Figure 8.4 Crack growth (750 broken beams) for different stiffnesses of the spring at the top of the model of Figure 8.3b: (a) $k = 0$ and (b) $k = \infty$.[497]

adhesive strength between the steel bar and the concrete was modelled by giving the first row of lattice elements in contact with the steel a low strength, viz. 0.5 MPa. It should be mentioned that the adhesive strength was selected rather arbitrarily. A parameter study (inverse analysis) would be needed to determine the adhesive strength. Such an analysis should be accompanied by detailed bond-slip experiments with emphasis on studying crack growth. The results from the lattice analysis suggest a strong dependency on the adopted boundary conditions. In this particular example the spring stiffness k is the most important variable, although, as we will show next, the adhesive strength between the steel and the concrete has quite some influence as well. For $k = 0$, an inclined crack is computed near the rib. However, the inclination of the crack is smaller than the 60° found by Goto[498] in his experiments. When the spring stiffness is set to infinite, no clear discrete crack develops from the rib,

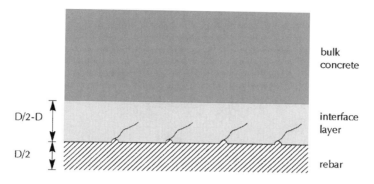

Figure 8.5 Bond-layer model. A layer with a thickness equal to half or one times the diameter of the reinforcing bar is given different material properties than the surrounding bulk concrete.

but rather a diffuse crack system is observed. The single spring is of course far from the real boundary conditions. Again, it would be better to model a complete tension-pull specimen, including the real boundary conditions. However, the computational effort would easily surpass the capacity of the largest computer available to date.

Another option is to scale down the bond-slip experiment.[490,491] The idea is to model the interface layer as a continuous, weak concrete layer between steel and bulk concrete as shown in Figure 8.5. The interface layer where cracks develop around the lugs at the rebar extends for about $0.5D$ to $1.0D$ (where D is the diameter of the rebar) into the concrete. This interfacial layer, where the lugs are located, and where the short inclined cracks may developed is given different smeared properties as the bulk concrete. Dragosavić and Groeneveld carried out small scale bond tests as shown in Figure 8.6. A rebar is situated in an aluminium ring filled with concrete. The thickness of the concrete layer is equal to the diameter of the rebar. The aluminium ring (thickness of 6.3 mm) should simulate the confining effect of the bulk concrete, which is normally much larger than the thin concrete layer used in these experiments. In order to ensure sufficient bonding between the concrete and the aluminum, a grove is created in the interior of the aluminium ring as shown in Figure 8.6a. The cross-section of the aluminium cylinder is shown in Figure 8.6b. The experiment has been simulated with the lattice model. A two-dimensional analysis was carried out.[497] For a ribbed rebar, the lattice is shown in Figure 8.6c. The concrete is connected to the rebar by means of a row of beam elements which either have a strength of 0.5 MPa, or alternatively a strength which varies between 1.25 and 10 MPa. In the latter case, simply the values corresponding to the particle structure of the concrete have been adopted. Examples of computed crack patterns are shown in Figure 8.7, both for a smooth and a ribbed reinforcing bar [(a,c) and (b,d), respectively] and for variable adhesive strength (Figures 8.7a,b) and low constant adhesive strength (0.5 MPa, Figures 8.7c,d).

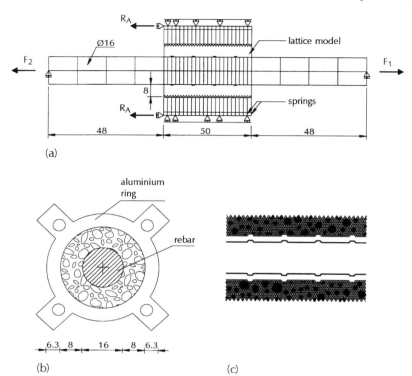

Figure 8.6 Detailed bond-slip experiment carried out by Dragosavić and Groeneveld.[490,491] A longitudinal section of the specimen is given in (a), the cross-section is shown in (b). The loads were distributed following $F_2 = 0.5F_1$ and $R_A + R_B = 0.5F_1$. The regular triangular lattice with particle overlay used in the analyses is shown in (c).[497]

The inclination of the internal cracks corresponds to the crack angles measured by Goto.[498] However, large differences in crack patterns emerge from the differences in adhesion between the rebar and the concrete. For a low interfacial strength (0.5 MPa), the row of beam elements in contact with the rebar is the first to fracture. Only limited cracking is found in the concrete layer surrounding a smooth rebar (Figure 8.7c). When a ribbed rebar is used, more diagonal cracking emerges as shown in Figure 8.7d. The growth of the inclined cracks seems to initiate at the ribs. Analyses with variable adhesive strength show the growth of a large amount of inclined cracks from the rebar. It was not important in this case whether the rebar was ribbed or smooth (Figures 8.7a,b). The experiments of Dragosavić and Groeneveld did not reveal the crack-growth sequence. Only bond stress-slip relations were measured. The maximum bond strength varied between 1 and 6 MPa. In the lattice analyses much lower values were found, which can perhaps be explained from the fact that no frictional restraint was taken into account. Moreover, the 3D experiment was

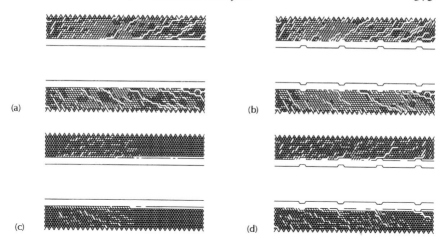

Figure 8.7 Crack patterns from the analyses of the small-scale bond test.[497] For (a) and (b) the adhesive strength varied between 1.25 and 10 MPa (i.e., the values from the concrete strength distribution, Equation 6.3), whereas for the analyses (c) and (d) a constant low adhesive strength of 0.5 MPa was assumed.

simulated by means of an approximate 2D analysis. Similar as for compressive simulations of plain concrete, the lattice model seems too simple. In spite of this, substantial insight in the fracturing around a rebar can be obtained from a parameter study. Again, this example shows that the numerical models, both at the meso- and macro-level, can be a strong supporting tool in experimental studies.

8.3 ANALYSIS OF ANCHOR PULL-OUT

A second interesting fracture problem is the pull-out of a steel anchor bolt from a concrete block. In order to demonstrate the effectiveness of fracture mechanics for the analysis of complex structural problems, a Round Robin analysis and test programme was organised by RILEM committee 90FMA.[489] Early information of anchor bolt pull-out was mainly limited to experimental data. Anchor pull-out is very difficult to address by means of analytical models because of the three-dimensional nature of the problem, see for example Eligehausen.[500] Fracture mechanics and numerical tools can be used for a better understanding of the phenomena, but again, the fracture mechanics models should be regarded as a supporting tool for the experiment. The Round Robin analysis of RILEM committee 90FMA revealed that the predictive qualities of most fracture mechanics models is very poor indeed. Experimental information of the pull-out problem analyzed was not available when the Round Robin was started, but became available after a first series of analyses was completed. A detailed account of the case is presented in Elfgren[501] and RILEM TC 90-FMA.[502]

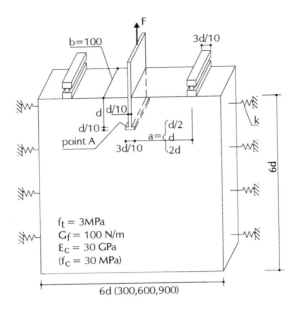

Figure 8.8 Round Robin analysis and tests of anchor bolts, after RILEM TC 90-FMA.[489]

Contributors could choose between a 2D or 3D variant of the anchor bolt problem. The two-dimensional geometry is shown in Figure 8.8. The three-dimensional pull-out analysis was reduced to an axi-symmetric analysis of an anchor embedded in a cylindrical block of concrete. Here, only the two-dimensional problem is considered, because this allows for easy study of crack growth, and because the number of participants was largest for this case. In the plane-stress analysis, a T-shaped steel anchor is embedded over a depth d in a concrete plate of 100 mm thickness. The distance from the centre line of the steel anchor to the centre of the supports was either $d/2$, d or $2d$. Three different embedment depths d should be considered, namely 50, 150 and 450 mm. Concrete and steel properties were specified. For concrete the tensile strength f_t was 3 MPa, the fracture energy G_f according to the Fictitious Crack Model was equal to 100 N/m, the Young's modulus was 30 GPa. A compressive strength f_c of 30 MPa was prescribed. Important in the analysis is the connection of the anchor to the surrounding concrete. Loads from the steel to the concrete were transmitted along the horizontal parts of the anchor only (designated with A in Figure 8.8). The shaft of the anchor was not bonded to the concrete and no frictional stresses could develop. The contributors were asked to supply a detailed description of the experimental and/or numerical/analytical methods used, the peak load, the displacement at point A at peak load, and the failure mechanism. If possible, the contributors were also asked to supply the full load-displacement diagram, as well as the crack patterns at peak load and at total collapse. In the default analysis, the confinement $K = 0$ (see Figure 8.8).

It was also possible to analyze the anchor bolt under a constant confinement K = ∞, but in that case only the embedded depth d = 150 mm would have to be considered.

In a meeting of the committee in Delft in 1991, all contributions were presented. In total 21 contributions were submitted. The models used varied from a simple LEFM based analysis,[503] to smeared crack analysis (among others Feenstra et al.[504] and Červenka et al.[505]), non-local microplane analysis,[506] and non-local continuum damage analysis (Clement and Mazars[502]). Some of the references mentioned are later improvements of earlier analyses and sometimes include simulations of different anchor bolt geometries as well. In an earlier — incomplete — 1990 report of the RILEM committee, the scatter of results was considerable. With the 1991 report,[502] which contained more contributions, the scatter was smaller but still substantial. Some experiments were performed, and a comparison could be made. The highest computed failure load for the geometry with $a = 2d$ and d = 150 mm ($K = 0$), was 865 kN/m, whereas the lowest value was 219 kN/m. Later experiments by Helbling et al.[507] gave peak loads of 384 kN/m. For the analysis with $K = 0$, $a = d$, and d = 150 mm, the highest failure load was 2365 kN/m, whereas the lowest computed failure load was equal to 350 kN/m. The experiments of Helbling et al.[507] gave a failure load of 619 kN/m for this case. Later experiments[508,509] gave average peak loads of 330 and 445 kN/m for the case $a = 2d$ and $a = d$ ($d = 150$ mm), respectively. It should be mentioned that also in the anchor bolt experiments considerable boundary condition effects may appear. Thus, an unrestricted comparison of the results is not possible. It should also be mentioned that some reservations exist concerning the highest values for the peak loads that were cited above. The second highest numbers were 710 and 1020 kN/m, respectively, and were therefore much closer to the experimental result.

A comparison of load-displacement diagrams from the analyses and experiments has been reproduced in Figure 8.9. This graph suggests widely varying load-displacement diagrams, and also substantial differences in crack growth predictions. The various analyses are marked 1 through 5, the experiment of Helbling is curve no. 6. Quite surprisingly, LEFM based analyses and fracture models in which an elastic/perfect brittle fracture law is used predict the horizontal crack growth found by Helbling et al. in experiments (which was later confirmed by others[508,509]). On the other hand, the computed failure loads are either too high (LEFM) or too low (lattice analysis). The computed load-displacement diagram from the LEFM analysis[503] is marked no. 1 in Figure 8.9. Smeared crack analyses and higher order continuum models generally show a better agreement in load-displacement response, but modelling curved or kinked crack growth is much more difficult (see also the analysis of the four-point-shear problem in Section 7.3.1). It is not my intention to give a full overview of the anchor-bolt analyses here. Instead, an example will be given of results obtained with a smeared crack model (Feenstra[510]), as well as some findings from lattice model analyses by Vervuurt.[509]

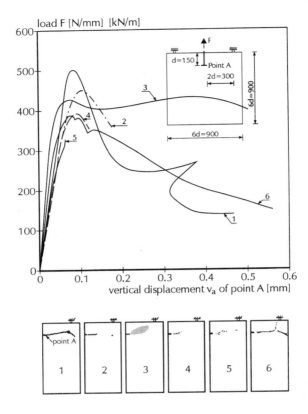

Figure 8.9 Preliminary results of the anchor bolt Round Robin organised by RILEM TC 90-FMA, after Elfgren.[501] (Reprinted with kind permission of Dr. Elfgren.)

First of all the smeared analysis of Feenstra is explained. The anchor bolt with an embedded length of 100 mm was analyzed (with $a = 2d$ and $K = 0$). In the analyses the material properties determined by Vervuurt were used ($f_c = 35$ MPa, $E_c = 37$ GPa, $f_t = 2.5$ MPa). An exponential softening curve was used (i.e., a simplified version of Equation 5.47). The finite element mesh is shown in Figure 8.10. The advantage of a non-orthogonal mesh is that mesh dependency, which is still a key-problem in smeared finite element analysis, can be avoided to some degree. It should be mentioned here that in the very first analyses of the anchor bolt, many researchers assumed that cracks would grow under an angle from the tip of the anchor directly to the supports. Many of the meshes used in these early analyses were therefore given a preferential direction, coinciding with the direction of the line connecting the tip of the anchor to the centre of the support. Quite surprisingly, the anchor pull-out experiments gave a horizontal crack, rather than an inclined crack in the direction of the supports. Vervuurt et al.[509] showed a clear dependency of crack growth on boundary conditions adopted in the experiment. In most cases, under freely

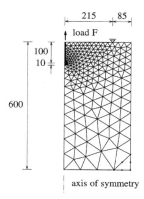

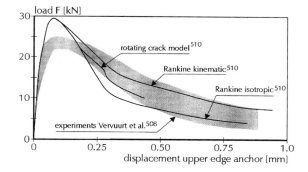

Figure 8.10 Pull-out of an anchor bolt. Finite element model used by Feenstra.[510] (Reprinted by kind permission of Dr. Feenstra.)

Figure 8.11 Comparison of load-displacement diagrams computed by Feenstra[510] and experiments performed by Vervuurt et al.[508] (Reprinted by kind permission of Dr. Feenstra.)

rotating supports, and when free lateral movement of the supports is allowed as well (which was again achieved by using pendulum bars, similar as in the four-point-shear experiments of Figure 4.7), non-symmetric crack growth was observed as will be shown later on. Feenstra assumed symmetry along the centre line of the anchor bolt, as can be seen in Figure 8.10. The computed load-displacement diagrams are compared to the range of experimental results obtained by Vervuurt et al. in Figure 8.11.

The correspondence is quite satisfactory. Also, as shown in Figure 8.12, crack growth in a horizontal direction from the tip of the anchor bolt and the subsequent growth of a kinked crack towards the supports is confirmed in the observations in the experiments by Helbling et al.[507] and Vervuurt et al.[508] In the earlier analyses by Feenstra et al.,[504] crack growth was in the direction of the support because the finite element mesh was aligned parallel to the line connecting the tip of the anchor and the support as discussed before.

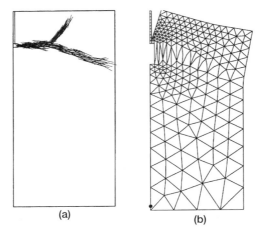

(a)   (b)

Figure 8.12 Final crack pattern (a) and total deformations (b) of the analysis of Feenstra.[510] (Reprinted by kind permission of Dr. Feenstra.)

LEFM and lattice model analyses showed the correct failure mode, but here contrary to the smeared crack analyses, the load-displacement diagrams, and also the "predicted" failure load are far from the experimental observations. The simulations of Vervuurt were accompanied by a large number of experiments, in line with the philosophy of this book, that experiment and model should form an inseparable pair of tools. An example of a lattice analysis of the anchor pull-out problem and a comparison to an experimental result is shown in Figure 8.13. The specimens were loaded between pendulum bars. This allows not only free rotation of the support points, but also free lateral translation.

Because of the heterogeneity of the concrete, when crack growth was observed at one tip of the anchor, this crack would therefore normally grow into the direction of the support. However, at the other side of the anchor, the second crack would grow towards the free opposite side of the specimen. This mechanism was visible in the specimens with small support span a. When a would increase, for example to $a = 2d$, cracks would nucleate and propagate into a horizontal direction from the tips at both sides of the anchor. Such behaviour corresponds with the findings by Helbling in his experiments. The mechanisms were reproduced quite satisfactory in the lattice analyses.[509] Thus, Figure 8.13 clearly demonstrates that non-symmetric behaviour is observed in the anchor test. It is easy to imagine that the symmetry in crack growth will depend to a large extent on the boundary conditions adopted in the experiment. As a matter of fact, the situation is completely comparable to the non-symmetric crack growth in uniaxial tensile tests. The source for all non-symmetry is the material itself. The heterogeneity of the concrete is sufficiently large to cause the asymmetry. One should wonder whether any symmetry may be

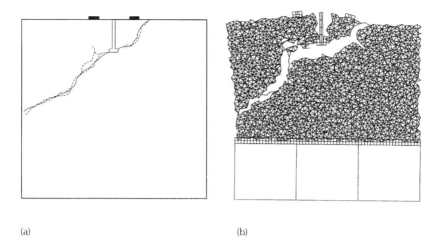

(a)   (b)

Figure 8.13 Crack growth observed in experiments with $a = d/2$ (a), and from lattice analyses (b).[509]

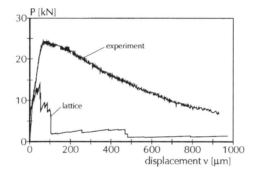

Figure 8.14 Load-displacement response for an experiment and simulation of an anchor with $d = 100$ mm and $a = 2d$.[509]

expected in experiments on concrete at all. If symmetry is observed, it may be complete coincidence, but it is more likely caused by stress-redistributions that can occur within the entire specimen-machine system. The example demonstrates again that the fracture behaviour of a structure cannot be separated from its environment.

A comparison between the computed and experimental load-displacement diagram of an unconfined test with span $a = 2d$ ($d = 100$ mm) is shown in Figure 8.14. It demonstrates that the computed load-displacement behaviour is far too brittle in the lattice analysis. A random lattice without particle overlay was used. Neglecting the details of the material structure, such as particle

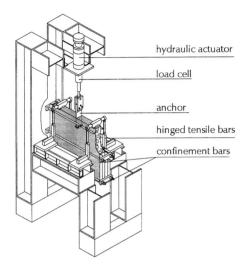

hydraulic actuator

load cell

anchor

hinged tensile bars

confinement bars

Figure 8.15 Schematic view of the test set-up for anchor bolt pull-out.[509]

distribution and porosity, the third dimension could explain the differences. Including the material structure in the analysis is of extreme importance. If it is left out, increased brittle behaviour follows from the computations as was shown clearly by Schlangen[469] in a number of analyses on isotropic meshes with varying fineness. Thus, one of the essential ingredients in lattice analyses is the heterogeneous material structure. It should be incorporated in an analysis up to the smallest structural detail. In that case a brittle fracture law might suffice to compute realistic structural behaviour.

When lateral confinement is applied to the sides of the anchor pull-out specimen, the fracture behaviour changes considerably. In the experiments lateral confinement was achieved by clamping the sides of the specimen between laterally connected steel bars as can be seen in Figure 8.15. The steel bars were given a small prestressing force of 1 kN, which prevented them from slipping off the specimen. During the test, however, no loading was applied in the lateral direction, but the bars would act as passive confinement only. The increase of confinement during the fracture process was monitored by means of strain gauges that were glued to the steel bars. A comparison between an experiment and a computation for a confined anchor-bolt is shown in Figure 8.16. The confinement from the horizontal steel bars was equal to $k = 500$ MPa, which was not the infinite confinement required in the Round Robin analysis. However, in the simulation the confinement was applied in exactly the same manner as in the experiments. Horizontal bars with the same stiffness as those used in the experiments were included in the finite element model. In Figure 8.16, the location of the horizontal confining bars is indicated. Because of the horizontal confinement, crack growth was forced into a horizontal direction.

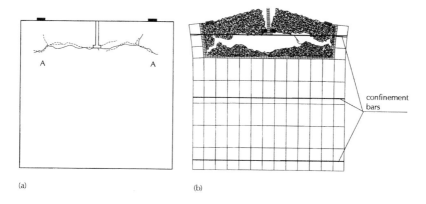

(a)   (b)

Figure 8.16 Experiment (a) and numerical lattice simulation of a laterally confined specimen with a support span $a = 2d$ ($d = 150$ mm).[509]

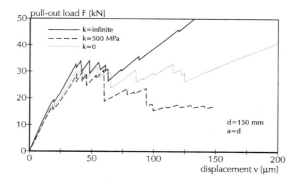

Figure 8.17 Effect of lateral confinement k on the load-displacement behaviour in the anchor pull-out problem.[509]

Only when the cracks reached a point directly under the supports (point A in Figure 8.16), would they branch off towards the support points. The confinement has an important effect on the load-displacement diagrams as shown in Figure 8.17. Increasing k from zero to infinity results in a change from softening to hardening behaviour. Due to crack growth, the passive confinement will become active at an earlier stage of loading when k is large. Cracks tend to drive the concrete away from the anchor, but this is restrained by the lateral confinement.

In three contributions to the Round Robin analysis, a purely brittle fracture law was assumed, namely an LEFM-based particle model,[503] a probabilistic approach with a brittle strength criterion,[511] and the lattice model. In each of these analyses the failure mechanism was predicted to a high degree of accuracy, as demonstrated above for the lattice analyses. In Figure 8.18 results of

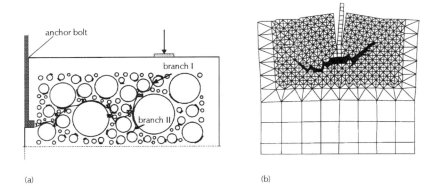

(a)   (b)

Figure 8.18 Crack patterns computed from the LEFM-based meso-level model of Wang et al.[503] (a), and the outcome of the probabilistic model of Rossi and Wu[511] (b). In both examples the unconfined 2D anchor pull-out problem was addressed. (Figure (b) is reprinted by kind permission of Balkema, Rotterdam, The Netherlands.)

both Wang et al.[503] and Rossi and Wu[511] are shown. All these results strongly suggest that a brittle analysis will suffice to determine, at least qualitatively, the failure mechanism. As a second step, one could perform a smeared crack analysis using the information gathered from the "brittle predictor analysis".

8.4 EVALUATION OF BRITTLENESS OF STRUCTURES

From Section 3.3.3 it may have become clear that the fracture energy of concrete may increase when (a) the concrete mixture becomes increasingly more heterogeneous, (b) when fibres are added to the concrete, and (c) when the strength of concrete increases. The fracture energy is, however, not a sufficient quantity for expressing the brittleness of a structure. In Chapters 4, 5 and 7 it was shown from different points of view that boundary conditions and size have a significant effect on the fracture energy of a specimen. Most pronounced seem the effects of the boundary conditions, although the size of the structure and the associated energy release during crack propagation may have a substantial effect on the post-peak behaviour as well.

So how is brittleness defined? In Equation 3.5 the characteristic length of a material l_{ch} was defined. This equation contains material properties only, and the characteristic size of the structure is neglected. The size of the structure has an important influence on brittleness, see for example the effect of specimen length on the tensile softening diagram (Figure 4.4), as well as the slenderness effect in uniaxial compression (Figure 3.14). Other structures, like reinforced concrete structures with low reinforcement percentages exhibit varying brittleness depending on the size of the structure.[512] A better "brittleness number" should therefore include the structural size. The number proposed in Elfgren[513]

(a)   (b)   (c)

Figure 8.19 Toughness increase by adding fibres to the concrete: randomly oriented low fibre volume (a), randomly oriented high fibre volume as in SIFCON (b), and uni-directional fibre composite (c).[227]

might be more suitable in this respect. Brittleness (as opposed to toughness) can, for example, be conveniently expressed by the quotient of stored elastic energy and fracture energy following

$$\frac{elastic\ energy}{fracture\ energy} = \frac{L^3 f_t^2 / E}{L^2 G_f} = \frac{L f_t^2}{E G_f} \qquad (8.1)$$

where L is a characteristic size of the structure, f_t is the tensile strength of the material, E is the Young's modulus and G_f is the fracture energy of the material in [J/m²] or [N/m]. A structure behaves as brittle if the stored elastic energy is much larger than the energy needed to fracture the structure. On the other hand ductility or toughness of the structure is improved when the fracture energy is larger than the stored elastic energy. Equation 8.1 clearly shows that for given material properties E, f_t, and G_f, a more brittle or ductile behaviour may be obtained by changing the size of the structure. Also, quite curiously, improving the tensile strength, which might be interpreted as a high performance property, might have a rather negative effect as the brittleness increases. The Young's modulus seems rather unimportant. For example, for composites like SIFCON,[222,225,228] the Young's modulus does not change by more than a factor 2 in comparison to plain concrete.[514] The fracture energy plays an important role. It can be improved by adding fibres to the concrete. For fibre volumes up to 3%, Ono and Ohgishi[515] reported fracture energies up to 48670 N/m, obtained from a four-point bend test (see also Figure 3.51). In that case fibres are randomly oriented in the concrete structure as shown in Figure 8.19a. In SIFCON, the larger aggregates are left out, and the fibre volumes may increase above 10%. In that case a material structure as depicted in Figure 8.19b is obtained, which leads to an even further increase of fracture energy.[225] The highest fibre densities can be obtained when the fibres are aligned as shown in Figure 8.19c. This can be done either by using continuous fibres,[172] or in SIFCON by using a special fibre-alignment method.[227] In the latter case, relatively short fibres are sprinkled in a mould using a specially designed

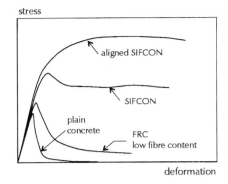

Figure 8.20 Stress-crack opening diagrams for the three fibre composites of Figure 8.19, in comparison to plain concrete behaviour.[227]

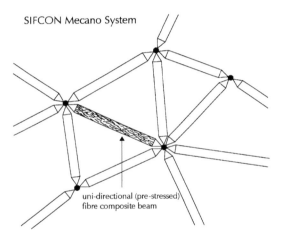

Figure 8.21 Truss-system containing uni-directional SIFCON bar elements.[227]

nozzle. Fibre volumes up to 21% were achieved. Of course, the properties of aligned composites improve in the direction of the fibres only. When load is applied perpendicular or under a small angle to the fibre direction, a very low strength is measured, not only in tension but also in shear.

The stress-crack opening diagrams for the three composites of Figure 8.19 are shown schematically in Figure 8.20. Clearly, the behaviour becomes more ductile when the subsequent improvements of Figure 8.19 are made. The above shows that for structural applications not only the improvement of fracture energy should be considered, but the other material properties are of concern too. Through a balanced design procedure, where new materials and structural design are combined, the most optimal result can be obtained. For example, in the uni-directional SIFCON used in a truss-system as shown in Figure 8.21, the deflection of the truss is inverse proportional to the Young's modulus of the material and the cross-sectional area of the elements, and proportional to the size of the elements and the applied load. The deflection will never change very much when a high performance material like SIFCON is used. The fracture strength and fracture energy will increase enormously, but the Young's modulus, which is mainly governing the deflection of a truss, will not increase by more than a factor of two. Obviously, not only fracture energy must be considered in the optimisation of structures, but the other mechanical properties are equally important.

Appendix

PRINCIPAL STRESSES AND INVARIANTS

The stress tensor for general three-dimensional states of stress contains nine stress components, i.e., three normal stresses σ_{ii} and six shear stresses σ_{ij} ($i \neq j$), following

$$\sigma_{ij} = \begin{vmatrix} S^1 \\ S^2 \\ S^3 \end{vmatrix} = \begin{vmatrix} \sigma_{11} & \sigma_{12} & \sigma_{13} \\ \sigma_{21} & \sigma_{22} & \sigma_{23} \\ \sigma_{31} & \sigma_{32} & \sigma_{33} \end{vmatrix} \tag{A.1}$$

S^1, S^2 and S^3 are the three stress vectors acting on the sides of a volume element in the directions of the three axes (1), (2) and (3) as indicated in Figure A1. Each stress vector can be divided in a normal stress component and two shear stress components as in Equation A.1 and Figure A1.

For the discussion in Sections 3.5 and 5.3 principal stresses and invariants must be defined. In order to obtain the principal stresses, we have to rotate the elementary material volume of Figure A1 such that all the shear stress components become zero. Thus, can we find area elements such that

$$\sigma_{ij} n_j = \sigma n_i \tag{A.3}$$

must be solved, which can be written as

$$\left(\sigma_{ij} - \sigma \delta_{ij} \right) n_j = 0 \tag{A.4}$$

In Equation A.4, δ_{ij} is the Kronecker delta, which has the property that it is equal to zero when $i \neq j$, and equal to one when $i = j$. The solution for Equation A.3 is found when the determinant of the coefficients vanishes, i.e., when

$$\left| \sigma_{ij} - \sigma \delta_{ij} \right| = 0 \tag{A.5}$$

This equation can be solved easily, and in expanded form it can be written as

$$\sigma^3 - I_1 \sigma^2 + I_2 \sigma - I_3 = 0 \tag{A.6}$$

389

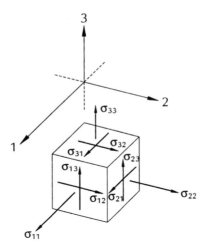

Figure A1. Stresses acting on a three-dimensional element.

where

$$I_1 = \sigma_{ii} = \sigma_{11} + \sigma_{22} + \sigma_{33} \tag{A.7}$$

$$I_2 = \frac{1}{2}\left(I_1^2 - \sigma_{ij}\sigma_{ji}\right) = \begin{vmatrix} \sigma_{11} & \sigma_{12} \\ \sigma_{21} & \sigma_{22} \end{vmatrix} + \begin{vmatrix} \sigma_{22} & \sigma_{23} \\ \sigma_{32} & \sigma_{33} \end{vmatrix} + \begin{vmatrix} \sigma_{11} & \sigma_{13} \\ \sigma_{31} & \sigma_{33} \end{vmatrix} \tag{A.8}$$

and

$$I_3 = \frac{1}{6}\left(2\sigma_{ij}\sigma_{jk}\sigma_{ki} - 3I_1\sigma_{ij}\sigma_{ji} + I_1^3\right) = \begin{vmatrix} \sigma_{11} & \sigma_{12} & \sigma_{13} \\ \sigma_{21} & \sigma_{22} & \sigma_{23} \\ \sigma_{31} & \sigma_{32} & \sigma_{33} \end{vmatrix} \tag{A.9}$$

Because the principal stresses σ_1, σ_2 and σ_3 must fulfil Equation A.6, I_1, I_2 and I_3 can be written as,

$$\begin{aligned} I_1 &= \sigma_1 + \sigma_2 + \sigma_3 \\ I_2 &= \sigma_1\sigma_2 + \sigma_2\sigma_3 + \sigma_3\sigma_1 \\ I_3 &= \sigma_1\sigma_2\sigma_3 \end{aligned} \tag{A.10}$$

The three roots σ_1, σ_2 and σ_3 fulfilling Equation A.6 are called the *principal stresses*, and their directions, the *principal directions*. The quantities I_1, I_2 and I_3 are the same for any combination of stresses, i.e., they do not change under a rotation of the coordinate system, and are called the invariants of the stress tensor σ_{ij}.

For the description of failure surfaces in Sections 3.5 and 5.3 the use of isotropic and deviatoric stress are quite convenient. The isotropic stress σ_0 is the average of the three normal stress components

$$\sigma_0 = \frac{1}{3}\sigma_{ii} = \frac{1}{3}\left(\sigma_{11} + \sigma_{22} + \sigma_{33}\right) \tag{A.11}$$

Note that $\sigma_0 = I_1/3$. The remaining stress components are called deviatoric stresses, and can be interpreted as the deviation of the stress state from a purely isotropic (or hydrostatic) state of stress. The deviatoric stresses can be written as

$$s_{ij} = \sigma_{ij} - \sigma_0\delta_{ij} \tag{A.12}$$

Similar as for the principal values of the stress tensor, the principal values of the deviatoric stress tensor can be found. The principal values are the three roots of the equation

$$s^3 - J_1 s^2 - J_2 s - J_3 = 0 \tag{A.13}$$

where J_1, J_2 and J_3 are the invariants of the deviatoric stress tensor. They can be written as follows:

$$
\begin{aligned}
J_1 &= s_{ii} = s_{11} + s_{22} + s_{33} = \left(\sigma_{11} - \sigma_0\right) + \left(\sigma_{22} - \sigma_0\right) + \left(\sigma_{33} - \sigma_0\right) = \\
&= \sigma_{11} + \sigma_{22} + \sigma_{33} - 3\sigma_0
\end{aligned} \tag{A.14}
$$

$$
\begin{aligned}
J_2 &= \frac{1}{2}s_{ij}s_{ji} = -\begin{vmatrix} s_{11} & s_{12} \\ s_{21} & s_{22} \end{vmatrix} - \begin{vmatrix} s_{22} & s_{23} \\ s_{32} & s_{33} \end{vmatrix} - \begin{vmatrix} s_{11} & s_{13} \\ s_{31} & s_{33} \end{vmatrix} = \\
&= \frac{1}{2}\left(s_1^2 + s_2^2 + s_3^2\right) = \\
&= \frac{1}{6}\left[\left(\sigma_x - \sigma_y\right)^2 + \left(\sigma_y - \sigma_z\right)^2 + \left(\sigma_z - \sigma_x\right)^2\right] + \tau_{xy} + \tau_{yz} + \tau_{zx}
\end{aligned} \tag{A.15}
$$

$$
J_3 = \frac{1}{3}s_{ij}s_{jk}s_{ki} = \begin{vmatrix} s_{11} & s_{12} & s_{13} \\ s_{21} & s_{22} & s_{23} \\ s_{31} & s_{32} & s_{33} \end{vmatrix} = \frac{1}{3}\left(s_1^3 + s_2^3 + s_3^3\right) = s_1 s_2 s_3 \tag{A.16}
$$

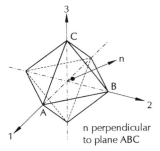

Figure A2. Definition of octahedral planes. All octahedral planes form a regular octahedron. The normal *n* to the plane ABC coincides with the space diagonal.

As can be seen the invariants of the deviatoric stress tensor are defined identical to the invariants of the stress tensor. Again, these invariants are not affected by coordinate rotations. J_2 and J_3 are quite convenient for describing limit surfaces.

Octahedral shear stresses are defined on planes that make equal angles with the axes of principal stress as shown in Figure A2. The normal to the octahedral plane in the first octant coincides with the space diagonal and has the form

$$n_i = \left[n_1 n_2 n_3 \right] = \frac{1}{\sqrt{3}} \left[1 \; 1 \; 1 \right] \tag{A.17}$$

The direction of the octahedral normal stress σ_{oct} coincides with the normal on the octahedral plane and can be expressed as

$$\sigma_{oct} = \frac{1}{3} \left(\sigma_1 + \sigma_2 + \sigma_3 \right) = \sigma_0 \tag{A.18}$$

The octahedral shear stress τ_{oct} is defined as

$$\tau_{oct}^2 = \frac{1}{9} \left[\left(\sigma_1 - \sigma_2 \right)^2 + \left(\sigma_2 - \sigma_3 \right)^2 - \left(\sigma_3 - \sigma_1 \right)^2 \right] = \left(\frac{2}{3} J_2 \right)^{1/2} \tag{A.19}$$

A geometric interpretation of the stress state at a point in stress space is given in Figure A3. The point $P(\sigma_1, \sigma_2, \sigma_3)$ has coordinates σ_1, σ_2 and σ_3. The line connecting point *P* with the origin *O* can be considered as the stress-vector *OP* as a representation of the state of stress. Alternatively, we can describe the state of stress by separating the hydrostatic and deviatoric components. The *hydrostatic axis* is the space diagonal $n = 1/\sqrt{3} \; |1 \; 1 \; 1|$ with the characteristic that $\sigma_1 = \sigma_2 = \sigma_3$. In every point on the hydrostatic axis, the deviatoric stress is equal to zero. The planes perpendicular to the hydrostatic axis are called the

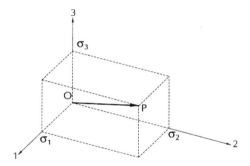

Figure A3. Stress point P in stress space.

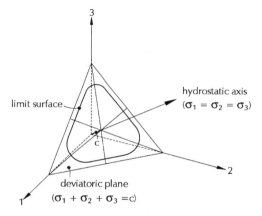

Figure A4. Definition of hydrostatic axis and deviatoric planes. When $c = 0$, the deviatoric plane passes through the origin and is called π-plane.

deviatoric planes, see Figure A4. These planes are characterised through the formula

$$\sigma_1 + \sigma_2 + \sigma_3 = c \tag{A.20}$$

When $c = 0$, the deviatoric plane passes through the origin O and is called the π-plane. Now, if we consider a point $P(\sigma_1, \sigma_2, \sigma_3)$ outside the hydrostatic axis, then the projection P' of P on the π-plane is given by the coordinates ($\sigma_1 - \sigma_0$, $\sigma_2 - \sigma_0$, $\sigma_3 - \sigma_0$), where σ_0 is defined through Equation A.11. The coordinates of P' are the principal deviatoric stresses. The distance r from the origin to P' can be computed following

$$r^2 = \left(\sigma_1 - \sigma_0\right)^2 + \left(\sigma_2 - \sigma_0\right)^2 + \left(\sigma_3 - \sigma_0\right)^2 \tag{A.21}$$

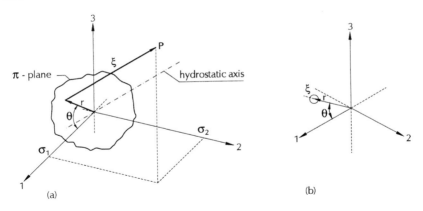

Figure A5. Two coordinate systems to describe a stress point P in stress space: cartesian coordinate system $(1,2,3)$ and cylindrical coordinate system (r, ξ, θ).

Substituting Equation A.11, and comparison to Equation A.15 yields

$$r^2 = \frac{1}{3}\left[\left(\sigma_1 - \sigma_2\right)^2 + \left(\sigma_2 - \sigma_3\right)^2 + \left(\sigma_3 - \sigma_1\right)\right] = 2J_2 \qquad \text{(A.22)}$$

From the above it can be concluded that the distance from a stress point P to the π-plane is equal to $\sqrt{3}\ \sigma_0$, and is determined by the first invariant of the principal stresses, whereas the distance from the stress point to the intersection with the hydrostatic axis is determined by J_2 following Equation A.22. Through Equations A.18 and A.19 we can make a link to the octahedral stresses. The stress point P can also be represented in terms of the octahedral stresses using these two equations. In fact, the stress point is represented in cylinder coordinates (r, ξ, θ), where the direction ξ (i.e., the axis of the cylinder) coincides with the hydrostatic axis, r is the distance from P' to the origin in the π-plane and θ is the angle of the vector OP' in the π-plane as shown in Figure A5. All three cylinder coordinates can be expressed in the invariants I_1, J_2 and J_3 following

$$\xi = \frac{I_1}{\sqrt{3}} = \sigma_{oct}\sqrt{3} \qquad \text{(A.23a)}$$

$$r = \sqrt{2J_2} = \tau_{oct}\sqrt{3} \qquad \text{(A.23b)}$$

$$\cos 3\theta = \sqrt{2}\ \frac{J_3}{\tau_{oct}^3} \qquad \text{(A.23c)}$$

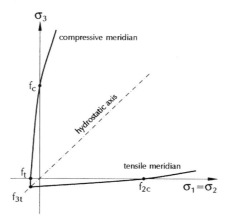

Figure A6. Failure surface shown in the rendulic plane. States of stress where $\sigma_1 = \sigma_2$ can be explored using a conventional triaxial cell.

The description of limit surfaces in cylinder coordinates is quite convenient because for concrete, rock and soils a circle symmetry exists around the hydrostatic axis. The coordinate system of Figure A5a is sometimes referred to as the Haigh-Westergaard coordinate system.

Experimental data are often shown in the so-called rendulic plane and/or deviatoric planes. When data are presented in deviatoric planes, we look to the limit surface in the direction of the hydrostatic axis (Figure A5b). The rendulic plane is the plane $\sigma_1 = \sigma_2$, see Figure A6. Triaxial data from cylinder tests as described in Section 3.5 (for example Figure 3.87) and Chapter 4 are most conveniently presented in a rendulic plane because the two intersections of the limit surface in the rendulic plane correspond to the set of stress-states that can be tested in a triaxial cell (see Chapter 4). The upper meridian in Figure A6 is the *compressive meridian*, i.e. stress states $\sigma_3 < \sigma_2 = \sigma_1$ (σ is negative in compression), whereas the lower meridian is the *tensile meridian* for stress states $\sigma_3 > \sigma_2 = \sigma_1$. Typical stress states, such as uniaxial compression (f_c), uniaxial tension (f_t), biaxial compression ($\sigma_1 = \sigma_2 = f_{2c}$) and triaxial tension are indicated in Figure A6.

REFERENCES

REFERENCES TO CHAPTER 1

1. Griffith, A.A. 1921, The phenomena of rupture and flow in solids, *Phil. Trans. Royal Soc. London*, Series A, 221, 163.
2. Dugdale, D.S 1960, Yielding of sheets containing slits, *J. Mech. Phys. Sol.*, 8, 100.
3. Barenblatt, G.I. 1962, The mathematical theory of equilibrium of cracks in brittle fracture, *Advances in Appl. Mech.*, 7, 55.
4. Dennett, D.C. 1991, *Consciousness Explained*, Penguin Books, London.
5. Hillerborg, A., Modéer, M. and Petersson, P.-E. 1976, Analysis of crack formation and crack growth in concrete by means of fracture mechanics and finite elements, *Cem. Conc. Res.*, 6, 773.
6. Vecchio, F. and Collins, M.P. 1981, Stress-strain characteristics of reinforced concrete in pure shear, in *Advanced Mechanics of Reinforced Concrete*, IABSE Colloquium Delft 1981, IABSE, Zürich, 34, 211.
7. Wittmann, F.H., Structure of concrete with respect to crack formation, in *Fracture Mechanics of Concrete*, Wittmann, F.H., Ed., Elsevier, London/New York, 1983, 43.
8. CUR 1969, Loading tests on a full size suspended beam and a model of this beam for a metro viaduct at Rotterdam, *Research Report 40*, CUR, Gouda, The Netherlands.

REFERENCES TO CHAPTER 2

9. Yunping Xi and Jennings, H.M. 1992, Relationships between microstructure and creep and shrinkage of cement paste, in *Materials Science of Concrete III*, Skalny, J. Ed., The American Ceramic Society Inc., Westerville, OH, 37.
10. Mindess, S. and Young, J.F. 1981, *Concrete*, Prentice Hall, Englewood Cliffs, N.J., 1981.
11. Reinhardt, H.W. 1985, *Beton als Constructiemateriaal—Eigenschappen en Duurzaamheid* (Concrete as Structural Material — Properties and Durability), Delftse Universitaire Pers, Delft.
12. Skalny, J. Ed. 1989, *Materials Science of Concrete I*, The American Ceramic Society, Westerville, OH.
13. Skalny, J. Ed. 1992, *Materials Science of Concrete II*, The American Ceramic Society, Westerville, OH.
14. Taylor, H.F.W. 1990, *Cement Chemistry*, Academic Press, London.
15. Klemm, 1991, in *Materials Science of Concrete II*, Skalny, J. Ed., The American Ceramic Society Inc., Westerville, OH.
16. Jensen, 1991, in *Materials Science of Concrete II*, Skalny, J. Ed., The American Ceramic Society Westerville, OH.
17. Scrivener, K. 1989, The microstructure of concrete, in *Materials Science of Concrete I*, Skalny, J. Ed., The American Ceramic Society Westerville, OH, 127.
18. Bogue, R.H. 1955, *The Chemistry of Portland Cement*, Reinhold, New York, 2nd ed.
19. US Department of the Interior 1981, *Concrete Manual*, 8th revised edition, United States Governement Printing Office, Washington, D.C.
20. Wittmann, F.H. 1977, Grundlagen eines Modells zur Beschreibung charakteristischer Eigenschaften des Betons (Fundamentals of a model for describing characteristic properties of concrete), *Deutscher Ausschuss für Stahlbeton*, Vol. 290, 45 (in German).
21. Van Breugel, K. 1992, Numerical simulation of hydration and microstructural development in hardening cement-based materials, *HERON*, 37(3), 1.
22. Bentz, D.P., Garboczi, E.J. and Stutzman, P.E. 1992, Computer modelling of the interfacial zone in concrete, in *Interfaces in Cementitious Composites*, Maso, J.C., Ed., E&FN Spon, London/New York, 107.

23. Gartner, E.M. and Gaidis, J.M. 1989, Hydration mechanisms I, in *Materials Science of Concrete I*, Skalny, J.P. Ed., The American Ceramic Society, Westerville, OH, 95.

24. Locher, F.W., Richartz, W. and Sprung, S. 1976, Erstarren von Zement (Hardening of cement), *Zement - Kalk - Gips*, 29(10), 435 (in German).

25. Odler, I. 1991, Strength of cement, Final report of RILEM-TC 68-MMH Mathematical Modelling of Cement Hydration Task Group 1, *Mater. Struct. (RILEM)*, 24, 143.

26. Birchall, J.D., Howard, A.J. and Kendall, K 1981, Flexural strength and porosity of cements, *Nature*, 320(289), 388.

27. Richard, P. and Cherezy, M.H. 1994, Reactive Powder Concretes with High Ductility and 200-800 MPa Compressive Strength, *ACI Spring Convention*, San Francisco.

28. Bache, H.H. 1995, Concrete and concrete technology in a broad perspective, in *Proceedings Nordic Symposium on Modern Design of Concrete Structures*, Aakjaer, K., Ed., Aalborg University, Department of Building Technology and Structural Engineering, 1.

29. Hansen, W. et al. 1986, SMDDHC, Boston, MRS fall symposium, Proceedings Vol. 85, 105.

30. Diamond, S. and Leeman, M.E. 1995, Pore size distributions in hardened cement paste by SEM image analysis, in *Microstructure of Cement-Based Systems/Bonding and Interfaces in Cementitious Materials*, Diamond, S., Mindess, S., Glasser, F.P., Roberts, L.W.m Skalny, J.P. and Wakeley, L.D., Eds., MRS, Pittsburgh, PA, Vol. 370, 217.

31. Wischers, G. and Richartz, W. 1982, Einfluss der Bestandteile und der Granulometrie des Zements auf das Gefüge des Zementsteins (Influence of composition and particle size on the structure of cement). *Beton*, 32, 10, 379 (in German).

32. Powers, T.C. 1958, The physical structure and engineering properties of concrete, *Research and Development Bulletin No. 90*, Portland Cement Association, Skokie, IL.

33. Feldman, R.F. and Sereda, P.J. 1968, A model for hydrated portland cement paste as deduced from sorption length change and mechanical properties, *Mater. Struct. (RILEM)*, 1, 509.

34. Hsu, T.T.C., Slate, F.O., Sturman, G.M. and Winter, G. 1963, Microcracking of plain concrete and the shape of the stress-strain curve, *J. Am. Conc. Inst.*, 60, 209.

35. Rehm, G. Diem, P. and Zimbelmann, R. 1977, Technische Möglichkeiten zur Erhöhung der Zugfestigkeit von Beton (Technical means to increase the tensile strength of concrete), *Deutscher Ausschuss für Stahlbeton*, Vol. 283 (in German).

36. Mindess, S. 1989, Interfaces in Concrete, in *Materials Science of Concrete I*, Skalny, J. Ed., The American Ceramic Society, Westerville, OH, 163.

37. VBT 1995, (NEN5950), Regulations for Concrete Technology—Requirements, Production and Inspection, NNI September 1995.

38. Aitcin, P.C. and Pinsonneault, P. 1981, Utilisation de la Poussiere de silice submicroscopique des usines de silicium et de ferrosilicium dans les beton, *Proceedings CANMET*, Ottawa 1981, 185.

39. Kelly, A. and Macmillan, N.H. 1986, *Strong Solids*, Oxford University Press.

40. ACI 1987, Silica fume in concrete, Preliminary report of ACI committee 226, *ACI Mater. J.*, 84(2), 158.

41. Kreijger, P.C. 1984, The skin of concrete: composition and properties, *Mater. Struct. (RILEM)*, 17(100), 275.

42. Hassanzadeh, M. 1995, Fracture mechanical properties of rocks and mortar/rock interfaces, in *Microstructure of Cement-Based Systems/Bonding and Interfaces in Cementitious Materials*, Diamond, S., Mindess, S., Glasser, F.P., Roberts, L.W., Skalny, J.P. and Wakeley, L.D., Eds., MRS, Pittsburgh, PA, 370, 377.

43. Maso, J.C. 1992, *Interfaces in Cementitious Composites*, RILEM Proceedings 18, E&FN Spon/Chapman & Hall, London/New York.

44. Diamond, S., Mindess, S., Glasser, F.P., Roberts, L.W.m Skalny, J.P. and Wakeley, L.D., Eds. 1995, *Microstructure of Cement-Based Systems/Bonding and Interfaces in Cementitious Materials*, Proceedings MRS Fall meeting, Boston, November 1994, MRS, Pittsburgh, PA, Vol. 370.

45. Zimbelmann, R. 1985, A contribution to the problem of cement-aggregate bond, *Cem. Conc. Res.*, 15(5), 801.
46. Scrivener, K. and Gartner, E.M. 1988, Microstructural gradients in cement paste around aggregate particles, in *Bonding in Cementitious Composites,* Mindess, S. and Shah, S.P., Eds., Materials Research Society, Pittsburgh, PA, Vol. 114, 77.
47. Garbozci, E.J. and Bentz, D.P. 1991, Digital simulation of the aggregate-cement paste interfacial zone in concrete, *J. Mat. Res.*, 6, 196.
48. Bentz, D.P., Garbozci, J. and Stutzman, P.E. 1993, Computer modelling of the interfacial zone in concrete, in *Interfaces in Cementitious Composites*, Maso, J.C., Ed., E&FN Spon/ Chapman & Hall, London/New York, 107.
49. Hsu, T.T.C.1963, Mathematical Analysis of Shrinkage Stresses in a Model of Hardened Concrete, *J. Am. Concr. Inst.*, 60, 371.
50. Torrenti, J.-M., Acker, P., Boulay, C. and Lejeune, D. 1988, Contraintes initiales dans le béton (Eigenstresses in concrete), *Bull. Liaison Labo. P. et Ch.*, 158, 39 (in French).
51. Acker, P., Boulay, C. and Rossi, P. 1987, On the importance of initial stresses in concrete and of the resulting mechanical effects, *Cem. Conc. Res.*, 17, 755.
52. Vervuurt, A. and Van Mier, J.G.M. 1995, Optical microscopy and digital image analysis of bond-cracks in cement-based materials, in *Microstructure of Cement-Based Systems/ Bonding and Interfaces in Cementitious Materials*, Diamond, S., Mindess, S., Glasser, F.P., Roberts, L.W., Skalny, J.P. and Wakeley, L.D., Eds., MRS, Pittsburgh, PA, 370, 337.
53. Vervuurt, A. 1997, Interface Fracture in Concrete, Ph.D. thesis, Delft University of Technology, Delft, The Netherlands (in preparation).
54. Vervuurt, A., Chiaia, B. and Van Mier, J.G.M. 1995, Damage evolution in different types of concrete, *HERON,* 40, 285.
55. Zhang, M.H. and Gjørv, O.E. 1990, Microstructure of the interfacial zone between light-weight aggregate and cement paste, *Cem. Conc. Res.*, 20, 610.
56. Sarkar, S., Aïtcin, P.-C. and Djellouli, H. 1990, Synergistic roles of slag and silica fume in very high strength concrete, *Cement, Concrete and Aggregates*, 12, 32.
57. Van Mier, J.G.M. 1991, Crack face bridging in normal, high stength and lytag concrete, in *Fracture Processes in Concrete, Rock and Ceramics*, Van Mier, J.G.M., Rots, J.G. and Bakker, A., Eds., Chapman & Hall/E&FN Spon, London/New York, 27.
58. Cai, H., Stevens Kalceff, M.A., Lawn, B.R. 1993, Deformation and fracture of mica-containing glass-ceramics in hertzian contacts, *J. Mater. Res.* (preprint).
59. Alexander, K.M., Wardlaw, J., Gilbert, D.J. 1965, Aggregate-cement bond, cement paste strength and the strength of concrete, in *Proceedings Int'l. Conference on 'The Structure of Concrete'*, Cement & Concrete Association, 59.
60. Mindess, S. and Alexander, M. 1995, Mechanical phenomena at cement/aggregate interfaces, in *Materials Science of Concrete IV*, Skalny, J.P. Ed., The American Ceramic Society, Westerville, OH, 263.
61. Alexander, M.G. 1993, Two experimental techniques for studying the effects of the interfacial zone between cement paste and rock, *Cem. & Conc. Res.*, 23, 567.
62. Alexander, M.G. and Mindess, S. 1995, Use of chevron-notched cylindrical specimens for paste/rock interface experiments, *Cem. Conc. Res.*, 25(2), 345.
63. Tschegg, E.K., Rotter, H.M., Roelfstra, P.E., Bourgund, U. and Jussel, P. 1995, Fracture mechanical behavior of aggregate-cement matrix interfaces, *J. Mater. Civil Engrg. (ASCE)*, 7(4), 199.
64. Taylor, M.A. and Broms, B.B. 1964, Shear bond strength between coarse aggregate and cement paste or mortar, *J. Am. Conc. Inst.*, 61, 939.
65. Mitsui, K., Zongjin Li, Lange, D.A. and Shah, S.P. 1993, A study of the paste-aggregate interface, in *Interfaces in Cementitious Composites*, Maso, J.C., Ed., E&FN Spon/Chapman & Hall, London/New York, 119.
66. Lee, K.M., Buyukozturk, O. and Oumera, A. 1992, Fracture analysis of mortar-aggregate interfaces in concrete, *J. Eng. Mech. (ASCE)*, 118(10), 2031.

67. RILEM 1985, TC-50 FMC Draft Recommendation, Determination of the fracture energy of mortar and concrete by means of three-point bend tests on notched beams, *Mater. Struct. (RILEM)*, 18(106), 285.
68. Tasdemir, M.A., Maji, A.K. and Shah, S.P. 1990, Crack propagation in concrete under compression, *J. Eng. Mech. (ASCE)*, 116(5), 1058.

REFERENCES TO CHAPTER 3

69. Müller, R.K. 1964, Der Einfluss der Messlänge auf die Ergebnisse bei Dehnmessungen an Beton (Influence of measuring length on strain measurements of concrete), *Beton*, 5, 204.
70. Stroeven, P. 1973, Some aspects of the micromechanics of concrete, Ph.D. thesis, Delft University of Technology, The Netherlands.
71. Huet, C. 1993, An integrated approach of concrete micromechanics, in *Micromechanics of Concrete and Cementitious Composites*, Huet, C., Ed., Presses Polytechniques et Universitaires Romandes, Lausanne, 117.
72. Hill, R. 1952, The elastic behaviour of a crystalline aggregate, *Proc. Phys. Soc. A.*, 65, 349.
73. Hansen, T.C. 1968, Theories of multi-phase materials applied to concrete, cement mortar and cement paste, in *Proceedings Int'l. Conference on Structure of Concrete*, Cement & Concrete Association, London, 16.
74. Dantu, P. 1958, Etude des contraintes dans les milieux hétérogenes. Application au béton, *Annales de l'Institut de Bâtiment et des Travaux Publics*, 11(121), 55.
75. Dantu, P. and Mandel, J. 1963, Contribution à l'étude théorique et expérimental du coefficient d'élasticité d'un milieu hétérogène mais statistiquement homogène, *Annales des Ponts et Chausées*, Paris, 133(2), 115.
76. Newman, K. 1968, The structure and properties of concrete, an introductory review, in *Proceedings Int'l. Conference on Structure of Concrete*, Cement & Concrete Association, London, viii. 77.
77. Hashin, Z. and Shtrikman, S. 1963, A variational approach to the theory of the elastic behaviour of multiphase materials, *J. Mech. Phys. Solids*, 11, 127.
78. Wittmann, F.H., Sadouki, H. and Steiger, T. 1993, Experimental and numerical study of effective properties of composite materials, in *Micromechanics of Concrete and Cementitious Composites*, Huet, C., Ed., Presses Polytechniques et Universitaires Romandes, Lausanne, 59.
79. Hughes, B.P. and Ash, J.E. 1970, Anisotropy and failure criteria for concrete, *Mater. Struct. (RILEM)*, 3, 371.
80. Van Mier, J.G.M. 1984, Strain-softening of concrete under multiaxial loading conditions, Ph.D. thesis, Eindhoven University of Technology, The Netherlands.
81. Wang, P.T., Shah, S.P. and Naaman, A.E. 1978, Stress-strain curves of normal and lightweight concrete in compression, *J. Am. Conc. Inst.*, 75, 603.
82. Cornelissen, H.A.W., Hordijk, D.A. and Reinhardt, H.W. 1986, Experimental determination of crack softening characteristics of normalweight and lightweight concrete, *HERON*, 31(2), 45.
83. Hordijk, D.A. and Salet, T. 1989, Experimental investigation into the tensile properties of foamed concrete, *Report 25.5-89-03/VFA*, Delft University of Technology, Department of Civil Engineering, Stevin Laboratory, Delft, The Netherlands.
84. Hordijk, D.A. 1991, Local approach to fatigue of concrete, Ph.D. thesis, Delft University of Technology, The Netherlands.
85. Mihashi, H. and Wittmann, F.H. 1980, Stochastic approach to study the influence of rate of loading on strength of concrete, *HERON*, 25(3).
86. Maher, A. and Darwin, D. 1982, Mortar constituent of concrete in compression, *J. Am. Conc. Inst.*, 79, 100.
87. Vile, G.W.D. 1968, The strength of concrete under short term static biaxial stress, in *Proceedings Int'l. Conference on 'The Structure of Concrete'*, Cement & Concrete Association, London, 275.

88. Gramberg, J. 1989, *A Non-Conventional View on Rock Mechanics and Fracture Mechanics*, Balkema, Rotterdam.

89. Vonk, R.A. 1992, Softening of concrete loaded in compression, Ph.D. thesis, Eindhoven University of Technology, The Netherlands.

90. Wischers, G. and Lusche, M. 1972, Einfluss der innerren Spannungsverteilung das Tragverhalten von druckbeanspruchtem Normal- und Leichtbeton (Influence of internal stress distribution on the behaviour of normalweight and lightweight concrete in compression), *Beton-technische Berichte*, 18, 137 (in German).

91. Gallagher, J.J., Friedman, M., Handin, J. and Sowers, G.M. 1974, Experimental studies relating to microfracture in sandstone, *Tectonophysics*, 21, 203.

92. Vonk, R.A., Rutten, H.S., Van Mier, J.G.M. and Fijneman, H.J. 1990, Size effect in softening of concrete loaded in compression, in *ECF8 'Fracture Behaviour and Design of Materials and Structures'*, Firrao, D., Ed., EMAS Publishers, Warley, U.K., 767.

93. Rüsch, H. 1960, Researches toward a general flexural theory for structural concrete, *J. Am. Conc. Inst.*, 57, 1.

94. Sturman, G.M., Shah, S.P. and Winter, G. 1965, Effects of flexural strain gradients on micro-cracking and stress-strain behaviour of concrete, *J. Am. Conc. Inst.*, 62, 805.

95. Hillerborg, A. 1990, Fracture mechanics concepts applied to moment capacity and rotational capacity of reinforced concrete beams, *Eng. Fract. Mech.*, 35, 233.

96. Markeset, G. 1993, Failure of concrete under compressive strain gradients, Ph.D. thesis, Norges Tekniske Høgskole, Trondheim, Norway.

97. Bazant, Z.P. 1989, Identification of strain-softening constitutive relation from uniaxial tests by series coupling model for localization, *Cem. Conc. Res.*, 19, 973.

98. Bieniawski, Z.I. 1967, Mechanism of brittle fracture of rock, Part I, II and III, *Int. J. Rock Mech. Min. Sci.*, 4, 395.

99. Hudson, J.A., Brown, E.T. and Fairhurst, C. 1972, Shape of the complete stress-strain curve for rock, in: *Stability of Rock Slopes, Proceedings 13th Int'l. Symposium on Rock Mechanics*, Cording, E.J., Ed., ASCE, New York, 773.

100. Kotsovos, M.D. 1983, Effect of testing techniques on the post-ultimate behaviour of concrete in compression, *Mater. Struct. (RILEM)*, 16, 3.

101. Vonk, R.A., Rutten, H.S., Van Mier, J.G.M. and Fijneman, H.J. 1989, Influence of boundary conditions on softening of concrete loaded in compression, in *Fracture of Concrete and Rock — Recent Developments*, Shah, S.P., Swartz, S.E. and Barr, B., Eds., Elsevier Applied Science, London/New York, 1989, 711.

102. Van Mier, J.G.M. and Vonk, R.A. 1991, Fracture of concrete under multiaxial stress — Recent developments, *Mater. Struct. (RILEM)*, 24, 61.

103. Van Vliet, M.R.A. and Van Mier, J.G.M. 1995, Softening behaviour of concrete under uniaxial compression, in *Proceedings FraMCoS-2 'Fracture Mechanics of Concrete Structures'*, Wittmann, F.H., Ed., AEDIFICATIO Publishers, Freiburg, 383.

104. Van Vliet, M.R.A. and Van Mier, J.G.M. 1995, Concrete under uniaxial compression, *Report TUD 25.5-95-9*, October 1995, Stevin Laboratory, Delft University of Technology.

105. Van Vliet, M.R.A. and Van Mier, J.G.M. 1996, Experimental investigation of concrete fracture under uniaxial compression, *Mech. Cohesive-frict. Mater.*, 1(1), 12.

106. Choi, S., Thienel, K-C. and Shah, S.P. 1994, Strain Softening of Concrete — RILEM Round Robin Test, *Research Report NSF-ACBM*, Northwestern University, Evanston, IL, July 1994.

107. König, G., Simsch, G. and Ulmer, M. 1994, Strain-softening of concrete, *Research Report*, Technische Hochschule Darmstadt, October 1994.

108. Lange-Kornbak, D. and Karihaloo, B.L. 1994, Strain softening of concrete under compression, *Research Report*, School of Civil and Mining Engineering, University of Sydney, November 1994.

109. Bascoul, A., Arnaus, M., Balayssac, J.P. and Turatsinze, A. 1994, Report on the round-robin test of RILEM committee 148ssc, *Research Report*, INSA Toulouse, December 1994.

110. Dasenbrock, D., Labuz, J. and French, C. 1994, Strain softening of concrete — Preliminary summary of test results, *Research Report*, Department of Civil Engineering, University of Minesota, September 1994.

111. Van Geel, H.J.G.M. 1994, Uniaxial strain-softening of concrete, *Report BKO-94.09*, Eindhoven University of Technology, The Netherlands.

112. Gobbi, M.E. and Ferrara, G. 1995, Strain Softening of Concrete under Compression, *Research report*, ENEL-CRIS, Milano, May 1995.

113. Zissopoulos, D., Pavlovic, M.N. and M.D. Kotsovos 1994, *Strain Softening of Concrete — RILEM Round Robin Test*, Imperial College, London, and National Technical University of Athens, Greece, August 1994.

114. Taerwe, L. 1993, Empirical analysis of the fracture process in high strength concrete loaded in uniaxial compression, in *Fracture and Damage of Concrete and Rock - FDCR2*, Rossmanith, H.P., Ed., Chapman & Hall/E&FN Spon, London/New York, 122.

115. Markeset, G. 1995, High Strength Concrete Phase 3E - SP4 - Comments on Size Dependence and Brittleness of HSC, *Research Report*, SINTEF Structures and Concrete, Trondheim, Norway, February 1995

116. Armer, G.S.T. and Grimer, F.J. 1989, On the use of full stress-strain characteristics in structural analysis, *Mater. Struct. (RILEM)*, 22, 48.

117. Neville, A.M. 1986, *Properties of Concrete*, Longman Scientific and Technical, Harlow, 3rd ed.

118. Holand, I. and Sellevold, E., Eds. 1993, *Utilization of High Strength Concrete*, Proceedings of the Symposium held in Lillehammer, Norway, June 20-23, 1993, Norwegian Concrete Association, Oslo.

119. Wischers, G. 1978, Aufnahme und Auswirkungen von Druckbeanspruchungen auf Beton (Behaviour of concrete under compressive stress), *Betontechnische Berichte*, 19, 31, (in German).

120. Rokugo, K., Ohno, S. and Koyanagi, W. 1985, Automatical measuring system of load-displacement curves including post-failure region of concrete specimens, in Pre-prints *Int'l. Conference on Fracture Mechanics of Concrete*, Wittmann, F.H., Ed., EPFL, Lausanne, 291.

121. Glavind, M. and Stang, H. 1991, Evaluation of the complete compressive stress-strain curve for high strength concrete, in *Fracture Processes in Concrete, Rock and Ceramics*, Van Mier, J.G.M., Rots, J.G. and Bakker, A., Eds., Chapman & Hall/E&FN Spon, London/New York, 749.

122. CEB 1990, *CEB-FIP Model Code 1990*, Thomas Telford.

123. Sargin, M. 1970, Stress-strain relationships for concrete and the analysis of structural concrete sections, *Solid Mechanics Division Studies No. 4*, University of Waterloo, Waterloo, Ontario, Canada.

124. Plauk, G. (ed.) 1982, *'Concrete Structures under Impact and Impulsive Loading'*, Introductory Report of the RILEM/CEB/IABSE/IASS Symposium held in Berlin, June 2-4, 1982, Bundesanstalt für Materialprüfung, Berlin.

125. Curbach, M. 1987, Festigkeitssteigerung von Beton bei hohen Belastungsgeschwindigkeiten (Strength increase of concrete under high loading rates), *Schriftenreihe des Instituts für Massivbau und Baustofftechnologie, Universität Karlsruhe*, Vol. 1 (in German).

126. Weerheijm, J. 1992, Concrete under impact tensile loading and lateral compression, Ph.D. thesis, Delft University of Technology, The Netherlands.

127. Reinhardt, H.W. 1991, Loading rate, temperature and humidity effects, in *Fracture Mechanics Test Methods for Concrete*, Shah, S.P. and Carpinteri, A., Eds., Chapman & Hall, London/New York, 199.

128. Reinhardt, H.W. 1982, Testing and monitoring techniques for impact and impulsive loading of concrete structures, in *Introductory Report of the RILEM/CEB/IABSE/IASS Symposium on 'Concrete Structures under Impact and Impulsive Loading'*, Plauk, G., Ed., Berlin, June 2-4, 1982, Bundesanstalt für Materialprüfung, Berlin, 65.

129. Ross, C.A. and Kuennen, S.T. 1989, Fracture of concrete at high strain-rates, in *Fracture of Concrete and Rock - Recent Developments*, Shah, S.P., Swartz, S.E. and Barr, B., Eds., Elsevier Applied Science, London/New York, 152.

130. Harsh, S., Shen, Z. and Darwin, D. 1990, Strain-rate sensitive behavior of cement paste and mortar in compression, *ACI Materials Journal*, 87, 508.

131. Reinhardt, H.W., Rossi, P. and Van Mier, J.G.M. 1990, Joint investigation of concrete at high loading rates", *Mater. Struct. (RILEM)*, 23(135), 213.

132. Han, N. and Walraven J.C. 1993, Sustained loading effects in high strength concrete, in *Proceedings Int'l. Symposium on Utilization of High Strength Concrete*, Holand, I. and Sellevold, E., Eds., Norwegian Concrete Association, Oslo, 1076.

133. Sinha, B.P., Gerstle, K.H. and Tulin, C.G. 1964, Stress-strain relations for concrete under cyclic loading, *J. Am. Conc. Inst.*, 61, 195.

134. Karsan, I.D. and Jirsa, J.O. 1969, Behavior of concrete under compressive loadings, *J. Struct. Div. (ASCE)*, 95, 2543.

135. Spooner, D.C. and Dougill, J.W. 1975, A quantitative assessment of damage sustained in concrete during compressive loading, *Mag. Conc. Res.*, 27, 151.

136. Bažant Z.P. and Kim, S.S. 1979, Plastic fracturing theory for concrete, *J. Eng. Mech. Div. (ASCE)*, 105, 407.

137. Dougill, J.W. and Rida, M.A.M. 1980, Further considerations of progressively fracturing solids, *J. Eng. Mech. Div. (ASCE)*, 106, 1021.

138. Dougill, J.W. 1983, Constitutive relations for concrete and rock: Applications and extensions of elasticity and plasticity theory, in *Preprints William Prager Symposium on 'Mechanics of Geomaterials: Rocks, Concretes, Soils'*, Bažant, Z.P., Ed., Northwestern University, Evanston, IL, 17.

139. Spooner, D.C., Pomeroy, C.D. and Dougill, J.W. 1976, Damage and energy dissipation in cement pastes in compression, *Mag. Conc. Res.*, 28, 21.

140. Bažant, Z.P. and Oh, B.-H. 1983, Crack band theory for fracture of concrete, *Mater. Struct. (RILEM)*, 16, 155.

141. Evans, R.H. and Marathe, H.S. 1968, Microcracking and stress-strain curves for concrete in tension, *Mater. Struct. (RILEM)*, 1, 61.

142. Petersson, P.-E. 1981, Crack growth and development of fracture zones in plain concrete and similar materials, *Report TVBM-1006*, Lund Institute of Technology, Sweden.

143. Mindess, S. 1991, Fracture process zone detection, in *Fracture Mechanics Test Methods for Concrete*, Shah, S.P. and Carpinteri, A., Eds., Chapman & Hall, London, Chap. 5.

144. Van Mier, J.G.M. 1990, Internal crack detection in single edge notched concrete plates subjected to uniform boundary displacement, in *Micromechanics of Failure of Quasi Brittle Materials*, Shah, S.P., Swartz, S.E. and Wang, M.L., Eds., Elsevier Applied Science Publishers, London/New York, 33.

145. Van Mier, J.G.M. 1990, Fracture process zone in concrete: A three dimensional growth process, in *ECF8 'Fracture Behaviour and Design of Materials and Structures'*, Firrao, D., Ed., EMAS Publishers, Warley, UK, 567.

146. Van Mier, J.G.M. 1991, Mode I fracture of concrete: Discontinuous crack growth and crack interface grain bridging, *Cem. Conc. Res.*, 21, 1.

147. Van Mier, J.G.M. 1992, Scaling in tensile and compressive fracture of concrete, in *Applications of Fracture Mechanics to Reinforced Concrete*, Carpinteri, A., Ed., Elsevier Applied Science, London/New York, Chap. 5.

148. Van Mier, J.G.M. and Nooru-Mohamed, M.B. 1990, Geometrical and structural aspects of concrete fracture, *Eng. Fract. Mech.*, 35, 617.

149. Bascoul, A., Kharchi, F. and Maso, J.C. 1989, Concerning the measurement of the fracture energy of a micro-concrete according to the crack growth in a three points bending test on Notched Beams, in *Fracture of Concrete and Rock*, Shah, S.P. and Swartz, S.E., Eds., Springer Verlag, New York, 396.

150. Swartz, S.E. and Refai, T. 1989, Cracked surface revealed by dye and its utility in determining fracture parameters, in *Fracture Toughness and Fracture Energy*, Mihashi, H., Takhashi, H. and Wittmann, F.H., Eds., Balkema, Rotterdam, 509.

151. Turatsinze, A. 1992, Caracterisation microstructurale de la fissuration des betons et mortiers en mode I propagation (Microstructural characterization of mode I cracking in concrete and mortar), Ph.D. thesis, University Paul Sabatier, Toulouse (in French).

152. Mazars, J. 1981, Mechanical damage and fracture of concrete structures, in *Proceedings ICF5 'Advances in Fracture Research'*, Francois, D., Ed., Vol. 4, Pergamon Press, Oxford, 1499.

153. Løland, K.E. 1980, Continuous damage model for load-response estimation of concrete, *Cem. Conc. Res.*, 10, 395.

154. Simha, K.R.Y., Fourney, W.L., Barker, D.B. and Dick, R.D. 1986, Dynamic photo-elastic investigation of two pressurised cracks approaching one another, *Eng. Fract. Mech.*, 23, 237.

155. Van Mier, J.G.M., Schlangen, E., Visser, J.H.M. and Vervuurt, A. 1992, Experimental and numerical analysis of cracking in concrete and sandstone, in *Topics in Applied Mechanics — Integration of Theory and Applications in Applied Mechanics*, Dijksman, J.F. and Nieuwstadt, F.T.M., Eds., Kluwer Academic Publishers, Dordrecht, 65.

156. Schlangen, E. 1993, Experimental and numerical analysis of fracture processes in concrete, Ph.D. thesis, Delft University of Technology, The Netherlands, 1993.

157. Sempere, J.-C. and Macdonald, K.C. 1986, Overlapping spreading centers: implications from crack growth simulation by the displacement discontinuity method, *Tectonics*, 5, 151.

158. Steinbrech, R.W., Dickerson, R.M. and Kleist, G. 1991, Characterization of the fracture behaviour of ceramics through analysis of crack propagation studies, in *Toughening Mechanisms in Quasi Brittle Materials*, Shah, S.P., Ed., NATO ASI series, Vol. E-195, Kluwer Academic Publishers, Dordrecht, 287.

159. Swanson, P.L., Fairbanks, C.L., Lawn, B.R., Mai, Y.-W. and Hockey, B.J. 1987, Crack-interface grain bridging as a fracture resistance mechanism in ceramics: I, experimental study on alumina, *J. Am. Ceram. Soc.*, 70, 279.

160. Zeyher, A. 1993, Simulated materials reveal microfractures, *Computers in Physics*, 7(4), 382.

161. Carpinteri, A. 1992, Fractal nature of material microstructure and size effects on apparent mechanical properties, *Internal Report 1/92*, Politecnico di Torino.

162. Schlangen, E. and Van Mier, J.G.M. 1991, Lattice model for numerical simulations of concrete fracture, in *Proceedings Int'l. Conference on Dam Fracture*, Saouma, V.E., Dungar, R. and Morris, D., Eds., University of Colorado, Boulder, Sept. 11-13, 1991, Electric Power Research Institute, Palo Alto, CA, 511.

163. Vonk, R.A., Rutten, H.S., Van Mier, J.G.M. and Fijneman, H.J. 1991, Micromechanical simulation of concrete softening, in *Fracture Processes in Concrete, Rock and Ceramics*, Van Mier, J.G.M., Rots, J.G. and Bakker, A., Eds., Chapman & Hall/E&FN Spon, London/ New York, 129.

164. Wang, J. and Huet, C. 1993, A numerical model for studying the influences of pre-existing microcracks and granular character on the fracture of concrete materials and structures, in *Micromechanics of Concrete and Cementitious Composites*, Huet, C., Ed., Presses Polytechniques et Universitaires Romandes, Lausanne, 229.

165. Duda, H. 1991, Grain-model for the determination of the stress-crack width relation, in *Analysis of Concrete Structures by Fracture Mechanics*, Elfgren, L. and Shah, S.P., Eds., Chapman & Hall, London/New York, 88.

166. Lange, D.A., Jennings, H.M. and Shah, S.P. 1993, Relationship between fracture surface roughness and fracture behaviour of cement paste and mortar, *J. Am. Ceram. Soc.*, 76(3), 589.

167. Nomura, N., Mihashi, H. and Izumi, M. 1991, Properties of fracture process zone and tension softening behaviour of concrete, in *Fracture Processes in Concrete, Rock and Ceramics*, Van Mier, J.G.M., Rots, J.G. and Bakker, A., Eds., Chapman & Hall/E&FN Spon, London/New York, 51.

168. Walraven, J.C., Den Uijl, J., Strobrand, J., Al-Zubi, N., Gijsbers, J. and Naaktgeboren, M. 1995, Structural lightweight concrete:recent research, *HERON*, 40(1), 5.

169. Carpinteri, A. 1985, Interpretation of Griffith instability as a bifurcation of the global equilibrium, in *Application of Fracture Mechanics to Cementitious Composites*, Shah, S.P., Ed., Martinus Nijhoff Publishers, Dordrecht, 287.

170. Barr, B. and Tokatly, Z.Y. 1992, Size effects in two compact test specimen geometries, in *Application of Fracture Mechanics to Reinforced Concrete*, Carpinteri, A., Ed., Elsevier Applied Science, London/New York, Chap. 4.

171. Brühwiler, E. 1988, Bruchmechanik von Staumauerbeton unter Quasi-Statischer und Erdbeben-dynamischer Belastung (Fracture mechanics of dam concrete under quasi-static and eartquake loading), Ph.D. thesis, Ecole Polytechnique Federale de Lausanne (in German).

172. Mobasher, B., Ouyang, C., Castro-Montero, A. and Shah, S.P. 1991, Tensile behaviour of fibre reinforced concrete, in *Fracture Processes in Concrete, Rock and Ceramics*, Van Mier, J.G.M., Rots, J.G. and Bakker, A., Eds., Chapman & Hall/E&FN Spon, London/New York, 269.

173. Aveston, J., Cooper, G.A. & Kelly, A. 1971, Single and multiple fracture, in *Proc. NPL Conference on the Properties of Fibre Composites*, National Physical Laboratory, London, England, 15.

174. Wang, Y., Li, V.C. and Backer, S. 1990, Experimental determination of tensile behavior of fiber reinforced cements, *ACI Mater. J.*, 87, 461.

175. Li, V.C. 1992, Performance driven design of fiber reinforced cementitious composites, in *Fibre Reinforced Cement & Concrete*, Swamy, R.N., Ed., Chapman & Hall/E&FN Spon, London/New York, 12.

176. Stang, H. 1995, Micromechanical parameters, fracture processes and application of fiber reinforced concrete, in *Fracture of Brittle Disordered Materials*, Baker, G and Karihaloo, B.L., Eds., Chapman & Hall/E&FN Spon, London/New York, 131.

177. Balaguru, P.N. and Shah, S.P. 1992, *Fibre Reinforced Cement Composites*, McGraw-Hill, New York.

178. Salet, T.A.M. 1990, Structural analysis of sandwich beams composed of reinforced concrete faces and a foamed concrete core, Ph.D. thesis, Eindhoven University of Technology, The Netherlands.

179. Bonzel, J. and Kadlecek, V. 1970, Einfluss der Nachbehandelung und des Feuchtig-keitszustands auf die Zugfestigkeit des Betons (Influence of curing and moisture content on the tensile strength of concrete) , *Beton*, 20, 303 and 351 (in German).

180. Fouré, B. 1985, Note sur la chute de résistance à la traction du béton léger consécutive à l'arrêt de la cure humide, *Ann. de l'Inst. Techn. du batiment et des Travaux Publics*, 432, 1 (in French).

181. Brameshuber, W. 1988, Bruchmechanische Eigenschaften von jungem Beton (Fracture mechanics properties of young concrete), Ph.D. thesis, Universität Karlsruhe (in German).

182. Van Mier, J.G.M. and Schlangen, E. 1989, On the stability of softening systems, In *Fracture of Concrete and Rock — Recent Developments*, Shah, S.P., Swartz, S.E. and Barr, B., Eds., Elsevier Applied Science Publishers, London/New York, 387.

183. Alvaredo, A.M. and Wittmann, F.H. 1992, Crack formation due to hygral gradients, in *Fracture Mechanics of Concrete Structures*, Bažant, Z.P., Ed., Elsevier Applied Science, London/New York, 960.

184. Van Mier, J.G.M. 1990, Mode I behavior of concrete: Influence of the rotational stiffness outside the crack zone, in *Analysis of Concrete Structures by Fracture Mechanics*, Elfgren, L. and Shah, S.P., Eds., Chapman & Hall, London/New York, 19.

185. Quenard, D. and Sallee, H. 1992, Water vapour adsorption and transfer in cement-based materials: a network simulation, *Mater. Struct. (RILEM)*, 25, 515.

186. Sadouhi, H. and van Mier, J.G.M. 1997, Simulation of hygral crack growth in concrete repair systems, and meso-level analysis of moisture flow in cement composites using a lattice type approach, accepted for publication in *Mater. Struct. (RILEM)*, and Analysis of hygral induced crack growth in multiphase materials, *HERON*, 41(4).

187. Reinhardt, H.W. 1984, Fracture mechanics of an elastic softening material like concrete, *HERON*, 29(2).

188. Cornelissen, H.A.W. 1984, Fatigue failure of concrete in tension, *HERON*, 29(4).

189. Reinhardt, H.W. and Cornelissen, H.A.W. 1984, Post-peak cyclic behavior of concrete in uniaxial tensile and alternating tensile and compressive loading, *Cem. Conc. Res.*, 14, 263.

190. Cornelissen, H.A.W., Hordijk, D.A. and Reinhardt, H.W. 1986, Experiments and theory for the application of fracture mechanics to normal and lightweight concrete, in *Fracture Toughness and Fracture Energy*, Wittmann, F.H., Ed., Elsevier, London/New York, 565.

191. Gylltoft, K. 1983, Fracture mechanics models for fatigue in concrete structures, Ph.D. thesis, Luleå University of Technology, Sweden.

192. Cornelissen, H.A.W. & Siemes, 1985, Plain concrete under sustained tensile or tensile and compressive fatigue loadings, in *Proceedings BOSS Conference*, Elsevier, 487.

193. Körmeling, H. 1986, Strain rate and temperature behaviour of steel fibre concrete in tension, Ph.D. thesis, Delft University of Technology, The Netherlands.

194. Reinhardt, H.W., Körmeling, H.A. and Zielinski, A.J. 1986, The split hopkinson bar, a versatile tool for impact testing of concrete, *Mater. Struct. (RILEM)*, 19, 55.

195. Reinhardt, H.W. 1982, Concrete under impact loading: tensile strength and bond, *HERON*, 27(3).

196. Rossi, P., Van Mier, J.G.M., Boulay, C. and LeMaou, F. 1992, The dynamic behaviour of concrete: influence of free water, *Mater. Struct. (RILEM)*, 25, 509.

197. Zielinski, A.J. 1982, Fracture of concrete and mortar under uniaxial impact tensile loading, Ph.D. thesis, Delft University of Technology, The Netherlands.

198. Bažant, Z.P. and Pfeiffer, P.A. 1986, Shear fracture tests of concrete, *Mater. Struct. (RILEM)*, 19, 111.

199. Davies, J. 1991, Numerical and experimental study of development of fracture path under mixed mode loading, in *Fracture Processes in Concrete, Rock and Ceramics*, Van Mier, J.G.M., Rots, J.G. and Bakker, A., Eds., Chapman & Hall Publishers, London/New York, pp. 717.

200. Davies, J. 1992, Macroscopic study of crack face bridging phenomenon in mixed-mode loading, in *Fracture Mechanics of Concrete Structures*, Bažant, Z.P., Ed., Elsevier Applied Science, London/New York, 713.

201. Barr, B., Hasso, A.B.D. and Khalifa, S.M.A. 1987, A study of mode II shear fracture of notched beams, cylinders and cones, in *Fracture of Concrete and Rock, SEM-RILEM Int'l. Conf.*, Houston, Shah, S.P. and Swartz, S.E., Eds., SEM, Bethel, CT, 370.

202. Carpinteri, A., Ferrara, G. and Melchiorri, G. 1989, Single edge notched specimen subjected to four point shear: an experimental investigation, in *Fracture of Concrete and Rock - Recent Developments*, Shah, S.P., Swartz, S.E. and Barr, B., Eds., Elsevier Applied Science Publishers, London/New York, 605.

203. Carpinteri, A., Ferrara, G., Melchiorri, G. and Valente, S. 1990, The four point shear test on single notched specimens: an experimental and numerical analysis, in *Fracture Behaviour and Design of Materials and Structures*, Firrao, D., Ed., EMAS Publishers, Warley, U.K., 667.

204. Hassanzadeh, M. and Hillerborg, A. 1989, Concrete properties in mixed mode fracture, in *Fracture Toughness and Fracture Energy*, Mihashi, H., Takahashi, H. and Wittmann, F.H., Eds., Balkema, Rotterdam, 565.

205. Hassanzadeh, M. 1992, Behaviour of fracture process zones in concrete influenced by simultaneously applied normal and shear displacements, Ph.D. thesis, Lund Institute of Technology, Report TVBM 1010, Sweden.

206. Reinhardt, H.W., Cornelissen, H.A.W. and Hordijk, D.A. 1989, Mixed mode fracture tests on concrete, in *Fracture of Concrete and Rock*, Shah, S.P. and Swartz, S.E., Eds., Springer Verlag, New York, 119.

207. Iosipescu, N. 1967, New accurate procedure for single shear testing of metals, *J. Materials*, 2(3), 537.

208. Swartz, S.E. and Taha, N.M. 1990, Mixed mode crack propagation and fracture in concrete, *Eng. Fract. Mech.*, 35, 137.

209. Van Mier, J.G.M., Schlangen, E. and Nooru-Mohamed, M.B. 1992, Shear fracture in cementitious composites, Part I: Experimental observations, part II: Numerical simulations, in *Proceedings FraMCoS 1*, Bažant, Z.P., Ed., Elsevier Applied Science Publishers, London/New York, 659.

210. Ingraffea, A.R. and Panthaki, M.J. 1986, Analysis of shear fracture tests of concrete beams, in *Finite Element Analysis of Reinforced Concrete Structures*, Meyer, C. and Okamura, H., Eds., American Society of Civil Engineers, New York, 151.

211. Carpinteri, A. and Swartz, S.E. 1991, Mixed-mode crack propagation in concrete, in *Fracture Mechanics Test Methods for Concrete*, Shah, S.P. and Carpinteri A., Eds., Chapman & Hall Publishers, London/New York, Chap. 3.

212. Schlangen, E. and Van Mier, J.G.M. 1993, Mixed-mode fracture propagation: A combined numerical and experimental study, in *Fracture and Damage of Concrete and Rock (FDCR-2)*, Rossmanith, H.P., Ed., Chapman & Hall/E&FN Spon, London/New York, 166.

213. Hassanzadeh, M., Hillerborg, A. and Zhou, F.P. 1987, Tests of material properties in mixed-mode I and II, in *Pre-Proceedings SEM-RILEM International Conference on Fracture of Concrete and Rock*, Shah, S.P. and Swartz, S.E., Eds., SEM, Bethel, CT, 353.

214. Nooru-Mohamed, M.B. 1992, Mixed mode fracture of concrete: an experimental approach, Ph.D. thesis, Delft University of Technology, The Netherlands.

215. Keuser, W. and Walraven, J.C. 1989, Fracture of plain concrete under mixed-mode conditions, in *Fracture of Concrete and Rock - Recent Developments*, Shah, S.P., Swartz, S.E. and Barr, B., Eds., Elsevier Applied Science Publishers, London/New York, 625.

216. Paulay, T. and Loeber, P.J. 1974, Shear transfer by aggregate interlock, *ACI Special Publication SP-42*, American Concrete Institute, Detroit, 1.

217. Laible. J.P., White, R.N. and Gergely, P. 1977, Experimental investigation of seismic shear transfer accross cracks in concrete nuclear containement vessels, *ACI Special Publication SP-53*, American Concrete Institute, Detroit, 203.

218. Walraven, J.C. 1980, Aggregate interlock: A theoretical and experimental analysis, Ph.D. thesis, Delft University of Technology, The Netherlands.

219. Millard, S.G. and Johnson, R.P. 1984, Shear transfer accross cracks in reinforced concrete due to aggregate interlock and dowel action, *Mag. of Concrete Res.*, 36(126), 9.

220. Van Mier, J.G.M. and Nooru-Mohamed, M.B. 1989, Fracture of concrete under tensile and shearlike loadings, in *Fracture Toughness and Fracture Energy - Test Methods for Concrete and Rock*, Mihashi, H., Takahashi, H. and Wittmann, F.H., Eds., Balkema, Rotterdam, 549.

221. Yin, W.S., Su, E.C.M., Mansur, M.A. and Hsu, T.T.C. 1990, Fibre reinforced concrete under biaxial compression, *Eng. Fract. Mech.*, 35, 261.

222. Lankard, D.R. and Newell, J.K. 1984, Preparation of highly reinforced steel fibre reinforced concrete composites, in *Fiber Reinforced Concrete (ACI-SP 81)*, American Concrete Institute, Detroit, 286.

223. Reinhardt, H.W. and Naaman, A.E., Eds. 1992, *High Performance Fiber Reinforced Cement Composites*, Chapman & Hall/E&FN Spon, London/New York.

224. Naaman, A.E. and Reinhardt, H.W. 1996, *High Performance Fiber Reinforced Cement Composites - 2*, Chapman & Hall/E&FN Spon, London/New York.

225. Naaman, A.E. 1992, SIFCON: Tailored properties for structural performance, in *High Performance Fiber Reinforced Cement Composites*, Reinhardt, H.W. and Naaman, A.E., Eds., Chapman & Hall/E&FN Spon, London/New York, 18.

226. Naaman, A.E., Paramasivan, P., Balazs, G., Bayasi, Z.M., Eibl, J., Erdelyi, L., Hassoun, N.,Krstulovic-Opara, N., Li, V.C. and Lohrman, G.1996, Reinforced and prestressed concrete using hpfrcc matrices, in *High Performance Fiber Reinforced Cement Composites - 2*, Chapman & Hall/E&FN Spon, London/New York, Chap. 8.

227. Van Mier, J.G.M. 1995, Fracture mechanics of concrete: Will applications start to emerge?, *HERON*, 40(2), 147.

228. Van Mier, J.G.M., Stang, H. and Ramakrishnan, V. 1996, Practical structural applications of FRC and HPFRCC, in *High Performance Fiber Reinforced Cement Composites - 2*, Chapman & Hall/E&FN Spon, London/New York, Chap. 12.

229. Van Mier, J.G.M. and Timmers, G. 1992, Shear fracture in slurry infiltrated fibre concrete (SIFCON), in *High Performance Fibre Reinforced Cement Composites*, Reinhardt, H.W. and Naaman, A.E., Eds., Chapman & Hall/E&FN Spon, London/New York, 348.

230. Van Mier, J.G.M. and Timmers G.1992, SIFCON subjected to shear: Effect of material anisotropy on strength and stiffness, in *Fibre Reinforced Cement and Concrete*, Swamy, R.N., Ed., Chapman & Hall, London/New York, 245.

231. Bažant, Z.P. and Pratt, P.C. 1988, Measurement of mode III fracture energy of concrete, *Nucl. Engrg. Design*, 106, 1.

232. Yacoub-Tokatly, Z. and Barr, B. 1989, Mode III fracture — A tentative test geometry, in *Fracture of Concrete and Rock — Recent Developments*, Shah, S.P., Swartz, S.E. and Barr, B., Eds., Elsevier Applied Science Publishers, London/New York, 596.

233. Xu Daoyuan and Reinhardt, H.W. 1989, Softening of concrete under torsional loading, in *Fracture of Concrete and Rock — Recent Developments*, Shah, S.P., Swartz, S.E. and Barr, B., Eds., Elsevier Applied Science Publishers, London/New York, 39.

234. Bažant, Z.P., Sener, S. and Pratt, P.C. 1988, Size effect tests of torsional failure of plain and reinforced concrete beams, *Mater. Struct. (RILEM)*, 21, 425.

235. Visser, J.H.M. 1997, Tensile hydraulic fracture of concrete and sandstone, Ph.D. thesis, Delft University of Technology, The Netherlands.

236. Kupfer, H. 1973, Das Verhalten des Betons unter Mehrachsiger Kurzzeitbelastung unter Besonderer Berücksichtigung des Zweiachsiger Beanspruchung (Behavior of concrete under multiaxial short term loading, with emphasis on biaxial loading), *Deutscher Ausschuss für Stahlbeton*, Vol. 229, Berlin, (in German).

237. Linse,D. and Stegbauer, A. 1976, Festigkeit und Verformungsverhalten von Leichtbeton, Gasbeton, Zementstein, und Gips unter Zweiachsigen Kurzzeitbelastung (Strength and eformation of lightweight concrete, foamed concrete, hardened cement paste and gypsum under short term biaxial loading), *Deutscher Ausschuss für Stahlbeton*, Vol.254, Berlin (in German).

238. Gerstle, K.H., Linse, D.L., Bertacchi, P., Kotsovos, M.D., Ko, H.-Y., Newman, J.B., Rossi, P., Schickert, G., Taylor, M.A., Traina, L.A., Zimmerman, R.M. and Bellotti, R. 1978, Strength of concrete under multiaxial stress states, in *Proceedings Douglas McHenry Int'l. Symposium on 'Concrete and Concrete Structures'*, ACI SP 55, American Concrete Institute, Detroit, MI, 103.

239. Kobayashi, S. and Koyanagi, W. 1972, Fracture criteria of cement paste, mortar and concrete subjected to multiaxial compressive stresses, in *Proceedings RILEM Symposium on 'The Deformation and the Rupture of Solids Subjected to Multiaxial Stress'*, Cannes, part I 'Concrete', paper I/9, 1972, 131.

240. Schickert, G. and Danssmann, J. 1984, Behaviour of concrete stressed by high hydrostatic compression, in *Proceedings RILEM/CEB Symposium on 'Concrete under Multiaxial Conditions'*, Vol. 2, Presses de l'Université Paul Sabatier, Toulouse, 69.

241. Bažant, Z.P., Bishop, F.C. and Chang, T.P. 1986, Confined compression tests of cement paste and concrete up to 300 ksi, *J. Am. Conc. Inst.*, 83, 553.

242. Richardt, F.E., Brandtzaeg. A. and Brown, R.L. 1929, The behaviour of plain and spirally reinforced ccncrete in compression, *University of Illinois Engineering Experiment Station*, Bulletin no. 190.

243. Newman, J.B. and Newman, K. 1972, The cracking and failure of concrete under combined stresses and its implications for structural design, in *Proceedings RILEM Symposium on 'The Deformation and the Rupture of Solids Subjected to Multiaxial Stress'*, Cannes, part I 'Concrete', paper I/10, 149.

244. Newman, J.B. 1979, Concrete under complex stress, in *Developments in Concrete Technology - 1*, Lydon, F.D., Ed., Applied Science Publishers, London, 151.

245. Kotsovos, M.D. and Newman, J.B. 1977, Behavior of concrete under multiaxial stress, *J. Am. Conc. Inst.*, 74, 443.

246. Kotsovos, M.D. 1979, Effect of stress-path on the behaviour of concrete under triaxial stress states, *J. Am. Conc. Inst.*, 76, 213.

247. Ahmad, S.H. and Shah, S.P. 1982, Complete triaxial stress-strain curves for concrete, *J. Struct. Div. (ASCE)*, 108, 728.

248. Jamet, P., Millard, A. and Nahas, G. 1984, Triaxial behaviour of a micro-concrete complete stress-strain curves for confining pressures ranging from 0 to 100 MPa, in *Proceedings RILEM/CEB Symposium on 'Concrete under Multiaxial Conditions'*, Vol. 1, Presses de l'Université Paul Sabatier, Toulouse, 133.

249. Visser, J.H.M. and Van Mier, J.G.M. 1994, Hydraulic fracturing in the tensile regime, in *Computer Methods and Advances in Geomechanics*, Siriwardane, H.J. and Zaman, M.M., Eds., Balkema, Rotterdam, 1647.

250. Willam, K.J. and Warnke, E.P. 1974, Constitutive model for the triaxial behaviour of concrete, in *Proceedings Colloquium on Concrete Structures Subjected to Triaxial Stresses, IABSE Report 19*, IABSE, Zürich.

251. Schickert, G. and Winkler, H. 1977, Versuchsergebnisse zur Festigkeit und Verformung von Beton bei mehraxialer Druckbeanspruchung (Strength and deformation of concrete under multiaxial compressive stress), *Deutscher Ausschuss für Stahlbeton*, Vol. 277, Berlin, (in German).

252. Podgorski, J. 1985, General failure criterion for isotropic media, *J. Eng. Mech. (ASCE)*, 111, 188.

253. Van Mier, J.G.M. 1986, Fracture of concrete under complex stress, *HERON*, 31(3).

254. Nelissen, L.J.M. 1972, Biaxial testing of normal concrete, *HERON*, 18(1).

255. Willam, K. Hurlbut, B. and Sture, S. 1985, Experimental constitutive and computational aspects of concrete fracture, in *Proc. Seminar on Finite Element Analysis of Reinforced Concrete Structures*, Ch. Meyer and H. Okamura, Eds., ASCE, New York, 149.

256. Paterson, M.S. 1978, *Experimental Rock Deformation - The Brittle Field*, Springer Verlag, New York.

257. Michelis, P.N. and Demiris, C.A. 1984, Conception et construction d'une vraie cellule triaxiale (Design and construction of a true triaxial cell), *C.R. Acad. Sc. Paris*, t.299, Série II, no.8, 375 (in French).

258. Michelis, P.N. 1985a, A true triaxial cell for low and high pressure experiments, *Int. J. Rock Mech. Min. Sci. Geomech. Abstr.*, 22, 183.

259. Michelis, P.N. 1985b, Polyaxial yielding of granular rock, *J. Eng. Mech. (ASCE)*, 111, 1049.

260. Kamp, W. and Cochram, M.J. 1991, Yield envelope investigation of felser sandstone with true triaxial compression tests including post failure behaviour, in *Fracture Processes in Concrete, Rock and Ceramics*, Van Mier, J.G.M., Rots, J.G. and Bakker, A., Eds., Chapman & Hall/E&FN Spon, London/New York, 761.

261. Bieniawski, Z.I. Denkhaus, H.G. and Vogler, U.W. 1969, Failure of fractured rock, *Int. J. Rock Mech. Min. Sci.*, 6, 323.

262. Hallbauer, D.K., Wagner, H. and Cook, N.G.W. 1973, Some observations concerning the microscopic and mechanical behaviour of quartzite specimens in stiff, triaxial compression tests, *Int. J. Rock Mech. Min. Sci. Geomech Abstr.*, 10, 713.

263. Bongers, J.P.W., Rutten, H.S., Van Mier, J.G.M. and Fijneman, H.J. 1994, Softening behaviour of concrete loaded in multiaxial compression, in *Computer Modelling of Concrete Structures*, Mang, H., Bicanic, N. and De Borst, R., Eds., Pineridge Press, Swansea, 303.

264. Smith, T.R. and Schulson, E.M. 1993, The brittle compressive failure of fresh-water columnar ice under biaxial loading, *Acta Metall. Mater.*, 41(1), 153.

265. Horii, H. and Nemat-Nasser, S. 1985, Compression-induced microcrack growth in brittle solids: axial splitting and shear failure, *J. Geophys. Res.*, 90, 3105.

266. Dyskin, A.V., Germanovich, L.N., Jewell, R.J., Joer, H., Krasinski, J.S., Lee, K.K., Roegiers, J.-C., Sahouryeh, E. and Ustinov, K.B. 1994, Study of 3D mechanisms of crack growth and interaction in uniaxial compression, *ISRM communications* (pre-print).
267. Cannon, N.P., Schulson, E.M., Smith, T.R. and Frost, H.J. 1990, Wing cracks and brittle compressive fracture, *Acta Metall. Mater.*, 38(10), 1955.

REFERENCES TO CHAPTER 4

268. Schorn, H. and Berger-Böcker, T. 1989, Test method for determining process zone position and fracture energy of concrete, *Experimental Techniques*, June 1989, 29-33.
269. Grimer, F.J. and Hewitt, R.E. 1971, The form of the stress-strain curve of concrete interpreted with a diphase concept of material behavior, in *Structure, Solid Mechanics and Engineering Design*, Te'eni, M., Ed., Wiley Interscience, 681.
270. Wawersik, W.R. and Fairhurst, C. 1970, A study of brittle rock fracture in laboratory compression experiments, *Int. J. Rock Mech. Min. Sci.*, 7, 561.
271. Gopalaratnam, V.S. and Shah, S.P. 1985, Softening response of plain concrete in direct tension, *J. Am. Conc. Inst.*, 82, 310.
272. Li, Z., Kulkarni, S.M. and Shah, S.P. 1993, New test method for obtaining softening response of unnotched concrete specimen under uniaxial tension, *Experimental Mechanics*, September 1993, 181.
273. Gjørv, O.E., Sorensen, S.I. and Arnesen, A. 1977, Notch sensitivity and fracture toughness of concrete, *Cem. Conc. Res.*, 7, 333.
274. Schlangen, E. and Van Mier, J.G.M. 1995, Crack propagation in sandstone: A combined experimental and numerical approach, *Rock Mech. Rock Eng.*, 28(2), 93.
275. Visser J.H.M. and Van Mier, J.G.M. 1992, Hydraulically driven fractures of concrete and rock: a new test cell, in *Proceedings FraMCoS1*, Bazant, Z.P., Ed., Elsevier Applied Science, London/New York, 1992, 512.
276. Linse, D. 1978, Lösung Versuchstechnischer Fragen bei der Ermittlung des Festigkeits und Verformungs verhaltens von Beton unter Dreiachsiger Belastung (Solutions for experimental problems in triaxial testing of concrete), *Deutscher Ausschuss für Stahlbeton*, Vol. 292, Berlin (in German).
277. Winkler, H. 1985, Grundsätzliche Untersuchungen zum Geräteeinfluss bei der mehraxialen Druckprüfung von Beton (Fundamental study on the influence of test-equipment on the multiaxial behavior of concrete), *Deutscher Ausschuss für Stahlbeton*, Vol. 366, Berlin (in German).
278. Meier, R.W., Ko, H.-Y. and Sture, S. 1985, Direct tensile loading apparatus combined with a cubical test cell for testing rocks and concrete, *Geotech. Testing J. (GTJODJ)*, 8, 71.
279. Sture, S. and Ko, H.-Y. 1978, Strain-softening of brittle geologic materials, *Int. J. Num. Anal. Methods in Geo. Mech.*, 2, 237.
280. Sture, S. and Desai, C.S. 1979, Fluid cushion truly triaxial or multiaxial testing device, *Geotech. Testing J. (GTJODJ)*, 2, 20.
281. Launay, P., Gachon, H. and Poitevin, P. 1970, Déformation et résistance ultime du béton sous étrainte triaxiale (Deformation and strength of concrete under multiaxial stress), *Annales de l'Institut Technique du Bâtiment et des Travaux Publics*, 269, 23 (in French).
282. Launay, P. and Gachon, H. 1971, Strain and ultimate strength of concrete under triaxial stress, in *Proceedings SMiRT-1*, paper H 1/3, Berlin, 23.
283. Mills, L.L. and Zimmerman, R.M. 1970, Compressive strength of plain concrete under multiaxial loading conditions, *J. Am. Conc. Inst.*, 67, 802.
284. Bertacchi, P. and Bellotti, R. 1972, Experimental research on deformation and failure of concrete under triaxial loads, in *Proceedings RILEM Symposium on 'The Deformation and the Rupture of Solids Subjected to Multiaxial Stress'*, Cannes, part I 'Concrete', paper I/3, 37.

285. Nojiri, Y., Kotani, K. and Abe, Y. 1984, Failure envelope of concrete subjected to multiaxial compressive stresses, in *Proceedings RILEM/CEB Symposium on 'Concrete under Multiaxial Conditions'*, Vol.1, Presses de l'Université Paul Sabatier, Toulouse, 141.

286. Erdei, C.K. 1980, Finite element analysis and tests with a new load transmitting medium to measure compressive strength of brittle materials, *Mater. Struct. (RILEM)*, 13, 83.

287. Labuz, J.F. and Bridell, J.M. 1991, Reducing frictional constraint in compression testing through lubrication, *Experimental Mechanics*, 1991.

288. Hilsdorf, H. 1965, Die bestimmung der Zweiachsiger Festigkeit des Betons (Determination of the biaxial strength of concrete), *Deutscher Ausschuss für Stahlbeton*, Vol. 173, Berlin (in German).

289. Schickert, G. 1973, On the influence of different load application techniques on the lateral strain and fracture of concrete specimens, *Cem. Conc. Res.*, 3, 487.

290. Schickert, G. 1980, Schwellenwerte beim Betondruckversuch (Threshold numbers for compressive testing of concrete), *Deutscher Ausschuss für Stahlbeton*, Vol. 312, Berlin (in German).

291. Guinea, G.V., Planas, J. and Elices, M. 1992, Measurement of the fracture energy using three-point bend tests: Part 1 - Influence of experimental procedures, *Mater. Struct. (RILEM)*, 25(148), 212.

292. Carpinteri, A. and Ferro, G. 1994, Size effects on tensile fracture properties: a unified explanation based on disorder and fractality of concrete microstructure, *Mater. Struct. (RILEM)*, 27(174), 563.

293. Reinhardt, H.W., Cornelissen, H.A.W. and Hordijk, D.A. 1986, Tensile tests and failure analysis of concrete, *J. Struct. Engrg. (ASCE)*, 112, 2462.

294. Van Mier, J.G.M., Schlangen, E. and Vervuurt, A. 1995, Lattice type fracture models for concrete, in *Continuum Models for Materials with Microstructure*, Mühlhaus, H.-B., Ed., John Wiley & Sons, Chap. 10.

295. Van Mier, J.G.M., Vervuurt, A. and Schlangen, E. 1994, Boundary and size effects in uniaxial tensile tests: a numerical and experimental study, in *Fracture and Damage of Quasibrittle Structures*, Bažant, Z.P., Bittnar, Z., Jirásek, M. and Mazars, J., Eds., E&FN Spon, London/New York, 289.

296. Daerga, P.A. 1992, Some experimental fracture mechanics studies in mode I of concrete and wood, *Licentiate thesis*, Luleå University of Technology, Sweden.

297. Budnik, J. 1985, Bruch- und Verformungsverhalten harzmodifizierter und faserverstärkter Betone bei einachsiger Zugbeanspruchung (Fracture and deformation of polymer modified and steel fibre concrete subjected to uniaxial tension), Ph.D. thesis, Ruhr University, Bochum (in German).

298. Hordijk, D.A., Reinhardt, H.W. and Cornelissen, H.A.W. 1987, Fracture mechanics parameters of concrete from uniaxial tensile tests as influenced by specimen length, in *Pre-Proceedings SEM/RILEM Conference on 'Fracture of Concrete and Rock'*, Shah, S.P. and Swartz, S.E., Eds., Society of Experimental Mechanics, Bethel, CT, 138.

299. Hurlbut, B. 1985, Experimental and computational strain-softening investigation of concrete, MS thesis, CEAE Department, University of Colorado, Boulder.

300. Carpinteri, A. and Ferro, G. 1992, Apparent tensile strength and fictitious fracture energy of concrete: A fractal geometry approach to related size effects, in *Fracture and Damage of Concrete and Rock (FDCR-2)*, Rossmanith, H.P., Ed., Chapman & Hall/E&FN Spon, London/ New York, 1993, 86.

301. Ferro, G. 1994, Effeti di scala sulla resistenza a trazione dei materiali (Scale effects on tensile properties of materials), Ph.D. thesis, Politecnico di Torino (in Italian).

302. Heilmann, H.G., Hilsdorf, H. and Finsterwalder, K. 1969, Festigkeit unde Verformung von Beton unter Zugspannungen (Strength and deformation of concrete subjected to uniaxial tension), *Deutscher Ausschuss für Stahlbeton*, Heft 203, Berlin (in German).

303. Carpinteri, A., Ferro, G. and Chiaia, B. 1995, Multifractal scaling law: An extensive application to nominal strength size effect of concrete structures, *Ati de Dipartimento No. 51*, Politecnico di Torino, 145 p.

304. Bažant, Z.P. 1984, Size effect in blunt fracture: concrete, rock, metal, *J. Eng. Mech. (ASCE)*, 110, 518.
305. Bažant, Z.P. 1986, Mechanics of distributed cracking, *Appl. Mech. Rev.*, 39(5), 675.
306. Weibull, W. 1939, A statistical theory of the strength of materials, *Proc. Royal Soc. Swedish Inst. Eng. Res.*, No. 159.
307. Arslan, A. and Ince, R. 1995, The neural network-based analysis of size effect in concrete fracture, in *Proceedings FraMCoS-2*, Wittmann, F.H., Ed., AEDIFICATIO Publishers, Freiburg, 693.
308. Leicester, R.H. 1973, Effect of size on the strength of structures, Forest Products Laboratory, *Division of Building Research Technological Paper No. 71*, CSIRO, Australia, 1.
309. Walsh, P.F. 1979, Fracture of plain concrete, *Indian Conc. J.*, 46, 469.
310. Van Vliet, M.R.A. and Van Mier, J.G.M. 1996, Size effect of concrete in uniaxial tension, in, *Localized Damage III*, Nisitani, H., Aliabadi, M.H., Nishida, S-I., and Cartwright, D.J., Eds., Computational Mechanics Publication, Southampton, 823.
311. Van Geel, H.J.G.M. 1995, Behaviour of concrete in plane-strain compression, *Report BKO95.19*, Eindhoven University of Technology, October 1995.
312. Reinhardt, H.W. and Hordijk, D.A. 1988, Various techniques for the assessment of the damage zone between two saw cuts, in *Strain Localization and Size Effect due to Cracking and Damage*, Mazars, J. and Bažant, Z.P., Eds., ENS de Cachan, Preprints, 1.
313. Jacobs, M.M.J., Hopman, P.C. and Molenaar, A.A.A. 1995, The crack growth mechanism in asphaltic mixes, *HERON*, 40(3), 181.
314. Karihaloo, B.L. and Nallathambi, P. 1991, Notched beam test: Mode I fracture toughness, in *Fracture Mechanics Test Methods for Concrete*, Shah, S.P. and Carpinteri, A., Eds., Chapman & Hall, London/New York, Chap.1.
315. Rossi, P., Brühwiler, E., Chhuy, S., Jenq, Y.-S. and Shah, S.P. 1991, Fracture properties of concrete as determined by means of wedge splitting tests and tapered double cantilever beam tests, in *Fracture Mechanics Test Methods for Concrete*, Shah, S.P. and Carpinteri, A., Eds., Chapman & Hall, London, Chap. 2.
316. Kobayashi, A.S., Hawkins, N.M., Barker, D.B. and Liaw, B.M. 1985, Fracture process zone of concrete, in *Applications of Fracture Mechanics to Cementitious Composites*, Shah, S.P., Ed., Martinus Nijhoff Publishers, Dordrecht, 25.
317. Desrues, J. 1983, On the application of stereo-photogrammetry to the measurement of large strains, *Rev. Franc. Mecan.* (Grenoble), 3 (in french).
318. Lierse, J. and Ringkamp, M. 1983, Investigations into cracked reinforced concrete structural elements with the aid of photo-elastic methods, in *Fracture Mechanics of Concrete*, Wittmann, F.H., Ed., Elsevier Applied Science Publishers, Amsterdam, 95.
319. Van Mier, J.G.M. 1988, Fracture study of concrete specimens subjected to combined tensile and shear loading, in *Proceedings GAMAC Int. Conf. on 'Measurement and Testing in Civil Engineering'*, Jullien, J.F., Ed., CAST/INSA Publishers, Villeurbanne, France, 337.
320. Raiss, M.E., Dougill, J.W. and Newman, J.B. 1989, Observation of the development of fracture process zones in concrete, in *Fracture of Concrete and Rock - Recent Developments*, Shah, S.P., Swartz, S.E. and Barr, B., Eds., Elsevier Applied Science, London/New York, 243.
321. Cedolin, L., Dei Poli, S. and Iori, I. 1987, Tensile behavior of concrete, *J. Eng. Mech. (ASCE)*, 113(3), 431.
322. Castro-Montero, A., Miller, R.A. and Shah, S.P. 1991, Study of the fracture process in mortar with laser holographic measurements, in *Toughening Mechanism in Quasi-Brittle Materials*, Shah, S.P., Ed., Kluwer Academic Publishers, Dordrecht, 249.
323. Maji, A.K. and Shah, S.P. 1991, Laser interferometry methods, in *Fracture Mechanics Test Methods for Concrete*, Shah, S.P. and Carpinteri, A., Eds., Chapman & Hall, London/New York, Chap. 6.
324. Torrenti, J.M., Desrues, J., Benaija, E.H. and Boulay, C. 1991, Stereophotogrammetry and localization in concrete under compression, *J. Eng. Mech.*, 117, 1455.

325. Torrenti, J.M., Benaija, E.H. and Boulay, C. 1992, Strain localization in concrete in compression: The influence of boundary conditions, in *Proceedings FraMCoS-1*, Bažant, Z.P., Ed., Elsevier Applied Science Publishers, London/New York, 281.

326. Sunderland, H., Tolou, A. and Huet, C. 1993, Multilevel numerical microscopy and tridimensional reconstruction of concrete microstructure, in *Micromechanics of Concrete and Cementitious Composites*, Huet, C., Ed., Presses Polytechniques et Univesitaires Romandes, Lausanne, 171.

327. Lange, D.A., Jennings, H.M. and Shah, S.P. 1993, Relationship between fracture surface roughness and fracture behavior of cement paste and mortar, *J. Am. Ceram. Soc.*, 76(3), 589.

328. Tait, R.B. and Garret, G.G. 1986, In situ double torsion fracture studies of cement mortar and cement paste inside a scanning electron microscope, *Cem. Conc. Res.*, 16, 143.

329. Krstulovic-Opera, N. 1993, Fracture process zone presence and behaviour in mortar specimens, *ACI Materials Journal*, 90, 618.

330. Hwang, C.L., Wang, M.L. and Miao, S. 1993, Proposed healing and consolidation mechanisms of rock salt revealed by ESEM, *Microscopy Res. Techn.*, 25, 456.

331. Ollivier, J.P. 1985, A nondestructive procedure to observe the microcracks of concrete by scanning electron microscopy, *Cem. Conc. Res.*, 15, 1055.

332. Turatsinze, A. 1992, Caracterisation microstructurale de la fissuration des betons et mortiers en mode I propagation (Microstructural characterization of mode I cracking in concrete and mortar), Ph.D. thesis, University Paul Sabatier, Toulouse (in French).

333. Vervuurt, A., Chiaia, B. and van Mier, J.G.M. 1996, Damage evolution in different types of concrete by means of splitting tests, *HERON*, 40(4), 285.

334. Goto, Y. 1971, Cracks formed in concrete around deformed tension bars, *J. Am. Conc. Inst.*, 68, 244.

335. Goudswaard, I. and Vonk, R.A. 1990, A detection method for internal cracks in concrete specimens, *Report TUE-BKO-89.12*, Eindhoven University of Technology (in Dutch).

336. Zheng, Z., Cook, N.G.W. and Myer, L.R. 1989, Stress induced microcrack geometry in unconfined and confined axial compressive tests, in *Rock Mechanics as a Guide for Efficient Utilization of Natural Resources*, Khair, A., Ed., Balkema, Rotterdam, 749.

337. Myer, L.R., Kemeny, J.M., Zheng, Z., Suarez, R., Ewy, R.T. and Cook, N.G.W. 1992, Extensile cracking in porous rock under differential compressive stress, *Appl. Mech. Rev. (ASME)*, 45, 263.

338. Nemati, K.M. 1994, Generation and interaction of compressive stress-induced microcracks in concrete, Ph.D. thesis, University of California at Berkeley, USA.

339. Fishbine, B.H., Macy, R.J., Ross, T.J. and Wang, M.L. 1990, SEM dynamic microscopy, in *Micromechanics of Failure of Quasi Brittle Materials*, Shah, S.P., Swartz, S.E. and Wang, M.L., Eds., Elsevier Applied Science Publishers, London/New York, 365.

340. Berthelot, J.M. and Robert, J.L. 1990, Damage evaluation of concrete test specimens related to failure analysis, *J. Eng. Mech.*, 116(3), 587.

341. Wissing, B. 1988, Acoustic emission of concrete, Masters thesis, Department of Civil Engineering, Delft University of Technology, (in Dutch).

342. Landis, E.N. and Shah, S.P. 1995, Experimental measurements of microfracture in cement-based materials, in *Proceedings FraMCoS-2*, Wittmann, F.H., Ed., AEDIFICATIO Publishers, Freiburg, 315.

343. Moczko, A.T., and Stroeven, P. 1994, Cracking behaviour resulting from a corroding bar in concrete, in *Proceedings ECF-10 Structural Integrity: Experiments, Models and Applications*, Schwalbe, K.-H. and Berger, C., Eds., EMAS Publishers, Warley, UK, 721.

344. Duda, H. 1991, Grain-model for the determination of the stress-crack width relation, in *Analysis of Concrete Structures by Fracture Mechanics*, Elfgren, L. and Shah, S.P., Eds., Chapman & Hall, London/New York, 88.

345. Labuz, J., Shah, S.P. and Dowding, C.H. 1985, Experimental analysis of crack propagation in granite, *Int. J. Rock Mech. Min. Sci. Geomech. Abstr.*, 22, 85.

346. Ohtsu, M. 1988, Source inversion of acoustic emission waveform, *Proc. JSCE*, 398/I-10, 71.

347. Suaris, W. and Van Mier, J.G.M. 1995, Acoustic emission source characterization in concrete under biaxial loading, *Mater. Struct. (RILEM)*, 28(182), 444.

348. Savic, M., Ziolkowski, A.M. and Cockram, M.J. 1991, Acoustic monitoring of laboratory scale dynamic processes under high pressure, in *Proc. 1991 SEG Meeting*, Houston, November 10-14.

349. Berthaud, Y. 1991, Damage measurements in concrete via an ultrasonic technique. Part I Experiments. Part II Modelling, *Cem. Conc. Res.*, 21, 73 and 219.

350. Schickert, G. 1985, Heutige Praxis der Zerstörungsfreie Prüfmethoden im Bauwesen (Modern applications of non-destructive testing in building engineering), *Proceedings of the 'ZfPBau-Symposium Zerstörungsfreie Prüfung im Bauwesen'*, Schickert, G. and Schnitger, D., Eds., BAM, Berlin, 91.

351. Knehans, R. and Steinbrech, R. 1982, Memory effect of crack resistance during slow crack growth in notched Al_2O_3 bend specimens, *J. Mater. Sci. Letters*, 1, 327.

352. Hu, X.Z. and Wittmann, F.H. 1989, Fracture process zone and K_r curve of hardened cement paste and mortar, in *Fracture of Concrete and Rock - Recent Developments*, Shah, S.P., Swartz, S.E. and Barr, B., Eds., Elsevier Applied Science, 307.

353. Hu, X.Z. 1990, Fracture process zone and strain-softening in cementitious materials, *Research Report 1*, Institute for Building Materials, ETH Zürich.

354. Foote, R.M.L., Mai, Y.W. and Cotterell, B. 1987, Process zone size and crack growth measurements in fibre cements, in *Fibre Reinforced Concrete Properties and Applications*, Shah, S.P. and Batson, G.B., Eds., SP-105, American Concrete Institure, Detroit, 55.

355. Otsuka, K. 1989, X-ray technique with contrast medium to detect fine cracks in reinforced concrete, in *Fracture Toughness and Fracture Energy - Test Methods for Concrete and Rock*, Mihahshi, H., Takahashi, H. and Wittmann, F.H., Eds., Balkema, Rotterdam, 521.

356. Luong, M.P. 1990, Infrared thermovision of damage processes in concrete and rock, *Eng. Fract. Mech.*, 35, 291.

357. Mendenhall, W. 1968, *Introduction to Linear Models and the Design and Analysis of Experiments*, Wadsworth.

358. Cochran, W.G. and Cox, G.M. 1957, *Experimental Designs*, John Wiley & Sons, New York, 2nd edition.

359. Imam, M., Vandewalle, L and Mortelmans, F. 1995, Are current concrete strength tests suitable for high strength concrete?, *Mater. Struct. (RILEM)*, 28(181), 384.

360. RILEM 1990a, TC 89-FMT Draft Recommendation, Determination of fracture parameters (K_{Ics} and $CTOD_c$) of plain concrete using three-point bend tests, *Mater. Struct. (RILEM)*, 23(138), 457.

361. RILEM 1990b, TC 89-FMT Draft Recommendation, Size-effect method for determining fracture energy and process zone size of concrete, *Mater. Struct. (RILEM)*, 23(138), 461.

362. Wittmann, F.H. 1992, *Numerical Models in Fracture Mechanics of Concrete*, Pre-Proceedings of the 1st Bolomey workshop, ETH Zürich, July 16-17, 1992, Appendix 2.

363. Ouchterlony, F. 1989, Fracture toughness testing of rock with core based specimens, the development of an ISRM standard, in *Fracture Toughness and Fracture Energy - Test Methods for Concrete and Rock*, Mihashi, H., Takahashi, H. and Wittmann, F.H., Eds., Balkema, Rotterdam, 231. (HFL 280.-)

364. Jenq, Y.S. and Shah, S.P. 1985, Two-parameter fracture model for concrete, *J. Engng. Mech. Div. (ASCE)*, 111(10), 1227.

365. Barr, B.I.G. and Swartz, S.E. 1996, The need for standard testing, in *Proceedings of the FraMCoS-2 Conference Workshop on "Numerical modelling and determination of fracture mechanics parameters,"* AEDIFICATIO Publishers, Freiburg, 1655.

366. Elices, M., Guinea, G.V. and Planas, J. 1992, Measurement of the fracture energy using three-point bend tests: Part 3 - Influence of cutting the P-δ tail, *Mater. Struct. (RILEM)*, 25(150), 327.

367. Planas, J., Elices, M. and Guinea, G.V. 1992, Measurement of the fracture energy using three-point bend tests: Part 2 - Influence of bulk energy dissipation, *Mater. Struct. (RILEM)*, 25, 305.

368. Elices, M. and Planas, J. 1996, Numerical modelling and determination of fracture mechanics parameters. Hillerborg Type Models, in *Proceedings of the FraMCoS-2 Conference Workshop on "Numerical modelling and determination of fracture mechanics parameters,"* AEDIFICATIO Publishers, Freiburg, 1611.

369. Hillerborg, A. 1985, The theoretical basis of a method to determine the fracture energy G_f of concrete, *Mater. Struct. (RILEM)*, 18, 291.

370. Tschegg, E.K. and Linsbauer, H.N. 1986, Prüfeinrichtung zur Ermittlung von bruchmechanischen Kennwerten (Testing procedure for determination of fracture mechanics parameters), *Patentschrift No. A-233/86*, Östereichisches Patentamt.

371. Linsbauer, H.N. and Tschegg, E.K. 1986, Fracture energy determination of concrete with cube-shaped specimens, *Zement und Beton*, 31, 38.

372. Roelfstra, P.E. and Wittmann, F.H. 1986, Numerical method to link strain softening with failure of concrete, In *Fracture Toughness and Fracture Energy*, Wittmann, F.H., Ed., Elsevier, London, 163.

373. Bažant, Z.P. and Li, Z. 1996, Zero-brittleness size effect method for one-size fracture test of concrete, *J. Eng. Mech. (ASCE)*, in press.

374. RILEM 30-TE Draft Recommendation 1983, Testing equipment, *Mater. Struct. (RILEM)*, 16(97).

375. Van Mier, J.G.M., Shah, S.P., Arnaud, M., Balayssac, J.P., Bascoul, A., Choi, S., Dasenbrock, D., Ferrara, G., French, C., Gobbi, M.E., Karihaloo, B.L., König, G., Kotsovos, M.D., Labuz, J., Lange-Kornbak, D., Markeset, G., Pavlovic, M.N., Simsch, G., Thienel, K-C., Turatsinze, A., Ulmer, M., Van Geel, H.J.G.M., Van Vliet, M.R.A., Zissopoulos, D. 1996, Strain-softening of concrete in uniaxial compression - Report of the round-robin test carried out by RILEM TC 148SSC, *Materials and Structures (RILEM)*, (to appear).

REFERENCES TO CHAPTER 5

376. Mühlhaus, H.B. (Ed.) 1995, *Continuum Models for Materials with Microstructure*, Wiley, Chichester.

377. Charmet, J.C., Roux, S. and Guyon, E., Eds. 1990, *Disorder and Fracture*, Plenum Press, New York.

378. Rashid, Y.R. 1968, Analysis of prestressed concrete pressure vessels, *Nuclear Eng. Design*, 7, 334.

379. Walraven, J.C. and Reinhardt, H.W. 1981, Concrete mechanics, Part A - Theory and experiments on the mechanical behavior of cracks in plain and reinforced concrete subjected to shear loading, *HERON*, 26(1a).

380. De Groot, A.K., Kusters, G.M.A. and Monnier, Th. 1981, Concrete mechanics, Part B - Numerical modelling of bond-slip behaviour, *HERON*, 26(1b).

381. Grootenboer, H.J., Leijten, S.F.C.H. and Blaauwendraad, J. 1981, Concrete mechanics, Part C -Numerical models for reinforced concrete structures in plane stress, *HERON*, 26(1c).

382. Van Mier, J.G.M., (Ed.) 1987, Examples of non-linear analysis of reinforced concrete structures with DIANA, *HERON*, 32(3).

383. De Borst, R. and Mühlhaus, H.-B. 1991, Continuum models for discontinuous media, in *Fracture Processes in Concrete, Rock and Ceramics*, Van Mier, J.G.M., Rots, J.G. and Bakker, A., Eds., Chapman & Hall/E&FN Spon, London/New York, 601.

384. Ngo, D. and Scordelis, A.C. 1967, Finite element analysis of reinforced concrete beams, *J. Am. Conc. Inst.*, 64, 152.

385. Petersson, P.-E. and Gustavsson, P.J. 1980, A model for calculation of crack growth in concrete-like materials, in *Numerical Methods in Fracture Mechanics*, Owen, D.R.J. and Luxmoore, A.R., Eds., Pineridge Press, Swansea, 707.

386. Ingraffea, A. and Saouma, V. 1984, Numerical modelling of discrete crack propagation in reinforced and plain concrete, in *Application of Fracture Mechanics to Concrete Structures*, Sih, G.C. and Di Tommaso, A., Eds., Martinus Nijhoff Publishers, Dordrecht, Chap. 4.

387. Carpinteri, A. and Valente, S. 1988, Size-scale transition from ductile to brittle failure: A dimensional analysis approach, in *Proceedings CNRS-NSF Workshop on Strain Localization and Size Effects due to Cracking and Damage*, Mazars, J. and Bazant, Z.P., Eds., Elsevier Applied Science, 477.

388. Valente, S. 1991, Influence of friction on cohesive crack propagation, in *Fracture Processes in Concrete, Rock and Ceramics*, Van Mier, J.G.M., Rots, J.G. and Bakker, A., Eds., Chapman & Hall/E&FN Spon, London/New York, 695.

389. Sousa,, J.O., Lutz, E.D. and Ingraffea, A.R. 1991, Three dimensional simulation of hydraulically driven fracture propagation in a rock mass, in *Fracture Processes in Concrete, Rock and Ceramics*, Van Mier, J.G.M., Rots, J.G. and Bakker, A., Eds., Chapman & Hall/E&FN Spon, London/New York, 545.

390. Elices, M. and Planas, J. 1989, Material models, in *Fracture Mechanics of Concrete Structures*, Elfgren, L., Ed., Chapman & Hall, London/New York, Chap. 3.

391. Hrennikoff, A. 1941, Solution of problems of elasticity by the framework method, *J. Appl. Mech.*, A169.

392. Hansen, A., Roux, S. and Herrmann, H.J. 1989, Rupture of central force lattices, *J. Phys. France*, 50, 733.

393. Herrmann, H.J., Hansen, H. and Roux, S. 1989, Fracture of disordered. elastic lattices in two dimensions, *Phys. Rev. B*, 39, 637.

394. Herrmann, H.J. and Roux, S. 1990, *Patterns and Scaling for the Fracture of Disordered Media*, Elsevier Applied Science Publishers B.V. (North Holland).

395. Herrmann, H.J. 1991, Patterns and scaling in fracture, in *Fracture Processes in Concrete, Rock and Ceramics*, Van Mier, J.G.M., Rots, J.G. and Bakker, A., Eds., Chapman & Hall/E&FN Spon, London/New York, 195.

396. Schlangen, E. and Van Mier, J.G.M. 1992, Experimental and numerical analysis of micromechanisms of fracture of cement-based composites, *Cem. Conc. Composites*, 14, 105.

397. Vervuurt, A., Schlangen, E. and Van Mier, J.G.M. 1993, A numerical and experimental analysis of the pull-out behaviour of steel anchors embedded in concrete, *Report 25.5-93-1/VFI*, Delft University of Technology, The Netherlands.

398. Vervuurt, A., Van Vliet, M.R.A., Van Mier, J.G.M. and Schlangen, E. 1995, Simulations of tensile fracture in concrete, in *Proceedings FraMCoS-2*, Wittmann, F.H., Ed., AEDIFICATIO Publishers, Freiburg, 353.

399. Mourkazel, C. and Herrmann, H.J. 1992, A vectorizable random lattice, *Preprint HLRZ 1/92*, KFA, Jülich.

400. Schlangen, E. and Van Mier, J.G.M. 1994, Fracture simulations in concrete and rock using a random lattice, in *Computer Methods and Advances in Geomaterials*, Siriwardane, H. and Zaman, M.M., Eds., Balkema, Rotterdam, 1641.

401. Schlangen, E. and Garboczi, E.J. 1996, New method for simulating fracture using an elastically uniform random geometry lattice, *Int. J. Eng. Sci.*, 34, 1131.

402. Jirásek, M. and Bažant, Z.P. 1995, Particle model for quasibrittle fracture and application to sea ice, *Int. J. Fract.*, 69, 201.

403. Burt, N.J. and Dougill, J.W. 1977, Progressive failure in a model heterogeneous medium, *J. Eng. Mech. Div. (ASCE)*, 103, 365.

404. Zubelewicz, A. and Bažant, Z.P. 1987, Interface element modeling of fracture in aggregate composites, *J. Eng. Mech. (ASCE)*, 113, 1619.

405. Bažant, Z.P., Tabbara, M.R., Kazemi, M.T. and Pijaudier-Cabot, G. 1990, Random particle model for fracture of aggregate or fiber composites, *J. Eng. Mech. (ASCE)*, 116, 1686.

406. Berg, A. and Svensson, U. 1991, Datorsimulering och analysis av brottförlopp i en heterogen materialstruktur (Simulation of fracture in heterogeneous materials), *Report no. TVSM-5050*, Lund Institute of Technology, Department of structural Engineering, Lund, Sweden (in Swedish).

407. Schorn, H. and Rode, U. 1989, 3D-Modelling of fracture process in concrete by numerical simulation, in *Fracture of Concrete and Rock*, Shah, S.P. and Swartz, S.E., Eds., Springer Verlag, New York, 220.

408. Murat, M., Anholt, M. and Wagner, H.D. 1992, Fracture behaviour of short-fiber reinforced materials, *J. Mater. Res.*, 7(11), 3121.

409. Termonia, Y. and Meakin, P. 1986, Formation of fractal cracks in a kinetic fracture model, *Nature*, 320, 429.

410. Schlangen, E. and Van Mier, J.G.M. 1992, Simple lattice model for numerical simulation of fracture of concrete materials and structures", *Mater. Struct. (RILEM)*, 25(153), 534.

411. Schlangen, E. 1995, Computational aspects of fracture simulations with lattice models, in *Proceedings FraMCoS-2*, Wittmann, F.H., Ed., AEDIFICATIO Publishers, Freiburg, 913.

412. Adley, M.D. and Sadd, M.H. 1992, Continuum models for materials with lattice-like microstructure, *Comp. Struct.*, 43, 13.

413. Beranek, W.J. and Hobbelman, G.J. 1994, Constitutive modelling of structural concrete as an assemblage of spheres, in *Computer Modelling of Concrete Structures'*, Mang, H., Bicanic, N. and De Borst, R., Eds., Pineridge Press, Swansea, 37.

414. Van Vliet, M.R.A. and Van Mier, J.G.M. 1996, Comparison of lattice type fracture models for concrete under biaxial loading regimes, in *Proceedings IUTAM Symposium on Size-Scale Effects in the Failure Mechanisms of Materials and Structures*, Torino, October 1994, Carpinteri, A., Ed., Chapman & Hall/E&FN Spon, 43.

415. Van Mier, J.G.M., Schlangen, E., Vervuurt, A. and Van Vliet, M.R.A. 1995, Damage analysis of brittle disordered materials: concrete and rock, in *Proceedings ICM-7*, The Hague, May 28 - June 2, 1995, Bakker, A., Ed., Delft University Press, 1995, 101.

416. Roelfstra, P.E., Sadouhi, H. and Wittmann, F.H. 1985, Le béton numérique (Numerical Concrete), *Mater. Struct. (RILEM)*, 18, 327 (in French).

417. Wang, J. 1994, Development and application of a micromechanics-based numerical approach for the study of crack propagation in concrete, Ph.D. thesis, EPFL Lausanne.

418. De Schutter, G. and Taerwe, L. 1993, Random particle model for concrete based on Delaunay triangulation, *Mater. Struct. (RILEM)*, 26, 67.

419. Chen, W.F. and Saleeb, A.T. 1982, *Constitutive Equations for Engineering Materials*, John Wiley & Sons, New York.

420. Chen, W.F. 1982, *Plasticity in Reinforced Concrete*, McGraw-Hill, New York, 1982.

421. Lade, P.V. 1982, Three-parameter failure criterion for concrete, *J. Eng. Mech. Div. (ASCE)*, 111, 188.

422. Ottosen, N.S. 1977, A failure criterion for concrete, *J. Eng. Mech. Div. (ASCE)*, 103, 527.

423. Podgorski, J. 1985, General failure criterion for isotropic media, *J. Eng. Mech. (ASCE)*, 111, 188.

424. CEB 1983, Concrete under multiaxial states of stress - Constitutive equations for practical design, *Comité Euro-Intrenational du Béton, EPFL-Lausanne, Bulletin d'Information* No. 156, June 1983.

425. Timoshenko, S.P. and Goodier, J.N. 1970, *Theory of Elasticity*, McGraw-Hill, 3rd edition.

426. Broek, D. 1982, *Elementary Engineering Fracture Mechanics*, Martinus Nijhoff Publishers, 3rd revised ed.

427. Ewalds, H.W. and Wanhill, R.J.H. 1986, *Fracture Mechanics*, VSSD, Delft.

428. Anderson, T.L. 1991, *Fracture Mechanics - Fundamentals and Applications*, CRC Press, Boca Raton, FL.

429. Moavenzadeh, F. and Kuguel, R. 1969, Fracture of Concrete, *J. Mater.*, 4(3), 497.

430. Tada, H., Paris, P.C. and Irwin, G.R. 1973, *The Stress Analysis of Cracks Handbook*, Del Research Corporation, Hellertown.

431. Marchand, N., Parks, D.M. and Pelloux, R.M. 1986, K_I-solutions for single edge notched specimens under fixed end displacements, *Int. J. Fracture*, 32, 53.

432. Murakami, Y., Ed. 1987, *Stress-Intensity Factor Handbook*, Pergamon Press.

433. Zaitsev, Y.W. and Wittmann, F.H. 1981, Simulation of crack propagation and failure of concrete, *Mater. Struct. (RILEM)*, 14, 357.

434. Costin, L.S. 1983, A microcrack model for the deformation and failure of brittle rock, *J. Geophys. Res.*, 88, 9485.

435. Kachanov, M. Tsukrov, I. and Shafiro, B. 1994, Effective elastic properties of solids with randomly located defects, in *Probabilities and Materials*, Breysse, D., Ed., NATO ASI series, Vol. E-269, Kluwer Academic Publishers, Dordrecht, 1994, 225.

436. Kemeny, J.M. and Cook, N.G.W. 1991, Micromechanics of deformation in rocks, in *Toughening Mechanisms in Quasi-Brittle Materials*, Shah, S.P., Ed., Kluwer Academic Publishers, Dordrecht, 155.

437. Nihei, K.T., Myer, L.R., Kemeny, J.M., Liu, Z. and Cook, N.G.W. 1994, Effects of heterogeneity and friction on the deformation and strength of rock, in *Fracture and Damage of Quasibrittle Structures*, Bažant, Z.P., Bittnar, Z., Jirásek, M. and Mazars, J., Eds., E&FN Spon, London/New York, 479.

438. Ortiz, M. 1988, Microcrack coalescence and macroscopic crack growth initiation in brittle solids, *Int. J. Solids Structures*, 24, 231.

439. Karihaloo, B.L. 1991, Tensile response of Quasi-Brittle Materials, in *Fracture Processes in Concrete, Rock and Ceramics*, Van Mier, J.G.M., Rots, J.G. and Bakker, A., Eds., Chapman & Hall/E&FN Spon, London/New York, 163.

440. Sammis, C.G. and Ashby, M.F. 1985, The failure of brittle porous solids under compressive stress states, *Acta. Metall. Mater.*, 34(3), 511.

441. Carrasquillo, R.L., Slate, F.O. and Nilson, A.H. 1981, Micro-cracking and behavior of high strength concrete subject to short term loading, *J. Am. Conc. Inst.*, 78, 179.

442. Attiogbe, E.K. and Darwin, D. 1985, Submicroscopic cracking of cement paste and mortar in compression, *Structural Engineering and Engineering Materials Report No. 16*, University of Kansas, Lawrence, Kansas.

443. Darwin, D. and Mohamed Nagib Abou-Zeid 1995, Application of automated image analysis to the study of cement paste microstructure, in *Microstructure of Cement-Based Systems, Proceedings of Symposium Va/Vb of the MRS Fall meeting 1994*, Boston, November 28 - December 2, 1994, Materials Research Society, Pittsburgh (in press).

444. Bažant, Z.P. 1987, Snapback instability at crack ligament tearing and its implication for fracture micromechanics, *Cem. Conc. Res.*, 17, 951.

445. Rice, J.R.1968, Mathmatical analysis in the mechanics of fracture, in *Fracture - An Advanced Treatise 2*, Liebowitz, H., Ed., Academic Press, New York, 191.

446. Horii, H., Hasegawa, A. and Nishino, F.1989, Fracture process and bridging zone model and influencing factors in fracture of concrete, in *Fracture of Concrete and Rock*, Shah, S.P. and Swartz, S.E., Eds., Springer Verlag, New York, 205.

447. Sih, G.C. (Ed.) 1972, *Mechanics of Fracture I. Methods of Analysis and Solutions of Crack Problems*, Noordhoff Int. Publ., Leyden.

448. Hillerborg, A. 1989, Stability problems in fracture mechanics testing, in *Fracture of Concrete and Rock - Recent Developments*, Shah, S.P., Swartz, S.E. and Barr, B., Eds., Elsevier Applied Science Publishers, London/New York, 369.

449. Foote, R.M.L., Mai, Y.-W. and Cotterell, B. 1986, Crack growth resistance curves in strain-softening materials, *J. Mech. Phys. Solids*, 34(6), 593.

450. Elfgren, L., Ed. 1989, *Fracture Mechanics of Concrete Structures - From Theory to Applications*, Chapman & Hall, London/New York.

451. Fanping Zhou 1988, Some aspects of tensile fracture behaviour and structural response of cementitious materials, *Report TVBM 1008*, Lund Institute of Technology, Sweden.

452. Rots, J.G. 1988, Computational modelling of concrete fracture, Ph.D. thesis, Delft University of Technology, The Netherlands.

453. Pamin, J. 1994, Gradient-dependent plasticity in numerical simulation of localization phenomena, Ph.D. thesis, Delft University of Technology, The Netherlands.

454. Hordijk, D.A., Van Mier, J.G.M. and Reinhardt, H.W. 1989, Material properties, in *Fracture Mechanics of Concrete Structures*, Elfgren, L., Ed., Chapman & Hall, London/New York, Chap.4.

455. Schechtman, D., Blech, I., Gratias, D. and Cahn, J.W. 1984, Metallic phase with long-range orientational order and no translational symmetry, *Phys. Rev. Lett.*, 53, 1951.

456. Penrose, R. 1989, *The Emperors New Mind - Concerning Computers, Minds and the Laws of Physics*, Oxford University Press.

457. Majumdar, A.J. and Laws, V. 1991, *Glass Fibre Reinforced Cement*, BSP Professional Books, Oxford.

458. Zeyher, A. 1993, Simulated materials reveal microfractures, *Computers in Physics*, 7(4), 382.

459. Breijsse, D. 1991, Understanding some aspects of damage development using hierarchic lattices, in *Fracture Processes in Concrete, Rock and Ceramics*, Van Mier, J.G.M., Rots, J.G. and Bakker, A., Eds., Chapman & Hall/E&FN Spon, London/New York, 241.

460. Beranek, W.J. and Hobbelman, G.J. 1995, 2D and 3D modelling of concrete as an assemblage of spheres. Revaluation of the failure criterion, in *Proceedings FraMCoS-2*, Wittmann, F.H., Ed., AEDIFICATIO Publishers, Freiburg, 965.

461. Arslan, A., Schlangen, E. and Van Mier, J.G.M. 1995, Effect of model fracture law and porosity on tensile softening of concrete, in *Proceedings FraMCoS-2*, Wittmann, F.H., Ed., AEDIFICATIO Publishers, Freiburg, 45.

REFERENCES TO CHAPTER 6

462. Bui, H.D., Tanaka, M., Bonnet, M., Maigre, H., Luzzato, E. and Reynier, Eds. 1994, *Inverse Problems in Engineering Mechanics*, Balkema, Rotterdam.

463. Wittmann, F.H, Roelfstra, P.E., Mihashi, H., Huang, Y.-Y., Zhang, X.-H. and Nomura, N. 1987, Influence of age of loading, water-cement ratio and rate of loading on fracture energy of concrete, *Mater. Struct. (RILEM)*, 20(116), 103.

464. Planas, J., Guinea, G.V. and Elices, M. 1993, Softening curves for concrete and structural materials, in *Fracture and Damage of Concrete and Rock - FDCR-2*, Rossmanith, H.P., Ed., Chapman & Hall/E&FN Spon, 3.

465. Roelfstra, P.E. 1989, Simulation of failure in computer generated structures, In *Fracture Toughness and Fracture Energy*, Mihashi, H., Takahashi, H. and Wittmann, F.H., Eds., Balkema, Rotterdam, 313.

466. Stankowski, T. 1990, Numerical simulation of progressive failure in particle composites, Ph.D. thesis, CEAE Department, University of Colorado, Boulder.

467. Schlangen, E. and Van Mier, J.G.M. 1992, Micromechanical analysis of fracture of concrete, *Int. J. Damage Mech.*, 1, 435.

468. Margoldová, J. and Van Mier, J.G.M. 1994, Simulation of compressive fracture in concrete, in *Proceedings ECF10 'Structural Integrity: Experiments, Models, Applications'*, Schwalbe, K.-H. and Berger C., Eds., EMAS Publishers, Warley, UK, pp. 1399.

469. Schlangen, E. 1995, Fracture simulations of brittle heterogeneous materials, in *Proceedings 10th ASCE-EMD Conference*, Sture, S. Ed., ASCE, New York, 130.

470. Schlangen, E. and Van Mier, J.G.M. 1993, Lattice model for simulating fracture of concrete, in *Numerical Models in Fracture Mechanics of Concrete*, Wittmann, F.H., Ed., Balkema, Rotterdam, 195.

471. Chiaia, B. and van Mier, J.G.M. 1996, Evaluation of damage in different concretes: fractal analysis and relation to toughness characteristics, in *Mechanisms and Mechanics of Damage and Failure (ECF11)*, Petit, J., de Fouquet, J., Henaff, G., Villechaise, P. and Dragon, A., Eds., EMAS Publishers, Warley, UK, 1871.

472. Huang, X. and Karihaloo, B.L. 1993, The role of microcracks and microvoids in the toughening of quasi-brittle materials, in *Fracture and Damage of Concrete and Rock - FDCR-2*, Rossmanith, H.P., Ed., E&FN Spon, London/New York, 34.

473. Arslan, A., Schlangen, E. and Van Mier, J.G.M. 1995, A study on effect of fracture law and porosity on concrete fracture under uniaxial tension, *Research report 25.5-95-8*, Delft University of Technology, Faculty of Civil Engineering, April 1995.

474. Chiaia, B., Vervuurt, A. and Van Mier, J.G.M. 1996, Lattice model evaluation of progressive failure in disordered particle composites, *Eng. Fract. Mech., Special Issue on Statistical Fracture*, Eds. M. Ostoja-Starzewski and H.J. Liebowitz (in press).

REFERENCES TO CHAPTER 7

475. Liu, Y.-Q., Hikosaka, H. and Bolander, J.E. 1995, Modelling compressive failure using rigid particle systems, in *Proceedings FraMCoS-2*, F.H. Wittmann, Ed., AEDIFICATIO Publishers, Freiburg, 375.

476. Bolander, J.E. and Kobayashi, Y. 1995, Size effect mechanisms in numerical concrete fracture, in *Proceedings FraMCoS-2*, F.H. Wittmann, Ed., AEDIFICATIO Publishers, Freiburg, 535.

477. Van Mier, J.G.M., Vervuurt, A. and Schlangen, E. 1994, Crack growth simulations in concrete and rock, in *Probabilities and Materials. Tests, Models and Applications*, Breysse, D., Ed., NATO-ASI Series E Applied Sciences, Vol. 269, Kluwer Academic Publishers, Dordrecht, 377.

478. Van Mier, J.G.M. and Vervuurt, A. 1995, Micromechanical analysis and experimental verification of boundary rotation effects in uniaxial tension tests on concrete, in *Fracture of Brittle Disordered Materials: Concrete, Rock and Ceramics*, Baker, G. and Karihaloo, B.L., Eds., E&FN Spon, London/New York, 406.

479. Rots, J.G. and De Borst, R. 1989, Analysis of concrete fracture in 'direct' tension, *Int. J. Solids Struct.*, 25, 1381.

480. Rots, J.G. and De Borst, R. 1987, Analysis of mixed-mode fracture in concrete, *J. Engng. Mech. (ASCE)*, 113, 1739.

481. Rots, J.G., Kusters, G.M.A. and Blaauwendraad, J. 1989, Strain-softening simulations of mixed-mode concrete fracture, in *Fracture of Concrete and Rock*, Shah, S.P. and Swartz, S.E., Eds., Springer Verlag, New York, 175.

482. Pamin, J. and De Borst, R. 1995, Numerical simulation of localization phenomena using gradient plasticity and finite elements, *HERON*, 40(1), 71-92.

483. Ohtsu, M. 1991, BEM analysis of crack propagation based on a discrete crack model of fracture mechanics, in *Fracture Processes in Concrete, Rock and Ceramics*, Van Mier, J.G.M., Rots, J.G. and Bakker, A., Eds., Chapman & Hall/E&FN Spon, London/New York, 571.

484. Arrea, M. and Ingraffea, A.R. 1982, Mixed mode crack propagation in mortar and concrete, *Report No. 81-13*, Department of Structural Engineering, School of Civil and Environmental Engineering, Cornell University, Ithaca, NY.

485. Rots, J.G., Nauta, P., Kusters, G.M.A. and Blaauwendraad, J. 1985, Smeared crack approach and fracture localization in concrete, *HERON*, 30(1).

486. De Borst, R. 1986, Non-linear analysis of frictional materials, Ph.D. thesis, Delft University of Technology, The Netherlands.

487. Nooru-Mohamed, M.B., Schlangen, E. and Van Mier, J.G.M. 1993, Experimental and numerical study on the behavior of concrete subjected to biaxial tension and shear, *Adv. Cement Based Mater.*, 1, 22.

488. Van Baars, S. 1995, Discrete element analysis of granular materials, *Communications on Hydraulic and Geotechnical Engineering, Report 95-4*, Delft University of Technology, Faculty of Civil Engineering, and *HERON*, 41(1996), 139.

REFERENCES TO CHAPTER 8

489. RILEM TC-90FMA 1990, Round-robin analysis of anchor bolts - Invitation, *Mater. Struct. (RILEM)*, 23(133), 78.

490. Dragosavić, M. and Groeneveld, H. 1988, Concrete mechanics - Local bond, part I: Physical behaviour and constitutive consequences, and Part II: Experimental research, *Research Report BI-87-18/19, TNO Building and Construction Research*, Rijswijk, The Netherlands.

491. Dragosavić, M. and Groeneveld, H. 1984, Bond model for concrete structures, in *Computer Aided Analysis and Design of Concrete Structures*, Damjanić, F. et al., Eds., Split, Pineridge Press, Swansea, 203.

492. Tepfers, R. 1979, Cracking of concrete cover along anchored deformed reinforcing bars, *Mag. of Conc. Res.*, 31, 3.

493. Ingraffea, A.R. Gerstle, W.H., Gergely, P. and Saouma, V.E. 1984, Fracture mechanics of bond in reinforced concrete, *J. Struct. Eng. (ASCE)*, 110, 871.

494. Rots, J.G. 1992, Simulation of bond and anchorage: usefulness of softening fracture mechanics, in *Applications of Fracture Mechanics to Reinforced Concrete*, Carpinteri, A., Ed., Elsevier Applied Science Publishers, London/New York, Chap. 11.

495. Mazars, J., Pijaudier-Cabot, G. and Clement, J.L. 1992, Analysis of steel-concrete bond with damage mechanics: Non-linear behaviour and size effect, in *Applications of Fracture Mechanics to Reinforced Concrete*, Carpinteri, A., Ed., Elsevier Applied Science Publishers, London/New York, Chap. 12.

496. Vos, E. 1983, Influence of loading rate and radial pressure on bond in reinforced concrete. A numerical and experimental approach, Ph.D. thesis, Delft University of Technology, The Netherlands.

497. Van Mier, J.G.M. and Vervuurt, A. 1995, Lattice model for analysing steel-concrete interface behaviour, in *Mechanics of Geomaterials Interfaces*, Selvadurai, A.P.S. and Boulon, M.J., Eds., Elsevier Science Publishers, Amsterdam, 201.

498. Goto, Y. 1971, Cracks formed in concrete around deformed tension bars, *J. Am. Conc. Inst.*, 68, 244.

499. Otsuka, K. 1989, X-ray Technique with contrast medium to detect fine cracks in reinforced concrete, in *Fracture Toughness and Fracture Energy - Test Methods for Concrete and Rock*, Mihashi, H., Takahashi, H. and Wittmann, F.H., Eds., Balkema, Rotterdam, 521.

500. Eligehausen, R. 1989, Anchorage to concrete, in *Fracture Mechanics of Concrete Structures - From Theory to Applications*, Elfgren, L., Ed., Chapman & Hall/E&FN Spon, London/New York, Chapt. 13.

501. Elfgren, L. 1992, Round robin analysis and tests of anchor bolts, in *Proceedings FraMCoS-1 Fracture Mechanics of Concrete Structures*, Bažant, Z.P., Ed., Elsevier Applied Science, London/New York, 865.

502. RILEM TC90-FMA 1991, Round robin analysis and tests of anchor bolts in fracture mechanics of concrete, *Report*, Elfgren, L., Ed., Division of Structural Engineering, Luleå University of Technology, Sweden.

503. Wang, J., Navi, P. and Huet, C. 1993, Finite element analysis of anchor bolt pull-out based on fracture mechanics, in *Fracture and Damage of Concrete and Rock - FDCR2*, Rossmanith, H.P., Ed., Chapman & Hall/E&FN Spon, London/New York, 559.

504. Feenstra, P.H., Rots, J.G. and De Borst, R. 1990, Round robin analysis of anchor bolts, Report 25.2-90-5-07, Delft University of Technology, Delft, The Netherlands.

505. Červenka, V., Pukl, R. and Eligehausen, R. 1991, Fracture analysis of concrete plane-stress pull-out tests, in *Fracture Processes in Concrete, Rock and Ceramics*, Van Mier, J.G.M., Rots, J.G. and Bakker, A., Chapman & Hall/E&FN Spon, London/New York, 899.

506. Ožbolt, J. and Eligehausen, R. 1993, Fastening elements in concrete structures - Numerical simulations, in *Fracture and Damage of Concrete and Rock - FDCR2*, Rossmanith, H.P., Ed., Chapman & Hall/E&FN Spon, London/New York, 527.

507. Helbling, A., Alvaredo, A.M. and Wittman F.H. 1991, Round-robin tests of anchor bolts, in *RILEM TC-90 FMA Fracture Mechanics of Concrete - Applications*, L. Elfgren, Ed., Luleå University of Technology, Sweden, 8:1.

508. Vervuurt, A., Schlangen, E. and Van Mier, J.G.M. 1993, A numerical and experimental analysis of the pull-out behaviour of steel anchors embedded in concrete, *Report 25.5-93-1/VFI*, Delft University of Technology, The Netherlands.

509. Vervuurt, A., Van Mier, J.G.M. and Schlangen, E. 1994, Analyses of anchor pull-out in concrete, *Mater. Struct. (RILEM)*, 27(169), 251.

510. Feenstra, P.H. 1993, Computational aspects of biaxial stress in plain and reinforced concrete, Ph.D. thesis, Delft University of Technology, Delft, The Netherlands.

511. Rossi, P. and Wu, X. 1993, A probabilistic model for material behaviour analysis and appraisal of the structure of concrete, in *Numerical Models in Fracture Mechanics of Concrete*, Wittmann, F.H., Ed., Balkema, Rotterdam, 207. (HFL 150.-)

512. Carpinteri, A. 1991, Fracture mechanics approach to reinforced concrete design, in *Fracture Processes in Concrete, Rock and Ceramics*, Van Mier, J.G.M., Rots, J.G. and Bakker, A., Eds., E&FN Spon, London/New York, 871.

513. Elfgren, L., Ed. 1989, *Fracture Mechanics of Concrete Structures*, Chapman & Hall/E&FN Spon, London/New York, page 399.

514. Naaman, A.E., Otter, D. and Najm, H. 1991, Elastic Modulus in Tension and Compression, *ACI Mater. J.*, 88, 603.

515. Ono, H. and Ohgishi, S. 1989, Fracture Toughness J_{Ic} and Fracture Energy G_f in the New Material Fibres Reinforced Concrete, in *Fracture Toughness and Fracture Energy*, Mihashi, H., Takahashi, H. and Wittmann, F.H., eds., Balkema, Rotterdam, 73.

INDEX

INDEX

A

Aardelite, 34
Acceloratory period (hydration), 23
Acoustic emission, 215, 230-233
Acoustic event, 230-231
Acoustic techniques, 215, 230-233; see also
    Crack detection techniques
Acoustic tomography, 232
Active restraint, see Loading platens
Actuator
    hydraulic, 159, 161, 174-176, 191, 198-
        200, 213-214, 345
    hysteresis in, 161, 245
    pneumatic, 159
Adhesion between steel and concrete, 367,
    369, 372, 374
Adhesive strength, 373-375
Advanced fracture experiments
    examples of, 166-177
Aggregate, 17, 33-38, 62
    alternative, 34, 38
    lightweight, 59-60, 340-342
    natural, 34, 39
Aggregate-cement bond strength, see
    Interfacial transition zone
Aggregate-cement bond tests, see Interfacial
    transition zone
Aggregate-cement interface, see Interfacial
    transition zone
Aggregate content, 59, 324
Aggregate grading, 35, 81; see also Particle
    distribution
Aggregate interlock models, 120-122, 301,
    367
Aggregate particles
    interaction between, 63-64, 133
Aggregate properties, 33-37
Aggregate roughness, 39, 63
Aggregate shape and texture, 35-36
    angular, 263
    circular, 261-263
    rounded, 263
    segment, 205
    smooth, 263
    spherical, 307
Aggregate size, 207, 209, 216, 261
    effect on tensile softening, 102-103, 106
    effect on tensile strength, 104
    related to crack band width, 277, 292, 298

related to size of lattice beams, 260, 321-
    323, 325, 340
Aggregate strength in lattice, see Aggregate
    zone
Aggregate types, 33-37, 102
Aggregate volume, 261-262, 324
Aggregate zone (lattice model), 262, 321-
    327, 330
    stiffness of, 324-325
    strength of, 324-326, 340, 358
Alkali-aggregate reaction, 18
Alkalinity, 23
Alkali oxides, 18
Alumina oxide, 18, 31
Analytical modelling, 7
Anchor bolt pull-out, 289, 367, 375-384
    effect of lateral confinement on, 382-383
Andesite, 49
Anhydrous material (cement), 20, 39-41
Anisotropy, 233
    effect on failure surface, 135
    effect on triaxial stress-strain curves, 148-
        149
    effect on uniaxial stress-strain curves, 66-
        67
    in concrete structure, 59, 66
    steel fibre, 114, 129-130
Argex, 34
Ascending branch of stress-strain curve, 56-
    57, 81, 87
Asymmetric deformations, 199
Asymmetric failure modes, 198-199, 355
Atomic bond, 303
Atomic lattice, 253, 259, 302-306
Atomic level, 253, 302-306
Atomic spacing, 303
Autoclaved aerated concrete, 243, 246
Axial cleavage crack, 138, 173
Axial deformations, see Deformations
Axial rigidity, see Loading platens
Axial strain, see Strain
Axial stress, see Stress

B

Bar element (lattice), 254, 301-302
Bar model, see Truss model
Basalt, 34-35, 49
Beam element (lattice), see Lattice beams
Beam length (lattice), 260, 321-325, 339-340